BACULOVIRUS EXPRESSION SYSTEMS AND BIOPESTICIDES

BACULOVIRUS EXPRESSION SYSTEMS AND BIOPESTICIDES

Editors
Michael L. Shuler
H. Alan Wood
Robert R. Granados
Daniel A. Hammer
School of Chemical Engineering
Cornell University
and
Boyce Thompson Institute
Ithaca, New York

WILEY-LISS
A JOHN WILEY & SONS, INC., PUBLICATION
New York • Chichester • Brisbane • Toronto • Singapore

Address All Inquiries to the Publisher
Wiley-Liss, Inc., 605 Third Avenue, New York, NY 10158-0012

Printed in the United States of America

While the authors, editors, and publisher believe that drug selection and dosage and the specification and usage of equipment and devices, as set forth in this book, are in accord with current recommendations and practice at the time of publication, they accept no legal responsibility for any errors or omissions, and make no warranty, express or implied, with respect to material contained herein. In view of ongoing research, equipment modifications, changes in governmental regulations and the constant flow of information relating to drug therapy, drug reactions, and the use of equipment and devices, the reader is urged to review and evaluate the information provided in the package insert or instructions for each drug, piece of equipment, or device for, among other things, any changes in the instructions or indication of dosage or usage and for added warnings and precautions.

Library of Congress Cataloging-in-Publication Data

Baculovirus expression systems and biopesticides / editors, Michael L. Shuler . . . [et al.].
p. cm.
Includes index.
ISBN 0-471-06580-3
1. Baculoviruses—Biotechnology. 2. Insect cell biotechnology. 3. Biological pest control agents. I. Shuler, Michael L., 1947–

[DNLM: 1. Baculoviridae—genetics. 2. Gene Expression. 3. Pest Control, Biological. 4. Cells, Cultured. QW 162 B131 1994]
QR398.5.B33 1994
576'.6484—dc20
DNLM/DLC
for Library of Congress 94-17194

The text of this book is printed on acid-free paper.

Contents

BIOREACTOR DESIGN AND SCALE-UP ISSUES

COMMERCIAL APPLICATION OF INSECT CELL CULTURE

Contributors

Jeffrey J. Chalmers, Department of Chemical Engineering, The Ohio State University, Columbus, OH 43210

Thomas R. Davis, Boyce Thompson Institute for Plant Research, Cornell University, Ithaca, NY 14853

Sandra L. Gould, Merck Research Laboratories, Rahway, NJ 07065

Robert R. Granados, Boyce Thompson Institute for Plant Research, Cornell University, Ithaca, NY 14853

Daniel A. Hammer, School of Chemical Engineering, Cornell University, Ithaca, NY 14853

George R. Hunt, Merck Research Laboratories, Rahway, NJ 07065

Deepak Jain, Biotechnology Division, R. W. Johnson Pharmaceutical Research Institute, Raritan, NJ 08869

Beth Junker, Merck Research Laboratories, Rahway, NJ 07065

Thomas A. Kost, Department of Molecular Biology, Glaxo, Inc. Research Institute, Research Triangle Park, NC 27709

Albert B. Lenny, Merck Research Laboratories, Rahway, NJ 07065

Verne A. Luckow, Molecular and Cellular Biology, Monsanto/Searle, Chesterfield, MO 63198

Kathryn E. Mazina, Merck Research Laboratories, Rahway, NJ 07065

Kevin A. McKenna, Boyce Thompson Institute for Plant Research, Cornell University, Ithaca, NY 14853

Laurie K. Overton, Department of Molecular Biology, Glaxo, Inc. Research Institute, Research Triangle Park, NC 27709

Oksana Palyha, Merck Research Laboratories, Rahway, NJ 07065

Kripashankar Ramasubramanyan, Merck Manufacturing Division, West Point, PA 19486

T. Craig Seamans, Merck Research Laboratories, Rahway, NJ 07065

Michael L. Shuler, School of Chemical Engineering, Cornell University, Ithaca, NY 14853

Melvin Silberklang, Enzon, Inc., Piscataway, NJ 08854

Ronald A. Taticek, School of Chemical Engineering, Cornell University, Ithaca, NY 14853

Michael R. Tota, Merck Research Laboratories, Rahway, NJ 07065

Shiping Wang, Merck Research Laboratories, Rahway, NJ 07065

Thomas J. Wickham, Genvec, Rockville, MD 20852

H. Alan Wood, Boyce Thompson Institute for Plant Research, Cornell University, Ithaca, NY 14853

Preface

The life cycle of the baculovirus is intrinsically fascinating. This unique life cycle also presents important technological opportunities. The possibility of using baculoviruses as alternatives to chemical pesticides has been historically a strong motivation for baculovirus research. The environmental and human health and safety issues have not been sufficiently strong to override economic barriers as highly cost effective chemical pesticides were introduced in the 1970s and early 1980s. However, the possibility of using baculovirus as an expression vector for production of heterologous proteins emerged just as interest in the baculovirus as a biological pesticide began to wane. This concept catalyzed an explosion of research on baculovirus and insect cell culture.

This book examines the current advances and remaining obstacles associated with the commercial scale development of the baculovirus expression vector system as well as the production of viral pesticides. In the first part, it develops an appreciation of the biology of the baculovirus system and how that biology can be exploited for developing expression vectors and for genetically improved viral pesticides. Furthermore, the importance of the host cell selection to system performance is emphasized. The second part of the book addresses basic issues in the development of appropriate commercial scale systems to produce proteins or viral pesticides. The last part provides industrial perspectives on the commercial application of insect cell cultures and baculovirus.

The topics discussed are interrelated through a focus on the transcriptional, translational, and post-translational regulatory factors governing foreign gene expression and virus replication. Although the individual chapters deal with specific issues, the reader is forced to appreciate that each issue is multi-dimensional, being influenced by a milieu of viral, cellular, and extracellular factors. These factors are represented by both biological and biophysical parameters, and the book aims to provide the reader with an integrated perspective on the interaction between these parameters.

It is this systems perspective that should aid the readers—students, professional scientists and engineers, or business persons—in developing a more complete understanding of the issues in commercial development of the insect cell-baculovirus system. We expect that this improved understanding will facilitate the commercial employment of this novel biological system.

Michael L. Shuler
H. Alan Wood
Robert R. Granados
Daniel A. Hammer

1

Overview of Baculovirus–Insect Culture System

Michael L. Shuler, Daniel A. Hammer, Robert R. Granados, and H. Alan Wood

School of Chemical Engineering (M.L.S., D.A.H.) and Boyce Thompson Institute for Plant Research (R.R.G., H.A.W.), Cornell University, Ithaca, New York 14853

PERSPECTIVE

Modern biology has made important promises to the human race. The complete fulfillment of these promises is still problematic. One tool that may help to solve some remaining problems is the insect cell–baculovirus system.

Genetic engineering promised the possibility of the inexpensive production of safe and biologically active proteins. This promise has not been fully achieved. All host–vector systems for production of proteins from recombinant DNA suffer particular disadvantages or constraints.

Another promise of modern science has been the replacement of chemical pesticides with environmentally safer biological pesticides. Current biologically produced insecticides have limited effectiveness and have been only partially successful.

The insect cell–baculovirus system has received rapid and wide acceptance as a laboratory or small-scale system for production of proteins from recombinant DNA. Also an "AIDS vaccine" produced by insect cell systems is in clinical trials. This system offers advantages of safety, high productivity, generally reliable production of active proteins, and a high degree of post-translational processing. Since the post-translational processing machinery of insect cells is not identical to that of mammalian cells, there may be some limitation on the types of proteins that are best produced in insect cell systems. However, products from insect cells typically elicit a strong immunogenic response, and this system appears particularly well-suited for use in vaccine production and possibly for antibodies.

In addition, the virus itself may be genetically engineered to improve its properties as a safe, specific, and effective insecticide against major crop pests and forest insect pests. The first tests in the country of a field-released genetically engineered baculovirus are being completed.

Baculovirus Expression Systems and Biopesticides, pages 1–11

For either production of proteins or viral pesticides, more efficient production systems will be required. Because these products may be required in large volumes and are likely to be cost-sensitive (e.g., animal vaccines and bioinsecticides), the success of the system will depend strongly on improved and novel engineering design.

In this chapter we will provide a brief overview of the key features of the insect cell–baculovirus system that provide the basis for genetic manipulation. Furthermore, we will identify some of the key limitations of this system that must be circumvented if the insect cell–baculovirus system is to become commercially successful. These topics will then be treated in much greater depth in the chapters that follow.

THE BACULOVIRUS SYSTEM

The baculoviruses are a diverse group of large viruses with double-stranded DNA that are pathogenic for invertebrates, primarily insects of the order *Lepidoptera* (i.e., moths and butterflies). The baculoviruses have a unique biphasic infection cycle that imparts unique opportunities for its use as an expression system for heterologous proteins and as an insecticide [1–3]. Figure 1 summarizes many aspects of the cellular life cycle of the baculovirus—the NPV (nuclear polyhedrosis virus) subtype—which have many virions occluded within numerous intranuclear crystals. Figure 2 depicts the baculovirus replication cycle at the organismal level.

Baculovirus particles exist in two forms, a polyhedra-derived (PDV) and budded form. The PDV particles are occluded within a protein crystal referred to as an *occlusion body* or *polyhedron* in the case of NPVs. The crystal matrix is a viral gene product called *polyhedrin.* Upon ingestion by a host insect larva, the alkaline nature in the midgut region results in dissolution of the crystal, releasing the PDV particles and allowing for infection of the midgut epithelial cells. Virus entry occurs by membrane fusion. The particles migrate into the nucleus through the nuclear pore where they uncoat and replicate.

Early in the infection cycle, the progeny nucleocapsids bud through the nuclear membrane into the cytoplasm. The membrane is removed in the cytoplasm, and another membrane surrounds the nucleocapsid as it buds through the plasma membrane. These enveloped nucleocapsids are referred to as *budded virus* (BV) or *nonoccluded virus* (NOV), and both terms are used interchangeably in this book. They are highly infectious to tissue culture cells, but are usually noninfectious if fed to insect larvae. The BVs are highly infectious when injected into the hemocoel (circulatory system).

Late in the replication cycle, progeny nucleocapsids become membrane bound within the nucleus. At this time, large amounts of polyhedrin protein are synthesized. The polyhedrin protein crystallizes around the membrane-bound particles, forming polyhedra. The PDV within the polyhedra are bounded by a membrane (virus envelope) that differs from that surrounding the budded virus. This accounts for the fact that the PDV is not very infectious to tissue culture cells. Not all

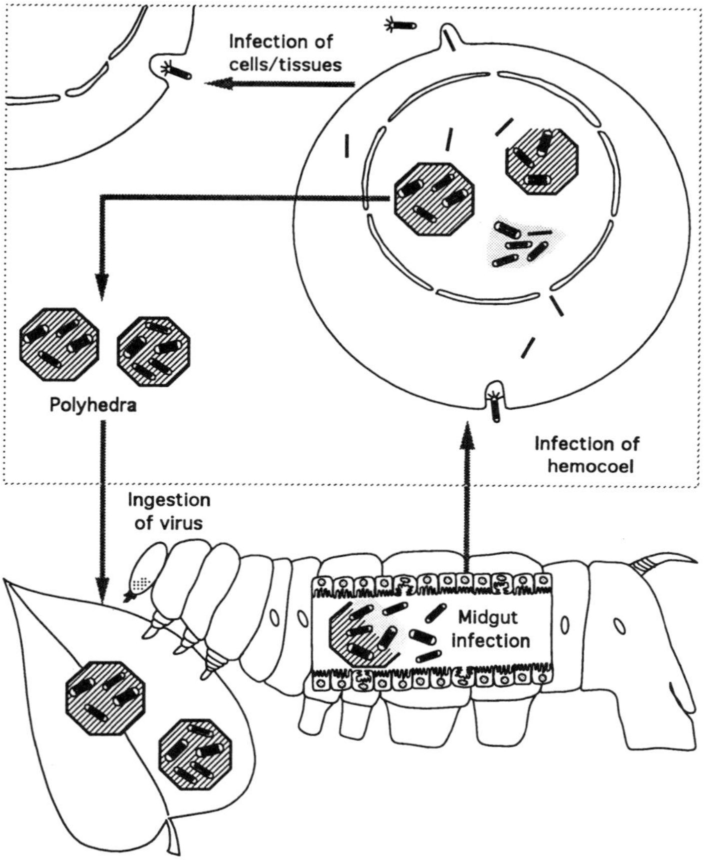

Fig. 1. The life cycle of the baculovirus is depicted. The occluded form of the virus is ingested and results in a primary infection. A nonoccluded form of virus is released into the hemocoel, resulting in secondary infection. At the late stages of the infection the occluded form is again produced and released. Details are discussed in the text.

baculoviruses are of the NPV type. Another type is the granulosis virus (GV), which is occluded in a protein matrix referred to as *capsule* or *granule*. The term *occlusion body* (OB) is commonly used to refer collectively to crystalline bodies produced by both NPVs and GVs. The occluded phenotypes of NPVs may be packaged as multiple (M) nucleocapsids within a single viral envelope (MNPV) or as single nucleocapsids per envelope (SNPV). All NPV species may have many enveloped virons occluded within a single occlusion body. In contrast, the occluded phenotype of GVs is usually packaged as single nucleocapsid per envelope and/or (or rarely two) virions per occlusion body.

In nature the virus is found in the soil or on leaves, and in the occluded form the

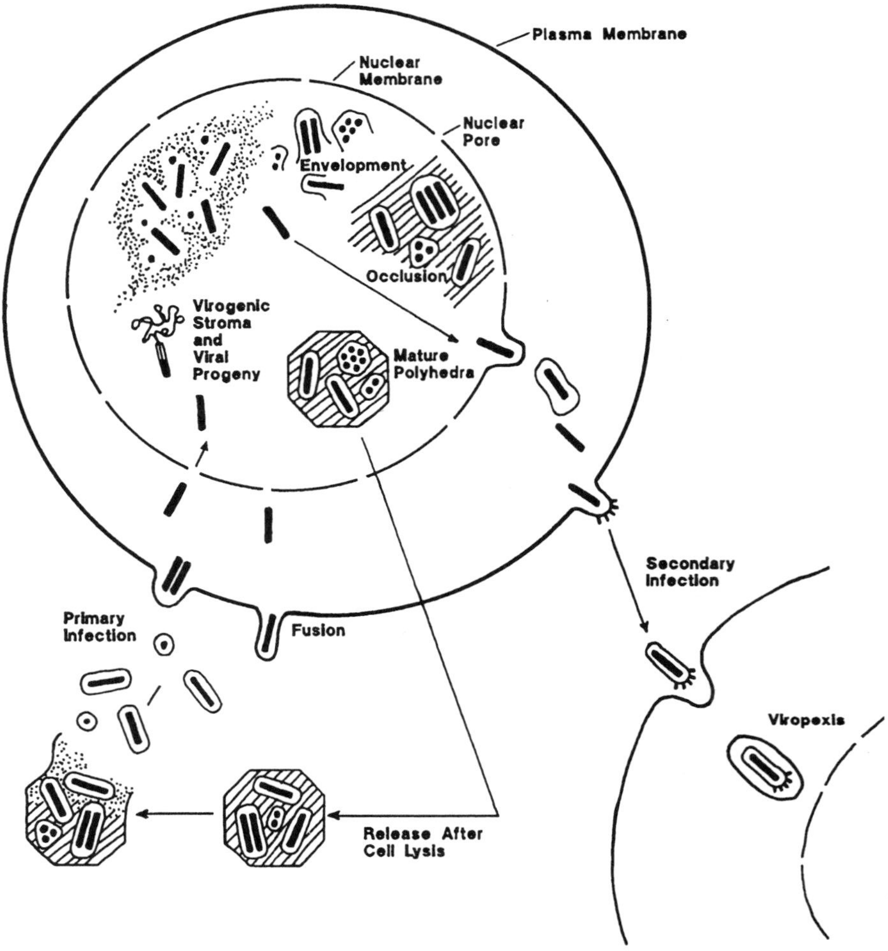

Fig. 2. The biphasic baculovirus replication cycle is depicted and described in the text.

virus is protected until ingestion by the insect. For use as an insecticide the occluded form is preferred. However, in the insect the nonoccluded virions are responsible for the secondary infection and are also the infective form in tissue culture. The protein matrix that encompasses virions in the occluded form is primarily polyhedrin, which is a 29 kd protein. The synthesis of polyhedrin is driven by a very strong late promoter. Because polyhedrin is unnecessary for infection in tissue culture, the polyhedrin gene loci make ideal cloning sites for genes for heterologous proteins. Another important hyperexpressed protein is the p10 protein. Its role in occlusion body formation is unclear; p10 mutants lack a properly associated polyhedron envelope, and the resulting occlusion body is very fragile. It is interesting to note that cells infected with p10 mutants fail to lyse late in the infection cycle [2]. The p10 locus has been used for cloning [4,5] and, as we will discuss later, offers an intriguing alternative to the much more commonly used polyhedrin cloning site.

In summary, the *Autographa californica* multiple nuclear polyhedrosis virus (AcMNPV) is well suited as an expression vector for cloned eukaryotic genes. The virus is nonpathogenic to vertebrates, and, because of its bacillus shape, it can encapsulate viral genomes with large pieces of additional, foreign DNA. In addition, the polyhedrin gene provides the following properties: 1) it is a nonessential gene that can be replaced with foreign DNA; 2) it has a very strong promoter; and 3) it provides a convenient phenotypic marker for selection of recombinant viruses.

Several strategies have been used in constructing AcMNPV expression vectors. Usually, the polyhedrin gene has been cloned and modified to interrupt gene expression and foreign DNA inserted. The foreign DNA can be inserted in frame to produce a fusion protein. Alternatively, the 3′ end of the open reading frame has been deleted and the foreign gene with its own initiation and termination condons inserted. Insect cells are then cotransfected with viral DNA and the plasmid containing the modified polyhedrin gene. Through recombination and allelic transplacement, the foreign gene is inserted into the viral DNA. The recombinant lacks a functional polyhedrin gene and therefore produces infectious viruses that produce foci lacking polyhedra. These are easily discernible by visual inspection from the polyhedra-containing wild-type infection foci.

These characteristics have made the baculovirus system very attractive to molecular biologists and biotechnologists.

ADVANTAGES AND LIMITATIONS FOR PROTEIN PRODUCTION

The baculovirus–insect cell system has quickly established itself as an important tool in biotechnology for the production of proteins (see Chapter 4) as well as a potential system for the production of safe, efficient insecticides (see Chapter 5). Luckow [6] lists nearly 200 proteins that have been successfully expressed in the baculovirus system. Luckow [6] states that

> one of the major advantages of these helper-independent invertebrate virus expression vectors over bacterial, yeast, and mammalian expression systems is the very abundant expression of recombinant proteins that are soluble and antigenically, immunogenically, and functionally similar to their authentic counterparts. Baculovirus-infected insect cells perform many of the post-translational modifications of higher eukaryotes including phosphorylation, glycosylation, correct signal peptide cleavage, proteolytic processing, palmitylation, and myristylation. In addition, the recombinant proteins are transported to their natural cellular location and can undergo oligomeric assembly, where appropriate, which makes this expression system a valuable tool for production of biologically active proteins.

Further advantages for the baculovirus–insect cell system reside in its relative safety. Baculoviruses are nonpathogenic to mammals, birds, fish, and reptiles. Insect cells can be subcultured continuously without being "transformed." In mammalian systems the host cell is often a transformed cell line that requires stringent purification procedures to ensure the removal of all nucleic acids, since it is feared that cancer-promoting substances could be transferred to recipients of the products [7]. Most mammalian cells contain endogenous viruses. Furthermore, the vectors used to engineer mammalian cells genetically are usually derived from primate viruses, and reversion to a pathogenic form is theoretically possible.

However, the baculovirus system, like any expression system, has disadvantages. Very high expression levels of active proteins can be obtained (100–800 mg/liter) in some cases (for 13 examples in which the expression level in the baculovirus system is 20–250 times greater than in mammalian cells, see Luckow [6]). However, secreted, glycosylated proteins often are expressed at much lower levels (1–10 mg/liter). These reduced expression levels are most likely related to rate limitations in the glycosylation and secretion pathways. These limitations are more severe late in the infection cycle [8]. In the case of the production of the papain precursor, it has been shown that low production levels are likely due to rate limitations in steps early in the secretery pathway and steps following N-glycosylation [9]. However, production could be improved fivefold with the honeybee melittin signal sequence in place of the natural plant signal sequence [9]. Another approach to improve low expression levels is to make a fusion protein with several N-terminal amino acids of the polyhedrin protein [10–13].

Careful analysis of glycoproteins from the few lines tested thus far has often shown differences with proteins in the natural hosts. However, in nearly all cases this alteration in glycosylation has had little, if any, effect on immunological properties or in vitro biological activity. No data are yet available on in vivo effects for therapeutic proteins made in insect cells. The glycosylation patterns in insect cells tend to be much less heterogeneous than the glycosylation patterns often observed in mammalian cells. Luckow [6] provides a specific example with a human erythropoietin.

Weiss et al. [14] point out that

> Chinese hamster ovary (CHO) cells, which are widely used for expression of foreign genes, do not always provide a solution when specific post-translational modifications are required for activity. For example, the vitamin K–dependent glycosylation of recombinant blood proteins does not occur in CHO cells. Considering this, CHO is probably not a desirable selection for production of blood or related proteins which require vitamin K–dependent mechanisms for activity. BEVS (baculovirus expression vector system) may provide a viable alternative vector system for expression of products where these types of modifications are required.

Slight alterations in glycosylation may influence antigenicity [15]. For example, for recombinant proteins intended to be used as vaccines, minor changes in structure may be desirable. Also, the high mannose forms of some proteins may be useful to target certain cells such as macrophages. The baculovirus expression system has been used to produce an HIV envelope protein, gp160, which gained federal approval for clinical trials "because its preparation stimulates a strong immune response in animals" [16]. In many cases viral proteins are produced at very high levels (100–500 mg/liter) [see refs. 5,6,17]. This high level of production greatly facilitates development of subunit vaccines.

The production of antibodies in insect cells has recently been reported [18]. The single chains were not secreted, but constituted 40% of Coomassie-stained protein. The glycosylated dual chain molecule was secreted at 25–30 mg/liter in an unoptimized system infected at 5×10^5 cells/ml. This level of production is equivalent to many hybridomas. The system allows production of both intact antibodies or biologically active antibody fragments in a nontumorigenic expression system.

While differences in post-translational processing with respect to mammalian cells may restrict the use of the baculovirus system for some proteins, the baculovirus system remains potentially very attractive for other proteins.

Another limitation on the baculovirus expression system is that it is lytic. The high expression levels are achieved as a burst. The economics of the process would be improved and perhaps less stress would be placed on the cell's post-translational machinery, if the period of active production could be extended. Jarvis et al [19] have obtained continuous expression and efficient processing of foreign gene products in stably transformed cells using an immediate early gene promoter (IE1). Although stable production is obtained, the level of β-galactosidase made decreased by more than 100-fold. However, with tissue plasminogen activator (TPA), the transformed cells secrete TPA more quickly with production levels slightly less than 50% of TPA production from the traditional baculovirus system.

Another difficulty in achieving the full potential of the baculovirus system is that specific productivity drops dramatically if cells at high density (above 2×10^6 cells/ml for SF9) are infected with virus [17,20]. However, the cells can grow to much higher densities ($1–2 \times 10^7$ cells/ml) [see refs. 20,21]. If productive infections at 6×10^6 cells/ml could be obtained, an approximate threefold increase in titers could be achieved, resulting in ≥ 1 g/liter concentrations for some protein products. The work of Caron et al. [17] demonstrates that medium changes could reverse this cell density effect. They conclude that "accumulation of some inhibitory metabolite contributes to the observed production decrease." However, other studies suggest that a type of "contact inhibition" may be present even in suspension cultures [22]. The dilution with fresh medium approach used by Caron et al. [17] does not differentiate between inhibition by a diffusable chemical species or a more intrinsic (cell-to-cell) type of inhibition.

Another effect of vital importance to expression of foreign proteins is the potential effect of multiplicity of infection (MOI). In many cases MOI has been reported to be of minor importance. However, King et al. [23] reported that 240 units/ml of choline acetyltransferase (CAT) were made at an MOI of 0.02 but only 7 units/ml at an MOI of 2. No explanation was available at the time, but recent observations corroborate the MOI effect and demonstrate that it is due to the formation of defective virus particles [see refs. 24,25], and this effect will occur with viruses subject to long periods of subculture.

In summary the baculovirus expression vector system shows much higher productivities for many proteins than mammalian cells show, can do many post-translational processing steps, and is inherently a safer system than transformed mammalian cells using modified mammalian virus vectors. Also, it is much easier to obtain active products more quickly in the baculovirus expression system than other expression systems, particularly CHO systems. For some protein products the baculovirus expression system will be potentially the system of choice. Its attractiveness would be further increased if the period of production could be extended, if productive infections at high cell densities could be achieved, and if more complete post-translational processing could be obtained through cell line selection, culture conditions, or genetic manipulation.

POTENTIAL AS AN INSECTICIDE

With the recent developments in serum-free medium formulations the commercial production of insecticides using tissue culture is potentially feasible if the virus can be "improved" through genetic engineering to increase speed of action, efficiency, and the number of species subject to its action. Much of the early work in insect cell culture was motivated by production of pesticides [see ref. 1]. Knowledge of the molecular biology and genetics of baculoviruses has accelerated greatly in recent years, and this database has made feasible the genetic improvement of these viruses. In particular, the introduction of new genes into the baculovirus genome, which could deliver some deleterious gene products in the insect cell larva, could greatly increase the speed of insect killing or baculovirus host range [3]. Recently the toxin gene of *Bacillus thuringiensis,* the juvenile hormone esterase gene of *Heliothis virescrens,* and the paralytic mite neurotoxin gene have been successfully cloned in the baculovirus system [26–28].

One primary concern with the use of a genetically altered baculovirus is whether regulatory approval can be obtained. No answer yet exists. However, the first project to allow the field release of a genetically engineered virus has been done with a genetically engineered baculovirus [3]. This work is beginning to build the framework for a commercially attractive product.

If the commercial production of an insecticide is undertaken, it will require very efficient, large-scale reactor systems.

ADVANTAGES OF IN VITRO PROPAGATION OF BACULOVIRUSES COMPARED WITH PRODUCTION IN INSECT LARVAE

There are various technical and biological advantages in using insect cell cultures instead of insect larvae for the production of baculovirus pesticides. Quality control procedures for certification of a virus product free of adventitious agents and/or contaminating insect debris are greatly facilitated by use of in vitro procedures. Insect cell lines can be readily tested to ascertain that they are free of contaminating viruses or other microbes, i.e., mycoplasma. Unlike sterile cultured cells, it is unpractical to rear thousands of living insects under absolute sterile conditions.

Insect cell lines that are free of adventitious agents can be easily maintained and stored in liquid nitrogen. During production, the growth conditions for cells and virus replication can be carefully controlled to yield an uncontaminated viral product. In addition to the presence of potential adventitious agents, insect-produced viruses may contain large amounts of larval hairs, chitin (cuticle), and proteins. Some of these contaminates, e.g. hairs, may be urtiatious and/or allergens that could be harmful to humans. Cell cultures would provide baculoviruses free of contaminating insect parts, and the harvesting of the occlusion body product would be relatively simple and inexpensive.

The efficiency of virus production in cultured cells is a major consideration. Cell lines with high susceptibility can be easily cloned, selected, and used to produce a virus product that exhibits high virulence and uniformity. These cell lines can be frozen and stored for long periods of time. In contrast, insect larval tissues are usually not uniformly susceptible and often do not support complete replication (e.g., intestine, Malpighian tubules, most hemocyte types, nerve tissue). Consequently, significant amounts of larval insect cellular mass may not contribute to the production of baculoviruses. Because of the diverse nature of susceptible tissues and/or organs with different physiological functions, the production of a synchronous, uniform viral product may not be possible in larvae.

Production facilities for cell culture-produced viral pesticides will depend on large-scale bioreactors that will not require special plant facilities, including quarantine areas for growth of uninfected cells. For insect-produced pesticides, unique facilities will be needed to rear hundreds of thousands of insect larvae (possibly more than one species, depending on the virus specificity). Special facilities will be required to separate insect rearing from virus infection areas. Special air filters, positive air-flow rooms, separation of personnel from virus and clean areas, and so forth, will all contribute to the high cost of production. Contamination of insect-rearing facilities will be a constant problem, since larvae are very sensitive to airborne baculovirus occlusion bodies. These viral occlusion bodies are not infectious to cultured insect cells and do not present a problem for in vitro production facilities.

SOME PROCESS AND SCALE-UP PROBLEMS

There are many articles that describe reactor systems or insect cell culture kinetics in bioreactors [e.g., refs. 14,17,29–33]. These are discussed in more detail in subsequent chapters.

Major problems in bioreactors and scale-up is the interaction of requirements for oxygen supply with the shear tolerance of insect cells. In particular, direct gas sparging and the resulting formation of gas bubbles is particularly problematic. Gas bubble size is a critical parameter. The bursting of a gas bubble results in tremendous shear forces on cells located at the bubble interface [34]. Chalmers discusses these effects in Chapter 9. The addition of agents such as methycellulose and Pluronic F-68 can greatly alter the shear tolerance of insect cells [35,36].

Another process problem relates to the efficient use of viral inocula. Strategies to obtain high rates of infection at the bench level often involve procedures (e.g., centrifugation and medium changes) that would be impracticable on a large scale. A more fundamental understanding of the kinetics of virus attachment, infection, and production is still needed. Such an understanding is not only important to batch processes, it is critical to the proper design of continuous culture processing strategies. So far the operation of continuous culture systems has proved problematic because of the formation of genetic variants of the virus that are of reduced productivity [37–39].

The lytic nature of the baculovirus system presents significant process limitations on both production and product recovery trains. Some progress has been made on nonlytic systems [19], and further advances can be anticipated. Appropriate bioreactors will need to be developed in conjunction with these developments.

Although insect cells can be cultured to fairly high cell densities ($1-2 \times 10^7$ cells/ml) [see ref. 20], the per cell productivity is greatly inhibited by high cell density ($> 2 \times 10^6$ cells/ml) [e.g., ref. 17]. The underlying cause of this problem must be understood, and a combination of appropriate medium development and reactor operation must be found to circumvent this limitation.

It is known that bioreactor operating strategies can alter glycosylation and other post-translational processing steps in mammalian cells [40]. Almost certainly the same will be true for insect cells.

Undoubtably other process and scale-up issues will appear as more experience is gained with the system. In this book we provide the background for the reader to understand more completely the nature of these problems and potential solutions.

REFERENCES

1. Granados, R.R., and Federici, B.A. The Biology of Baculoviruses. Volume 1. Biological Properties and Molecular Biology. CRC Press, Boca Raton, FL, 1986.
2. Blissard, G.W., and Rohrmann, G.F. Annu. Rev. Entomol. 35:127–155, 1990.
3. Wood, H.A., and Granados, R.R. Annu. Rev. Microbiol. 45:69–87, 1991.
4. Vlak, J.M., Klinkenberg, F.A., Zaal, K.J.M., Usmany, M., Klinge-Roode, E.C., Geervliet, J.B.F., Roosien, J., and van Lent, J.W.M. J. Gen. Virol. 69:765–776, 1988.
5. Vialard, J., Lalumiere, M., Vernet, T., Briedis, D., Alkhatib, G., Henning, D., Levin, D. and Richardson, C. J. Virol. 64:37–50, 1990.
6. Luckow, V.A. In: Recombinant DNA Technology and Applications (Ho, C., Prokop, A., and Bajpai, R., Eds.). McGraw-Hill, New York, pp. 97–149, 1990.
7. Ramabhadran, T.V. Trends Biotechnol. 5:175–178, 1987.
8. Jarvis, D.L., and Summers, M.D. Mol. Cell Biol. 9:214–223, 1989.
9. Vernet, T., Tessier, D.C., Richardson, C., Laliberte, F., Khouri, H.E., Bell, A.W., Storer, A.C., and Thomas, D.Y. J. Biol. Chem. 265:16661–16666, 1990.
10. Carbonell, L.F., Hodge, M.R., Tomalski, M.D., and Miller, L.K. Gene 73:409–418, 1988.
11. Lanford, R.E., Luckow, V., Kennedy, R.C., Dreesman, G.R., Notvall, L., and Summers, M.D. J. Virol. 63:1549–1557, 1989.
12. Luckow, V.A., and Summers, M.D. Virology 167:56–71, 1988.

13. Sekine, H., Fuse, A., Inaba, N., Takamizawa, H., and Simizu, B. Virology 170:92–98, 1989.
14. Weiss, S.A., Belisle, B.W., DeGiovanni, A., Godwin, G., Kohler, J., and Summers, M.D. Dev. Biol. Standard, 70:271–279, 1989.
15. Parekh, R.B., Dwek, R.A., Edge, C.J., and Rademacher, T.W. Trends Biotechnol. 7:117–122, 1989.
16. Barnes, D.M. Science 237:973, 1987.
17. Caron, A.W., Archambault, J., and Massie, B. Biotechnol. Bioeng. 36:1133–1140, 1990.
18. zu Putlitz, J., Kubasek, W.L., Duchene, M., Marget, M., von Specht, B.-U., and Domdey, H. Bio/Technology 8:651–654, 1990.
19. Jarvis, D.L., Fleming, J.G.W., Kovacs, G.R., Summers, M.D., and Guarino, L.A. Bio/Technology 8:950–955, 1990.
20. Weiss, S.A., Gorfien, S., Fike, R., DiSorbo, D., and Jayme, D. In: Biotechnology: The Science and the Business. Ninth Australian Biotechnology Conference, Gold Coast, Queensland, Australia (Sept. 24–27), Gold Coast, Queensland, Australia, pp. 220–231, 1990.
21. Vaughn, J.L. AIChE Annual Meeting (Nov. 11–16), Chicago, IL, 1990.
22. Wood, H.A., Johnston, L.B., and Burand, J.P. Virology 119:245–254, 1982.
23. King, G., Kuzio, J., Daugulis, A., Faulkner, P., Allen, B., Wu, J., and Goosen, M. Biotechnol. Bioeng. 38:1091–1099, 1991.
24. Kool, M., Voncken, J.W., van Lier, F.L.J., Tramper, J., and Vlak, J.M. Virology 183:739–746, 1991.
25. Wickham, T.J., Davis, T., Granados, R.R., Hammer, D.A., Shuler, M.L., and Wood, H.A. Biotechnol. Lett. 13:483–488, 1991.
26. Hammock, B.D., Bonning, B.C., Possee, R.D., Hanzlik, T.N., and Maeda, S. Nature 344:458–461, 1990.
27. Merryweather, A.T., Weyer, U., Harris, M.P.G., Hurst, M., Booth, T., and Possee, R.D. J. Gen. Virol. 71:1535–1544, 1990.
28. Tomalski, M.D., and Miller, L.K. Nature 352:82–85, 1991.
29. Maiorella, B., Inlow, D., Shauger, A., and Harano, D. Bio/Technology 6:1406–1410, 1988.
30. Wu, J., King, G., Daugulis, A.J., Faulkner, P., Bone, D.H., and Goosen, M.F.A. Appl. Microbiol. Biotechnol. 32:249–255, 1989.
31. Shuler, M.L., Cho, T., Wickham, T., Ogonah, O., Kool, M., Hammer, D.A., Granados, R.R., and Wood, H.A. Ann. N.Y. Acad. Sci. 589:399–422, 1990.
32. Stavroulakis, D.A., Kalogerakis, N., Behie, L.A., and Iatrou, K. Biotechnol. Bioeng. 38:116–126, 1991.
33. King, G.A., Daugulis, A.J., Faulkner, D., and Goosen, M.F.A. Biotechnol. Prog. 8:567–571, 1992.
34. Chalmers, J.J., and Bavarian, F. Biotechnol. Prog. 7:151–158, 1991.
35. Goldblum, S., Bae, Y.-K., Hink, W.F., and Chalmers, J. Biotechnol. Prog. 6:383–390, 1990.
36. Murhammer, D.W., and Goochee, C.F. Biotechnol. Prog. 6:391–397, 1990.
37. Kompier, R., Tramper, J., and Vlak, J.M. Biotechnol. Lett. 10:849–854, 1988.
38. de Gooijer, C.D., van Lier, F.L.J., van den End, E.J., Vlak, J.M., and Tramper, J. Appl. Microbiol. Biotechnol. 30:497–501, 1989.
39. van Lier, F.L.J., van den End, E.J., de Gooijer, C.D., Vlak, J.M., and Tramper, J. Appl. Microbiol. Biotechnol. 30:43–47, 1990.
40. Goochee, C.F., and Monica, T. Bio/Technology 87:421–427, 1990.

2

Insect Cell Culture Methods and Their Use in Virus Research

Robert R. Granados and Kevin A. McKenna

Boyce Thompson Institute for Plant Research, Cornell University, Ithaca, New York 14853

INTRODUCTION

Ever since the pioneering work of Goldschmidt [1] and Trager [2], insect cell culture has grown at an accelerated rate. With the many advancements that occurred during the 1960s and 1970s with regard to the establishment of new cell lines, and with the development of new media formulations in the 1980s, it was not surprising that insect cell culture research would attract considerable attention in the 1990s. The eloquent descriptions of the life cycles of insect-specific baculoviruses by Granados and Federici [3] and Doerfler and Bohm [4], which led to the inception of the baculovirus expression vector systems (BEVS) [5–8] decidedly advanced insect cell culture as a force in the field of biotechnology. As a viable alternative to mammalian cell culture and chemical pesticides as sources of pharmaceuticals and biological control agents, insect cell culture offers a natural base system from which to compete. The capacity of cultured insect cells to attain high densities and to produce copious amounts of virus and recombinant proteins makes this sytem attractive to both the researcher and the industrialist. This chapter addresses the utility of insect cell culture and the methods employed for use in baculovirus research.

INSECT CELL LINES

Sources of Cell Lines

The establishment of hundreds of invertebrate cell lines over the past 30 years has provided researchers with many choices in cell culture methods, baculovirus manip-

Baculovirus Expression Systems and Biopesticides, pages 13–39

ulation, and recombinant protein production options. Cell lines from over 94 species, covering 7 orders within Arthropoda, have been catalogued by Hink [9–13] and graphically portrayed by Lynn [14] (see Fig. 1). These cells originate from all stages of insect development, including eggs, embryos, larvae, and adults, and from specific tissues such as hemocytes, ovaries, imaginal discs, testes, midgut, and fat body.

There are various methods reported for the establishment of new insect cell lines some of which have been described in a collection of papers compiled by Mitsuhashi [15] and others [16–19]. Briefly, the tissue of choice is aseptically removed

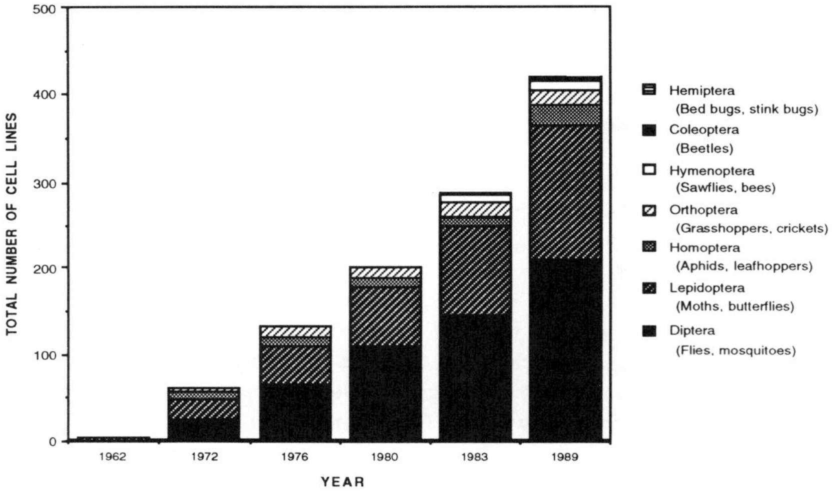

	1962	1972	1976	1980	1983	1989
Hemiptera	0	0	1	1	1	1
Coleoptera	0	0	0	0	0	3
Hymenoptera	0	0	0	0	10	11
Orthoptera	0	6	12	12	16	16
Homoptera	0	8	11	11	11	25
Lepidoptera	4	23	44	69	104	155
Diptera	0	25	65	109	146	209
Total	4	62	133	202	288	420

Fig. 1. Total number of insect cell lines as compiled by W.F. Hink [9–13] (Reproduced from D.E. Lynn [14], with permission of the publishers.]

from the insect and treated with an enzyme (e.g., trypsin) to separate the cells. Next, both the treated tissue and the cells are placed in a nutrient-rich medium. As nutrients are used up and toxic cell waste accumulates, the old medium is replaced with fresh. This procedure continues every 5–14 days until cell division is observed. When stable growth patterns are established, the cells can be maintained indefinitely by routine subculture.

The selection of a particular cell line based on the tissue of origin is a critical factor when determining research objectives. Generally, most cell lines are established from organs or whole embryos. It should be noted, however, that differences exist among cells derived from the same organ. Investigators have demonstrated that virus susceptibility varies greatly in some cell lines derived by identical methods [20–25]. For example, Granados et al. [25] demonstrated that some granulosis viruses (GV) will only infect cells of specific origin, whereas other viruses such as the *Autographa californica* multiple nuclear polyhedrosis virus (AcMNPV) will infect cells of different origin as well as those of different insects [26–28].

Continuous insect cell lines, that is, those cell lines that are beyond the primary culture stage, are maintained in many laboratories under stationary or suspension conditions in both serum-containing and serum-free media. Comprehensive lists of cells, their sources of origin, karyology, media type, and the investigator who established the line are available in the literature [12,13]. In addition, some cell lines are available through the American Type Culture Collection (12301 Parklawn Drive, Rockville, MD 20852-1776).

Growth Characteristics

Insect cells will grow in a variety of cell culture vessels. Some examples of vessel types include glass spinner or shaker flasks for suspension cultures and roller bottles or polystyrene culture flasks for anchorage-dependent cells. To maintain optimal growth, however, the cells must be subcultured at regular intervals, usually two to three times a week. This is accomplished by transferring a known concentration of cells (usually $0.2–1 \times 10^6$ cells/ml) to a new vessel containing fresh medium. Cells are counted either with a hemocytometer or electronically, as with a Coulter Counter (Coulter Electronics, Inc., Hialeah, FL). Attached cells, or anchorage-dependent cells, are dislodged for subculture by gentle agitation, pipetting, or scraping with rubber or Teflon scrapers or with proteolytic enzymes such as trypsin (0.25%), pronase (0.001%–0.01%), papain (5.0%), dispase (1 U/ml) [29], or versene preparations [30]. Although insect cells are not substrate dependent, they will grow attached to many surfaces, forming confluent monolayers, or suspended, as under bioreactor conditions for large-scale production [31–36]. Summers and Smith [37] describe several methods for the subculture of *Spodoptera frugiperda* (SF9) cells in both anchored and suspension cultures. Other publications are available that address maintenance of cell cultures and baculovirus infection procedures for a variety of lepidopteran cell lines [8,38–43].

For optimum cell growth, the physiochemical environment must be controlled. The parameters for lepidopteran cells, for example, include temperatures between

25° and 30°C, pH 6.2–6.6, and osmotic pressures of 300–380 mOsm [30,44]. The physiological conditions, which may be kept relatively constant, are more difficult to control because of the undefined nature of supplements in the various media. While some insect cells can tolerate broad ranges of growth conditions, other cells require more specific environments for optimal growth, as demonstrated by Sohi [45].

Following subculture, a lag phase occurs whereby cells re-adapt to fresh culture conditions, as indicated by little or no growth. This period can last from a few hours to several days, depending on the cell line. Usually, after a day or so, the cells begin to divide rapidly. This is the logarithmic growth phase, and it continues for 2–10 days. It is during this phase that the cells are the most physiologically active and are most suitable for cryopreservation or baculovirus infection studies. At approximately mid to late log phase, the cells are ready for subculture again. Typically, insect cell doubling times range from 16 to 48 hours and, depending on the culture method employed, can reach maximum cell densities of 10^6–10^7 cells/ml. Figure 2 depicts a typical insect cell growth curve for the *Trichoplusia ni*–derived BTI-TN5B1-4 cell line in TNM-FH medium.

Insect cells exhibit different growth characteristics in serum-free media. Although Hsu et al. [46] and Wilkie et al. [47] saw no gross morphological differences in cell lines adapted to serum-free media, Ogonah et al. [48] have reported cell size differences in SF9 cells in serum-containing medium (12 μm) versus serum-free

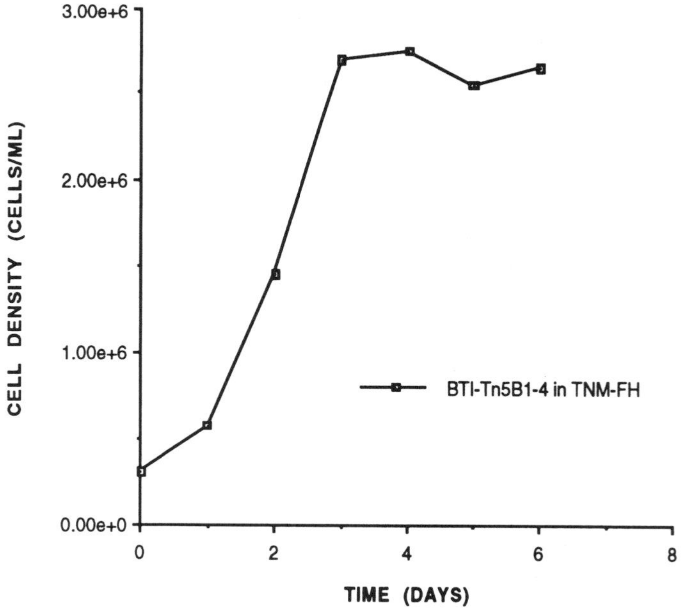

Fig. 2. A typical growth curve for the BTI-TN5B1-4 cell line in TNM-FH medium over a 6 day period.

medium (14 μm). We have also observed morphological differences in cells adapted to serum-free media; for example, a suspended cell line, BTI-EAA, became attached following adaptation to serum-free medium. Decreased cell growth rates have also been reported in cell lines adapted to serum-free media [46,49–52]. In contrast, many have found cell doubling times to be shorter or equal to those in serum-containing media [47,53–60].

Insect cells are more sensitive to lower temperatures in serum-free media [35]. This can be a problem during cryopreservation procedures. In addition, insect cells required lipid supplements in serum-free media in order to achieve maximum cell growth [67–69]. Other characteristics of insect cells in serum-free media include their sensitivity to antibiotics and the tendency for the cells to be easily damaged during manipulation [63].

Cryopreservation

Insect cells can be preserved under liquid nitrogen conditions for years and still retain character and viability. Keeping back-up stocks of cells in this way ensures continuation of research activities in the event cells are lost. The protocol chosen for cryopreservation must be considered when cells have been maintained in different types of media. A simple and effective method for freezing cells in serum-containing medium is described by Summers and Smith [37]. Weiss et al. [34] describe an alternative method for freezing cells optimized for roller bottle cell culture systems, but suitable for use in suspension and stationary systems also. As mentioned earlier, cells in serum-free media are easily damaged by freezing in liquid nitrogen due to their sensitivity to cold temperatures and increased fragility. Viability of recovered cells is decreased following cryopreseration in serum-free media compared with serum-containing media [633]. Cells attach stronger to vessel surfaces under serum-free conditions and are therefore more likely to be damaged when removed from a substrate. Consequently, great care must be taken when preparing for long-term storage of cells in serum-free media.

Goodwin [35] found that cells frozen in serum-free media without lipids could not be recovered, whereas the same cells frozen in serum-free media supplemented with lipids were successfully frozen and recovered. This result was supported by Vaughn and Fan [64] following successful recovery of cells frozen in serum-free media containing the lipid supplements CPSR-1 and CPSR-3 (Sigma, St. Louis, MO). In these experiments, they observed no differences in viability or morphology when compared with cells frozen in serum-containing media. A suggested method for preserving SF9 and other invertebrate cell lines in serum-free media is described by the manufacturers of serum-free SF900 II (GIBCO/BRL, Grand Island, NY) medium, which closely resembles that described by Weiss et al. [34], whereby cells are frozen in a mixture of 50% fresh medium, 50% cell-free conditioned medium, and 7.5% dimethylsulfoxide (DMSO).

It is important for the survival of the cells during long-term storage that they are frozen while in their optimum metabolic state, i.e., during mid-log phase, and at viabilities of > 90%. Typically, a known concentration of cells is pelleted by centrifugation at 1,000*g* for 10 minutes. The cells are then resuspended in a volume of freezing medium that consists of fresh and/or conditioned medium plus DMSO or glycerol at final concentrations of 5%–10%. Next, the cell suspension is dispensed into appropriately labeled cryovials and slowly cooled at 1°C/min or placed at −20°C for 1 hour. The frozen cells are then incubated at −70 to −90°C for 24 hours before being transferred to liquid nitrogen for indefinite storage.

Cells are recovered from liquid nitrogen by rapidly thawing cryovials in a 37°C water bath followed by dilution of the preservative (DMSO or glycerol) with fresh medium at a ratio of 1:5. The cells are allowed to re-adapt to either suspension or stationary conditions before subculture is attempted. Cell viability following cryopreservation will vary depending on the particular cell line, the type of medium used, and the procedures that are followed. As a general rule, all cell lines should be frozen at regular intervals to preserve their integrity and ensure a continuous stock.

Characterization

Traditionally, insect cell lines have been characterized based on morphology, karyology, serology, and isozyme analyses. However, not all insect cell lines can be characterized by all of these methods. For instance, cell lines from the orders Diptera, Hymenoptera, and Coleoptera are commonly characterized by karyotyping [65–70]. Species of Lepidoptera, on the other hand, are not as easily characterized by this method [71–73]. Lepidopteran cells, during metaphase, exhibit many microchromosomes, making it difficult to determine differences between cell lines. Cell morphology, as a means of characterizing insect cells, is not recommended, since changes can occur in the same cell line under different culture conditions and in different media [74]. In addition, many cell lines are morphologically heterologous. Serological methods, which rely on cell surface markers, are also limited in their ability to characterize broad ranges of insect cell lines [75–77]. These methods have inherent disadvantages due to variation in reagent preparation, lack of reference sera, and the production and storage of the antisera.

Isozymes, which are two or more forms of a single enzyme, have proven to be the most specific method for characterizing insect cell lines [70,75,78–84]. This relatively inexpensive and easy technique enables investigators to set up and run samples quickly. The resulting enzyme profile, or zymogram, identifies cell lines with high specificity [85]. For example, Mitsuhashi [86] demonstrated a simple and rapid method for analyzing as many as 19 isozymes from 19 different cell lines. This system, called APIZYM, was used to characterize different cell lines based on variable activity of the same enzyme between them. Harvey and Sohi [80], using a series of enzyme combinations, have demonstrated that cell lines from the same species can be differentiated. Using isoelectric focusing, Maskos and Miltenburger [83] reported success in distinguishing between different genera. Likewise, McIntosh and Ignoffo [84], using a single enzyme and the same method, were able to differentiate between five cell lines from *Heliothis zea* but not between two cell lines from *Heliothis virescens*.

One disadvantage to isozyme analysis, however, is that not all enzymes will identify different cell lines from the same species. For this reason, one cannot depend solely on isozyme analysis; it must be used in conjunction with other characterization methods. It is important to identify specific markers for all new cell lines, as well as existing ones, so that accidental cross-contamination is avoided and protection of proprietary cell lines can be ensured.

Adventitious Agents

Contamination of insect cells by adventitious microorganisms can be a serious problem in any lab that maintains several different cell lines. These agents are introduced into cultures from a variety of sources, the most notable being vertebrate sera [87–90]. The most commonly used vertebrate sera, in insect cell culture, is fetal bovine serum (FBS). Several investigators have demonstrated the presence of mycoplasmas, animal viruses, and bacteriophages in animal sera [91–93]. In addition, Price and Gregory [94] were able to show variability from lot to lot of FBS when analyzed for biochemical and growth-promoting properties. Viruses, such as porcine viruses, can also be introduced by trypsin, a compound used during the establishment of primary cell cultures and during subculture [92,95–99]. Vaughn [100] suggests that viral contaminants from these sources are rarely persistent in insect cell cultures, since incubation temperatures of 26°–28°C are generally lower than optimum for these vertebrate viruses.

Mycoplasmal contamination is also of great concern to cell culturists. These pleomorphic organisms, which belong to the order Mycoplasmatales, are small prokaryotic cells with no cell walls that vary in size from 10 to 100 μm. Mycoplasma and mycoplasma-like organisms, such as *Acholeplasma,* have been observed in cell lines maintained in medium containing FBS [88]. Surveys performed on cell lines maintained in different laboratories around the country have revealed alarming rates of contamination caused by these organisms [88,101]. It is often difficult to detect a mycoplasma contamination problem, since the cytopathogenic effect (CPE) is low. Again, this appears to be due to differences in temperature optimums, i.e., 25°C for insect cells, 30°C for mycoplasma.

Some microorganisms are transmitted directly to cell cultures from the insects themselves. These inherent microorganisms, such as the *Drosophila* X virus and the picornaviruses, are passed on from one generation to the next on the surface of the eggs. Unless the eggs are thoroughly cleaned with sodium hypochloride, contamination of newly established cell lines through primary explants will result [100,102,103]. Still other contaminants arise from organisms that are symbiotic with the host. These "rickettsia-like" organisms, which are also pleomorphic but have cell walls, have been detected by electron microscopy [104–111]. Interestingly, very few persistent infections of lepidopteran cell lines have been reported. Granados et al. [112], however, have reported a persistent baculovirus-like infection of unknown origin in a *H. zea* cell line, IMC-HZ-1. Here, the cells continued to grow for some time despite the presence of the virus.

It is of critical importance, then, that continuous cell cultures be monitored routinely for the presence of adventitious agents. For an excellent guide to good quality control, in particular, sera analysis prior to use, the reader is referred to a publication by Weiss and Vaughn [113].

CELL LINES FOR VIRUS RESEARCH

Numerous insect-associated viruses have been grown in insect cell cultures. The major groups of insect viruses include the arboviruses, phytoarboviruses, and insect-pathogenic viruses. Earlier reviews of the growth of these virus types in invertebrate cell lines are recommended to the reader [114–116]. Considerable interest is currently directed to the study of insect viruses that have potential for development as viral pesticides in agriculture or for viruses that may serve as vectors for expression of recombinant proteins. These virus types include the cytoplasmic polyhedrosis viruses (CPV), entomopoxviruses (EPV), and baculoviruses. Current advances in the study of CPVs and EPVs have recently been reviewed [117,118]. Only the viruses belonging to the family Baculoviridae will be considered in this chapter. Two genera, the nuclear polyhedrosis viruses (NPV) and the granulosis viruses (GV), have been assigned to the subfamily Eubaculovirinae [119]. Two subgenera have been established within the NPV genus, one consisting of viruses with multiple nucleocapsids per envelope (MNPV) and the other comprising viruses with a single nucleocapsid per envelope (SNPV). The nonoccluded baculoviruses (NOB) have been assigned to the subfamily Nudibaculovirinae.

In Vitro Host Range of Baculoviruses

MNPVs. The most widely studied baculovirus of this genus is AcMNPV. In addition to its value for use as a recombinant vector for expression of foreign genes, it is currently being considered for registration as a viral pesticide for agricultural purposes. This virus has the largest in vivo host range of any of the known NPVs [120,121] and infects over 35 insect species. Similarly, AcMNPV has been grown in numerous insect cell lines representing species from over six families. A cell line derived from *T. ni,* TN368, was used for many of the early studies with AcMNPV [122]. However, during the past decade, studies on the molecular biology of this virus and its use for foreign gene expression have primarily utilized the *S. frugiperda* cell line SF21 and its clonal variant SF9 [5,123]. Several studies have demonstrated that existing or novel insect cell lines could differ in their ability to replicate AcMNPV and/or in the production of recombinant proteins. In general, the studies identified cell lines that were superior to SF9 and SF21 [51,60,124,125]. In particular, one cell line established in our laboratory (Granados, unpublished data) from *T. ni* eggs and designated as BTI-TN5B1-4 was found to be superior to SF9 and SF21 in expression of recombinant proteins in stationary growth conditions [125–127]. The TN5B1-4 cells grow in serum-containing (TNM-FH) or serum-free media (EX-CELL 400 and SF900) and are highly susceptible to AcMNPV (Fig. 3). These studies have demonstrated that future development of novel cell lines from different insect species may result in high-yielding cell lines that can be grown as anchorage or suspension cultures in different types of media.

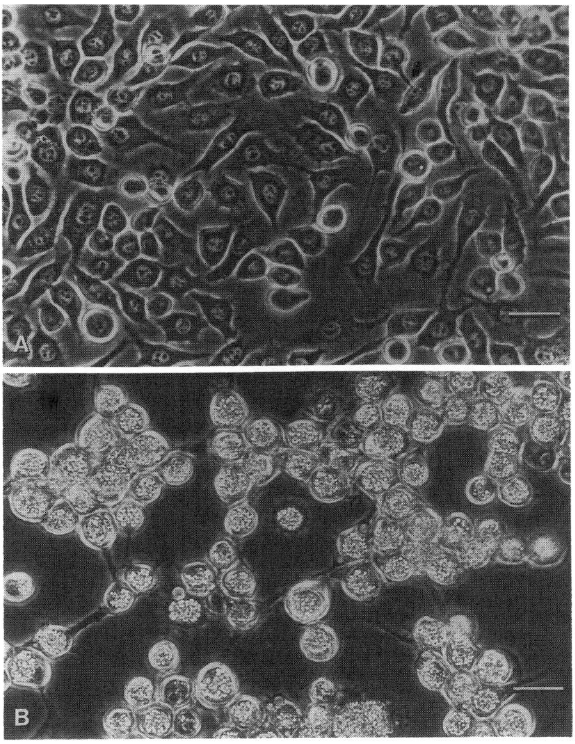

Fig. 3. Healthy BTI-TN5B1-4 cells in TNM-FH medium (A). BTI-TN5B1-4 cells at 48 hours postinoculation with AcMNPV (B). Cell nuclei are packed with occlusion bodies. Bar = 25 μm.

Recently, a newly discovered MNPV from the celery looper *Syngrapha falcifera* (SyfMNPV) was reported to have a wide in vivo host range, including 30 species from 10 families in the order Lepidoptera. In addition, eight cell lines from different insect species were susceptible to SyfMNPV infection [128]. This report suggested that SyfMNPV might have a wide in vitro host spectrum similar to that of AcMNPV. Some MNPVs have a very restricted in vitro host range. The gypsy moth *Lymantria dispar* MNPV (LdMNPV), which has a restricted host range, is primarily a pathogen of *Lymantria dispar* larvae and is only known to replicate in cell cultures from that species. The viral and/or host factors that determine the in vitro host range of baculoviruses are not known, but several mechanisms that may govern host range have been suggested (see below).

Other MNPVs that readily grow in cell cultures include *Orgyia pseudotsugata* MNPV (OpMNPV), *Anticarsia gemmatalis* MNPV (AgMNPV), and several *Spodoptera* spp. MNPVs. A partial listing of MNPVs that have been cultivated in different insect cell lines is presented in Table 1.

TABLE 1. Common Insect Cell Lines Susceptible to Multinucleocapsid Nuclear Polyhedrosis Viruses

Baculoviruses family/species[a]	Susceptible cell line designation and/or species of origin
Saturnidae	
Antheraea pernyi NPV	*Lymantria dispar*: Lymantridae
Noctuidae	
Autographa californica NPV	BTI-EAA
	Estigmene acrea: Arctidae
	IZD-MB503
	Mamestra brassicae: Noctuidae
	Manduca sexta: Sphingidae
	IPLB-SF1254
	Spodoptera littoralis: Noctuidae
	TN368, IPLB-TN-R, BTI-TN5B1-4
	Trichoplusia ni: Noctuidae
	UIV-SL-573, SPC-S1-48, SPC-S1-52
	Spodoptera littoralis: Noctuidae
	NIAS-LeSe-11
	Leucania separata: Noctuidae
	IPRI-MD-108
	Malacosoma disstria: Lasiocampidae
	IPLB-LD64BA
	Lymantria dispar: Lymantridae
	UCR-SE-1
	Spodoptera exigua: Noctuidae
Spodoptera frugiperda NPV	IPLB-SF21
	Spodoptera frudiperda: noctuidae
Spodoptera exempta NPV	UIV-SL-573
	Spodoptera littoralis: Noctuidae
	IPLB-SF21
	Spodoptera frugiperda: Noctuidae
Spodoptera littoralis NPV	IPLB-SF21
	Spodoptera frugiperda: Noctuidae
	UIV-SL-573, SPC-S1-52
	Spodoptera littoralis: Noctuidae
Spodoptera exigua NPV	UCR-SE-1
	Spodoptera exigua: Noctuidae
	IPLB-SF21
	Spodoptera frugiperda: Noctuidae
Trichoplusia ni NPV	IPLB-SF21
	Spodoptera frugiperda: Noctuidae
	TN368
	Trichoplusia ni: Noctuidae
Anticarsia gemmatalis NPV	TN368
	Trichoplusia ni: Noctuidae
	Hz-AM
	Heliothis zea: Noctuidae

TABLE 1. (*Continued*)

Baculoviruses family/species[a]	Susceptible cell line designation and/or species of origin
	IPLB-SF *Spodoptera frugiperda*: Noctuidae
	IAL-PiD *Plodis interpunctella*: Pyralidae
Xestia c-nigrum NPV	IPLB-SF21 AEII *Spodoptera littoralis*: Noctuidae
	CLS-79 *Spodoptera littoralis*: Noctuidae
Bombycidae	
Bombyx mori NPV	BM-5 *Bombyx mori*: Bombycidae
Lymantridae	
Orgyia pseudotsugata NPV	IPRI-01-12 *Orgyia leucostigma*: Lymantridae
Pyralidae	
Galleria mellonella NPV	TN368 *Trichoplusia ni*: Noctuidae
Choristoneura fumiferana NPV	IPRI-CF-124 *Choristoneura fumiferana*: Tortricidae
	IPRI-108 *Malacosoma disstria*: Lasiocampidae
Choristoneura murinana NPV	IZD-Cp58 *Cydia pomonella*: Olethreutidae
Geometridae	
Lambdina Fiscellaria somniaria NPV	IPRI-66, IPRI-108

[a]Baculoviruses nomenclature based on insect species where originally described. Data from Granados and Hashimoto [125].

SNPVs. Prior to 1984, the *Helicoverpa* (*Heliothis*) *zea* SNPV (HzSNPV) was the only SNPV to have been grown in an established cell line [21,130]. This status incorrectly led many virologists to believe that SNPVs were more difficult to grow in cell culture than MNPVs. At least three other SNPVs, including *Heliothis armigera* SNPV (HaSNPV), *Orgyia leucostigma* SNPV (OlSNPV), and *T. ni* SNPV (TnSNPV), have now been grown in cell culture (Table 2), and this suggests that this genus of baculoviruses can be readily grown in vitro. The HzSNPV has been the most widely studied virus of this genus, and several cell lines are available for virus replication. The first report of HzSNPV replication in vitro utilized the *H. zea* cell lines IPLB 1075, originally established by Goodwin et al. [131]. Although this cell line could support the growth of HzSNPV, most investigators reported some difficulty in routinely obtaining productive infections [132]. Cell line infection rates of

TABLE 2. In Vitro Host Range of Single-Nucleocapsid Nuclear Polyhedrosis Viruses

Baculoviruses family/species[a]	Susceptible cell line designation and/or species of origin
Noctuidae	
Heliothis zea NPV	IMC-HZ-1
	Heliothis zea: Noctuidae
	BCIRL-HV-AMI
	Heliothis virescens: Noctuidae
	BCIRL-HZ-AM1, BCIRL-HZ-AM2, BCIRL-HZ-AM3
	Heliothis zea: Noctuidae
Heliothis zea NPV	SIE-HA, SIE-HAH
	Heliothis armigera: Noctuidae
Orgyia leucostigma NPV	IPRI-01
	Orygia leucostigma: Lymantridae
Trichoplusia ni NPV	BTI-TN5B1-4
	Trichoplusia ni: Noctuidae

[a]Data from Granados and Hashimoto [129].

50%–90% can be variable, and the low infection rates have limited the use of this cell line for viral studies. Clonal cell strains from IPLB 1075 were isolated that varied in their abilities to support replication of HzSNPV [133]. Recently, new cell lines and/or cell clones were established [134] that exhibited good growth characteristics and improved susceptibility to HzSNPV. Several *Heliothis* cell lines are now available that may be used to study the replicative cycle of HzSNPV in greater detail.

Bombyx mori NPV (BmNPV) is considered the type species for the SNPV subgenus [119] and can be grown in *B. mori* cell lines [134]. However, in 1974, Raghow and Grace [135] reported that a *B. mori* NPV of the MNPV morphotype could be cultivated in a *B. mori* cell line. Furthermore, there are conflicting ultrastructure reports claiming that the *B. mori* NPV is an SNPV [136] and an MNPV [137]. The status and classification of the BmNPV remains to be determined.

GVs. Prior to 1984, attempts to grow GVs in primary organ cultures or established cell lines had met with minimal or no success [138]. In larvae, GVs replicate mainly in the midgut and fat body tissues. Most established insect cell lines are derived from hemocytes, larval homogenates, and ovarian cells and thus may not be appropriate cell substrates for GV growth, i.e., they lack viral receptors or host enzymes required for replication. Lepidopteran cell lines from midgut tissue have not been reported, and recently established cell lines from fat body have not been tested for GV susceptibility. In 1984, Miltenburger and coworkers [18,19] in Germany reported the first successful in vitro replication of *Cydia pomonella* (codling moth) GV (CpGV) in cell lines derived from *C. pomonella* eggs. From a total of 200 established *C. pomonella* cell lines, 81 were screened for CpGV growth and 9 were found susceptible. Cell line IZD-Cp-3300 was the most susceptible cell line, but only 20% of cells were infected at 8 days postinoculation. These *C. pomonella* cell lines have not been developed as efficient, reproducible in vitro systems for GV studies.

Granados et al. [25] established 36 new cell lines from *T. ni* eggs using the procedure reported by Miltenburger et al. [18] and were able to demonstrate infection of 15 of these cell lines by *T. ni* GV (TnGV). The percentage of infected cells in these 15 cell lines ranged from 1% to 50%. However, susceptibility was not stable; TnGV infection of these lines either decreased or was lost within 20–25 passages from the initial primary culture. Two important observations from these studies are 1) none of the floating cells were susceptible to TnGV, which confirms a similar finding reported by Miltenburger et al. [18]; and 2) these new cell lines, which represent many different embryonic cell types, may change their viral susceptibility properties upon passage. Loss of TnGV susceptibility appeared to correlate with faster cell growth. Indeed, the four susceptible *T. ni* cell lines had a slower rate of growth. The experimental approach developed by Miltenburger et al. [18] and Naser et al. [19], and confirmed by Granados et al. [25], suggests that new cell lines from different insect species may be developed for the growth of other GVs. In a similar study, Dwyer et al. [17] established 250 embryonic cell lines from *Pieris rapae* (imported cabbageworm), and 150 were tested for susceptibility to *P. rapae* GV (PrGV). They found that several primary cell lines contained susceptible cells that were subsequently lost after further subculturing.

Winstanley and Crook [139] have reinvestigated the development of new *C. pomonella* cell lines for the cultivation of CpGV. They were successful in obtaining susceptible cell lines from a group of 16 cell lines that were selected for virus susceptibility testing. In general, the cells grow slowly, with doubling times of 48 hours, and exhibit prolonged virus infection times of 8–28 days. Occlusion bodies (OB) were seen at 12 days postinoculation, and the proportion of OB-containing cells increased to about 60% by 17 days. These new cell lines have maintained their susceptibility for several years. This would suggest that they can be developed into reliable in vitro systems for the study of GVs.

NOBs. Nonoccluded baculoviruses have been reported from a variety of arthropod hosts, including insects, mites, and crustacea [140]. Only two NOBs (*Oryctes* and HZ-1) have been grown in insect cell culture (Table 3). *Oryctes* baculovirus was first reported to grow in primary cell cultures derived from the beetle *O. rhinocerus* [141]. Subsequently, the *Oryctes* virus was grown in a susceptible established cell line from the beetle, *Heteronychus arator;* and in vitro studies on *Oryctes* virus morphogenesis, infectivity of cell associated and budded virus, and synthesis of viral proteins were reported [142].

The HZ-1 NOB is a virus that persistently infects the *H. zea* IMC-HZ-1 cell line, but is also able to replicate in several other lepidopteran cell lines [112]. Studies

TABLE 3. In Vitro Range of Nonoccluded Baculoviruss

Baculoviruses family/species[a]	Susceptible cell line designation and/or species of origin
Noctuidae	
Heliothis zea (Hz-1)	IM-HZ-1
	Heliothis zea: Noctuidae
	TN368
	Trichoplusia ni: Noctuidae
	IPLB-652
	Lymantria dispar: Lymantridae
	IPLB-1075
	Spodoptera frugiperda: Noctuidae
	Mamestra brassicae: Noctuidae
Scarabaeidae	
Oryctes rhinocerus	DSIR-HA-1179
	Heteronychus arator: Scarabaeidae

[a]Data from Granados and Hashimoto [129].

on the molecular events of HZ-1 replication in cell cultures suggest that this virus–cell culture system may be ideal for the study of the molecular mechanisms of baculovirus persistence in vitro [143].

Replication of Baculoviruses

The budded virus (BV) phenotype from insect hemolymph or infected cell cultures is highly infectious to tissues in the hemolymph and in cell culture. In the preliminary phase, infection of cultured cells begins with the adsorption of BV to receptor(s) on the cell plasma membrane and entry into cells by adsorptive endocytosis [144]. The viral envelope is lost upon release of the nucleocapsid from the endosome, and the uncoated nucleocapsids traverse the cytoplasm and either enter the nucleus or discharge the DNA at the nuclear pores [145]. The process of entry and uncoating of the genome takes between 1 and 2 hours. This is followed by an eclipse period of 6–8 hours during which a densely stained virogenic stroma forms in the central area of the nucleus. Transcription and DNA replication occur during this time, and unenveloped nucleocapsids are seen in and around the virogenic stroma within the first 10 hours postinfection. During this early phase of infection, progeny nucleocapsids "bud" through the nuclear envelope, traverse the cytosol, and progeny BV are released by budding at the plasma membrane. The titer of BV normally increases until 36–48 hours after infection (Fig. 4).

In the second phase of replication, the budding of virus at the plasma membrane appears to be reduced [146], and de novo membrane proliferation occurs within the nucleus. Within 18 hours postinfection, nucleocapsids are enveloped by these de novo membranes and then become occluded within the matrix of proteinaceous OBs. Lysis of infected cells normally occurs between 48 and 72 hours postinfection.

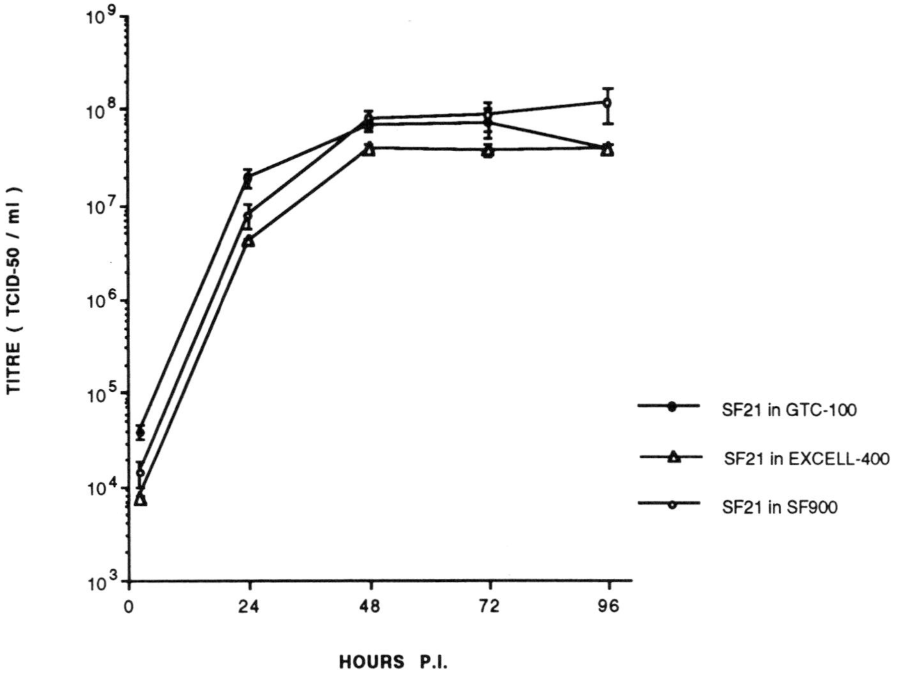

Fig. 4. Growth curve of wild-type AcMNPV budded virus in IPLB-SF21 cells in serum-containing (GTC-100) and serum-free (EX-CELL 400 and SF900) media.

Inoculation of cell cultures is normally carried out with highly infectious BV but occluded virus (OV) purified from OBs can also be used, although infection rates are greatly reduced. The BV and OV phenotypes have distinct morphological and biological characteristics [147], and these differences are related to the nature of the viral envelope. Lepidopteran cell lines can also be readily transfected with baculoviral and plasmid DNAs, and several transfection protocols have been developed [4]. Titration of baculovirus infections in cell culture is usually carried out by plaque assay using an agarose overlay or by the dilution end point assay ($TCID_{50}$) [115]. These titration assays are based on the production of OBs by the infected cells or by their lysis. An alternate immunological assay, which does not depend on OB production or cell lysis, was developed by Volkman and Goldsmith [24].

Studies on the host specificity of baculoviruses suggest that, in general, these viruses have a narrow host range in vivo and in vitro. However, there are a few examples (e.g., AcMNPV and SyfMNPV; see above) of baculoviruses with a moderately broad host range. Very little, however, is known about the factors that control the host specificity of baculoviruses. In one of the first studies to examine the nature of baculovirus specificity [148], it was reported that a recombinant AcMNPV was able to express a bacterial gene (chloramphenical acetyltransferase) in both permissive and nonpermissive insect cells. Although complete replication was not observed, this work demonstrated the ability of nonpermissive insect cells to allow the expression of foreign genes encoded in a recombinant baculovirus. Similar results were also reported using wild-type baculoviruses in semipermissive insect cell cultures [149]. Recently, Kondo and Maeda [150] reported on the in vitro and in vivo host range expansion by recombination of genomes of BmNPV and AcMNPV. They identified a "helper function" mechanism of AcMNPV for BmNPV replication in nonpermissive cells. Four recombinant baculoviruses were isolated that exhibited DNA restriction patterns of both BmNPV and AcMNPV, and these recombinant viruses showed an expanded in vitro host range compared with the parental isolates. These studies presented the first data to indicate that certain regions in the viral genome are responsible for determination of host specificity. There are numerous examples from almost all animal groups of persistent infections in cell culture. There are at least three examples of baculoviruses that may persistently infect insect cell cultures. The HZ-1-NOB was the first baculovirus demonstrated to persistently infect a cell line (IMC-HZ-1) [112]. Subsequently, both the SfMNPV and AcMNPV were established as persistent infections in IPLB-SF21 cells [151]. The establishment of persistent infections appears to be a complex interaction between the virus and host cells. For the HZ-1-NOB, Burand and Wood [152] demonstrated that the generation of defective interfering particles (DIPs) were required for the establishment of persistent infections. The generation of DIPs upon passage of baculoviruses can have an important impact on scale-up considerations.

Modification of Baculoviruses Passaged In Vitro

FP/MP Mutants. Serial passage of baculoviruses in cultured insect cells may result in the accumulation of spontaneous FP (few polyhedra) mutants [153]. The mutant FP phenotype is characterized by an increase of infectious extracellular BV and a decrease in the formation of OBs. Due to the production of few OBs, FP mutants are readily distinguished from the wild-type virus (MP, many polyhedral occlusion bodies) by plaque assay in insect cell cultures. This production of FP mutants can also be achieved by serial infection passage of BV in insect larvae; however, these mutants are not favored during serial per os feeding of OBs to susceptible larvae.

Analysis of FP mutants have shown the insertion of host cell DNA in the virus genome at a specific FP locus [154]. The BVs from FP mutants are highly infectious in cell culture, and, for all practical purposes, they appear to serve as excellent vectors for recombinant proteins. The stability of foreign genes in baculoviruses has not been systematically studied; therefore, it is recommended that any potential negative effect on gene expression can be minimized by limiting the number of serial passages [5].

Defective Interfering Particles. In 1983, Burand et al. [155] showed that defective virus particles of HZ-1-NOB could be generated following serial passage in cell culture and were required for the establishment of persistent infections. The defective particles were shown to interfere with infection and replication of standard virus particles. Therefore, it was not surprising that DIPs were identified in AcMNPV following continuous serial undiluted passaging of the virus [156,157]. The DIPs interfered with recombinant β-galactosidase and BV production in several insect cell lines. As predicted, these particles were smaller in size than the standard virus, contained large deletions in the genome, and required a helper virus for replication. It was reported that high MOI favored the selection of DIPs and that low MOI infections would decrease the percentage of DIPs in the infected cell cultures. Clearly, the occurrence of DIPs will affect the use of recombinant baculoviruses as expression vectors (see Chapter 7) and their use for viral pesticide production. Novel batch reactor system configurations must be designed to circumvent or minimize the impact of DIPs on gene expression.

MEDIA FOR CELL CULTURE

Serum-Containing

All insect cell media contain a basal element that was originally identified from insect hemolymph by Wyatt [158] and later modified by Grace [159] for *Antherea eucalypti* cultures. The basic nutrients include salts, vitamins, amino acids, sugars, organic acids, and trace elements. For years, this basic formulation was supplemented with insect hemolymph. As large quantities of insect hemolymph became increasingly more difficult to obtain, alternative sources with growth-promoting properties were sought. Vertebrate sera, in particular FBS, quickly replaced insect hemolymph [160–162]. FBS, although chemically undefined, is known to have growth-promoting properties; however, the actual compounds responsible have not yet been identified [163–165]. In addition to growth-promoting properties, FBS can enhance adherence of cells to flask walls [166] and protect cells from lysis [167].

The most commonly used serum-containing insect media are Grace's [159], IPL-41 [168]; TC100 [8,56,169]; and TNM-FH [122]. These types of media are usually supplemented with FBS at concentrations ranging from 5% to 20%. Serum-containing medium, or *complete medium* as it is often called, is usually supplemented with other undefined additives such as yeast extract and protein hydrolysates. Grace's medium, supplemented with yeastolate (3 g/liter), a primary source of vitamins and purines, and lactalbumin hydrolysate (3 g/liter), an additional source of amino acids, comprise TNM-FH medium. TNM-FH was originally formulated to establish primary cultures of *T. ni* cells [122]. Serum-containing media are routinely used in labs to maintain stock cultures of insect cells and for the propagation of baculoviruses [10,11,41,44,170]. For example, Broussard and Summers [171] reported that, by reducing the amount of FBS in TNM-FH from 10% to 0.5% following infection with either a recombinant or wild-type AcMNPV, the yields of polyhedra and recombinant protein were increased by 50% to 300%. Quelle et al. [172] also reported higher levels of erythropoietin in infected cells in TNM-FH containing 4% FBS and 20% lower concentrations of yeastolate and lactalbumin hydrolysate.

There are disadvantages however, associated with serum-containing media. The composition of serum, for instance, varies from batch to batch. This variability can alter experimental results with cell lines that are particularly sensitive to some serum component(s). Another consideration is the possibility of contamination from the serum. As discussed earlier, mycoplasma and other adventitious agents have been introduced into cultures from vertebrate sera [87–89,101]. Vertebrate sera, especially FBS, is very expensive and sometimes limited in availability. These situations can become major limiting factors if large-scale insect cell production is the goal. Finally, downstream processing of protein products from cells grown in serum-containing medium becomes exceedingly difficult due to the high protein content of FBS.

Serum-Free

Along with advances in insect cell culture technology, the need for the development of defined, more cost-effective media formulations became apparent. Following the breakthrough from hemolymph to vertebrate sera as the primary nutrient source for the cells, attempts were made to develop the first serum-free medium. Landureau and Jolles [173], Hsu et al. [46], Kitamura et al. [174], and Goodwin et al. [175] all had preliminary successes in adapting insect cells to serum-free media. Gardiner and Stockdale [56] questioned whether FBS was essential for cell growth. They began adjusting and/or omitting various vertebrate supplements, including FBS, and found that they could support the growth of a variety of cell lines over prolonged periods. Vail et al. [176] reported passaging *T. ni* (TN368) cells a minimum of 44 times in their serum-free medium. They accomplished this by omitting whole egg ultrafiltrate, crystalline bovine albumin, and FBS from TNM-FH medium. Hink et al. [50] also succeeded in subculturing TN368 cells 185 times in a serum-free medium containing Grace's basal components, yeastolate, lactalbumin hydrolysate, and bactotryptose.

The progress made in formulating serum-free media and the development of completely defined media such as those by Becker and Landureau [177] and Wilkie et al. [47] have enabled investigators to look closely at the actual nutritional requirements of insect cells in culture. By passaging cells from the German cockroach *Blatella germanica* for over 2 years in their serum-free medium formulation, Brooks et al. [53] and Brooks and Tsang [178] were able to determine that cholesterol was a vital component in membrane and ecdysone synthesis. Goodwin and Adams [179] were able to show that glycerol, glutamine, and folic acid played important roles in the survival of their *H. zea* and *L. dispar* cell lines. Mitsuhashi [180] adapted four lepidopteran and two dipteran cell lines to his serum-free version of Mitsuhashi and Maramorosch (MM) [161] medium. Cells were cultured indefinitely in this protein- and lipid-free medium.

Röder [55] maintained *S. frugiperda* (SF), *S. littoralis* (Sl), and *M. brassicae* (Mb) cells in a modification of Gardiner and Stockdale's BML-TC/10 medium [56] after substituting 1% egg yolk emulsion for FBS. More recently Léry and Fédèire [52] described a serum-free medium based on Grace's formulation, whereby an egg yolk emulsion at concentrations ranging from 0.01% to 0.05% was used in place of

FBS. In addition, calcium chloride, maltose, and β-alanine were omitted; and polyamines, trace metals, and higher concentrations of vitamins were added to support the growth of four lepidopteran cell lines.

Serum-free medium is available commercially, albeit mostly in liquid form. The specific formulations are considered proprietary; however, these media will support the growth of many insect cell lines from several different orders. Some examples are SF900 (GIBCO/BRL, Grand Island, NY), EX-CELL 400 (JRH Biosciences, Lenexa, KS), and HyQ (HyClone, Logan, UT).

Since the cost of producing large quantities of cells and virus is reduced with serum-free media, interest in scale-up systems has increased. Maiorella et al. [58] have reported the development of a protein/serum-free medium, capable of supporting commercial-scale production levels of cells and virus, in which a pluronic polyol-lipid emulsion is substituted for FBS. Likewise, Inlow et al. [59] describe a serum-free formulation, ISFM, that is based on IPL-41 [169] and is supplemented with yeastolate ultrafiltrate, α-tocopherol acetate, cholesterol, Tween 80, and a lipid emulsion consisting of cod liver oil fatty acid esters. Other investigators have recently reported successes in the development of serum-free media for scale-up applications [181,182]. In all cases, lipid emulsions were substituted for FBS.

Vaughn and Louloudes [61] have reviewed the lipid requirements of insect cells grown in culture. They conclude that lipids, in particular polyunsaturated fats, play important roles in stimulating cell growth. Goodwin [183] has also demonstrated that lipids in the medium are essential for membrane-to-membrane communication. Replacement of vertebrate serum with commercially available lipid supplements has been reported. Hink [60] evaluated Ex-Cyte VLE (Miles, Kankakee, IL), a low-protein, water-soluble lipid supplement, in place of FBS and found SF9 and TN368 cells grew to higher viable densities when compared with serum-containing medium. Vaughn and Fan [64] tested two commercially available serum replacements, CPSR-1 and CPSR-3 (Sigma, St. Louis, MO) in two *S. frugiperda* cell lines (SF21AE and SF21AE-CL-15) and found that they were suitable for insect cell culture despite the fact that they were originally developed for mammalian hybridoma cell culture. In addition, a synthetic serum replacement, Nuserum (Collaborative Research, Lexington, MA), was shown to be effective in the establishment of cell lines from a hymenopteran parasitic wasp [184].

The need for suitable, low-cost serum-free media is recognized as a major priority by cell culturists worldwide. All would agree that the prohibitive cost of establishing large-scale cell, virus, and protein production systems in serum-containing medium is probably the single most limiting factor impeding progress in this field. When costs of labor, serum replacements, and miscellaneous items such as filters are considered, \$5–\$8/liter of medium represents a reasonable goal.

Recently, Mitsuhashi [185] described an imaginative and inexpensive serum-free medium that should provide the impetus to others to address this issue of new media formulation. Mitsuhashi's medium, MTCM-1601, is simple, autoclavable, and utilizes diluted seawater and table sugar as sources of inorganic elements and energy, respectively. Although several cell lines were adapted to this medium, their growth was slower than that of the same control cells in MM [161] medium. Nevertheless, the fact that such novel formulations can support insect cell growth in culture is most encouraging. A current publication by O'Reilly et al. [186] discusses formulations and applications of serum-free medium for insect cell culture.

Growth of Baculovirus and Expression of Recombinant Protein in Serum-Free Medium

Although many factors have been demonstrated to play important roles in the infection of insect cells in vitro [26,47,62,182,185–193], the composition of the culture medium, in particular serum-free medium, has attracted much attention. This attention comes about due to the desired end product results. That is, a recombinant protein, produced by a cell infected with a recombinant virus, needs to be extracted not only in great amounts but also in a relatively pure form. Serum-containing medium impedes the downstream purification of recombinant proteins, whereas serum-free medium does not.

Watanabe [194] and others [171,195,196] reported that the production of nuclear polyhedrosis virus is dependent on healthy cell growth and the concentration of serum in the culture medium. More recently, however, investigators have shown that baculovirus infection and recombinant protein expression occurs equally as well, and in some cases better, in cells cultured in serum-free medium [55,60,62,176,179,197–199]. In early infection studies, Vail et al. [176] produced AcMNPV in TN368 cells in serum-free medium. The results compared favorably with virus produced in serum-containing medium and showed normal infectivity to *T. ni* larvae. Goodwin and Adams [179], after infecting *L. dispar* and *H. zea* cells in serum-free medium with *L. dispar*– and *H. zea*–specific NPV, found partial viral replication in the *L. dispar* cells but no replication in the *H. zea* cells. With modifications to the medium, such as increases in folic acid, glycerol, lipids, and the addition of fresh glutamine, they were able to improve viral replication significantly in these cell lines. Other investigators, such as Wilkie et al. [47] and Röder [55], reported successful NPV replication in *S. frugiperda* cells in serum-free medium.

Recently, Wang et al. [51] compared β-galactosidase expression and OB and BV production in three lepidopteran cell lines, SF21, SF9, and TN368, in serum-free and serum-containing media. No difference was reported in BV and OB production in all three cell lines in either media; however, β-galactosidase expression appeared lower under serum-free than under serum-containing conditions. Hink [60] also reported equal or better recombinant protein production in cells grown in a lipid/pluronic supplemented serum-free medium when compared with serum-containing medium. Inlow et al. [59] developed a protein-free medium, ISFM, that supported the growth of SF9 cells and the propagation of wild-type and recombinant AcMNPV. Their results showed similar cell growth and virus titers to that of serum-containing cultures.

In the area of large-scale insect cell and baculovirus production, greater cell densities and recombinant protein production was attained by Weiss et al. [199] in serum-free cultures of SF9 and TN368 cells than in serum-containing cultures.

Belisle et al. [200] also reported large-scale serum-free production of wild-type AcMNPV in SF9 cells. In addition, Godwin et al. [201] demonstrated comparable cell densities and equal or better BV and OB production in three different cell lines, SF9, TN368, and Ld (undesignated), in a new biopesticide serum-free medium. In yet another comparative study, Weiss et al. [202] reported equal production of AcMNPV-BV and AcMNPV-OBs in four different cell lines, SF9, TN368, Ld6524, and LdElta, in serum-free medium. In a 21L airlift bioreactor, Maiorella et al. [58] were able to produce human macrophage colony-stimulating factor (M-CSF) in SF9 cells in a serum/protein-free medium equal to that in serum-containing medium.

CONCLUSIONS

It is clear that the rapid growth of insect cell culture technologies, including the development of serum-free media and the evaluation of scale-up procedures, have the potential to make important contributions in the fields of medicine, agriculture, and pharmacology. The susceptibility of cultured insect cells to baculoviruses, their ability to express foreign genes in large quantities, and their capacity to modify post-translational products make insect cells and the baculovirus expression vectors (BEV) ideal systems with unlimited applications. Similarly, the development of cost-effective cell culture methods will eventually make possible the commercial production of baculovirus biopesticides. Where work is still left to be done, however, is in the establishment of new cell lines from specific insect tissue and from orders other than Lepidoptera and Diptera. In addition, the development of low-cost, broad cell range serum-free media will greatly enhance the expanding technologies originating from insect cell culture.

REFERENCES

1. Goldschmidt, R. Proc. Natl. Acad. Sci. U.S.A. 1:220–222, 1915.
2. Trager, W. J. Exp. Med. 61:501–513, 1935.
3. Granados, R.R., and Federici, B.A. (eds.). In: Biology of Baculoviruses, CRC Press, Boca Raton, FL, 1986.
4. Doerfler, W., and Bohm, P. (eds.). In: The Molecular Biology of Baculoviruses. Current Topics In Microbiology and Immunology. Volume 131. Springer-Verlag, Berlin, 1986.
5. Smith, G.E., Summers, M.D., and Fraser, M.J. Mol. Cell. Biol. 3:2156–2165, 1983.
6. Summers, M.D., and Smith, G.E. In: Banbury Report 22319–351. Cold Spring Harbor Laboratory, Cold Spring Harbor, NY, 1985.
7. Pennock, G.D., Shoemaker, C., and Miller, L.K. Mol. Cell. Biol. 4:399–406, 1984.
8. Miller, D.W., Safer, P., and Miller, L.K. In: Genetic Engineering (Setlow, J.K., and Hollaender, A. eds.). Volume 8, Principles and Methods. Plenum, New York, 1986.
9. Hink, W.F. In: Invertebrate Tissue Culture (Vago, C., ed.). Academic Press, New York, pp. 363–386, 1972.
10. Hink, W.F. In: Invertebrate Tissue Culture Research Applications (Maramorosch, K., ed.). Academic Press, New York, pp. 319–369, 1976.

11. Hink, W.F. In: Invertebrate Systems In Vitro (Kurstak, E., Maramorosch, K., and Dübendorfer, A. eds.). Elsevier/North Holland Biomedical Press, Amsterdam, The Netherlands, pp. 533–578, 1980.
12. Hink, W.F., and Bezanson, D.R. In: Techniques in the Life Sciences (Kurstak, E., ed.). Elsevier Scientific Publishers, County Clare, Ireland, Ireland, pp. 1–30, 1985.
13. Hink, W.F., and Hall, R.L. In: Invertebrate Cell System Applications (Mitsuhashi, J., ed.), Volume 2. CRC Press, Boca Raton, FL, pp. 269–293, 1989.
14. Lynn, D.E. In: Eighth International Conference on Invertebrate and Fish Tissue Culture (Fraser, M.J., ed.). Tissue Culture Association, Columbia, MD, pp. 2–6, 1991.
15. Mitsuhashi, J. (ed.) In: Invertebrate Cell System Applications. Volume 2. CRC Press, Boca Raton, FL, pp. 187–213, 1989.
16. Lynn, D.E. J. Tissue Culture Methods 12:23–29, 1989.
17. Dwyer, K.G., Webb, S.E., Shelton, A.M., and Granados, R.R. J. Invertebr. Pathol. 52:268–274, 1988.
18. Miltenburger, H.G., Naser, W.L., and Harvey, J.P. Z. Naturforsch. 39:993–1002, 1984.
19. Naser, W.L., Miltenburger, H.G., Harvey, J.P., Huber, J., and Huger, A.M. FEMS Microbiol. Lett. 24:117–121, 1984.
20. Goodwin, R.H., Tompkins, G.J., and McCawley, P. In Vitro 14:485–494, 1978.
21. McIntosh, A.H., and Ignoffo, C.M. J. Invertebr. Pathol. 37:258–264, 1981.
22. Sohi, S.S., Percy, J., Cunningham, J.C., and Arif, B.M. Can. J. Microbiol. 27:1133–1139, 1981.
23. Sohi, S.S., Percy, J., Arif, B.M., and Cunningham, J.C. Intervirology 21:50–60, 1984.
24. Volkman, L.E., and Goldsmith, P.A. Appl. Environ. Microbiol. 44:227–233, 1982.
25. Granados, R.R., Derkson, A.C.G., and Dwyer, K.D. Virology 152:472–476, 1986.
26. Lynn, D.E., and Hink, W.F. J. Invertebr. Pathol. 35:234–240, 1980.
27. McIntosh, A.H., Ignoffo, C.M., and Andrews, P.L. Intervirology 23:150–156, 1985.
28. McIntosh, A.H., and Ignoffo, C.M. J. Invertebr. Pathol. 54:97–104, 1989.
29. Roberts, P.L. FEMS Microbiol. Lett. 29:189–191, 1985.
30. Hink, W.F. Adv. Appl. Microbiol. 15:157–214, 1972.
31. Vaughn, J.L. J. Invertebr. Pathol. 28:233–237, 1976.
32. Weiss, S.A., Peplow, D., Smith, G.C., Vaughn, J.L., and Dougherty, E. In: Techniques in the Life Sciences (Kurstak, C., ed.). Volume C1, Cell Biology Techniques in Setting Up and Maintenance of Tissue and Cell Cultures. Elsevier Scientific, New York, pp. C110/1–C110/16, 1985.
33. Weiss, S.A., and Vaughn, J.L. In: The Biology of Baculoviruses (Granados, R.R., and Federici, B.A., eds.). Volume II, Practical Applications for Insect Control. CRC Press, Boca Raton, FL, pp. 64–87, 1986.
34. Weiss, S.A., Smith, G.C., Kalter, S.S., and Vaughn, J.L. In Vitro 17:495–502, 1981.
35. Goodwin, R.H. In: Techniques in the Life Sciences (Kurstak, C., ed.). Volume C1, Cell Biology Techniques in Setting Up and Maintenance of Tissue Cell Cultures. Elsevier Scientific, New York, pp. C109/1–C109/28, 1985.
36. House, W., Shearer, M., and Maroudas. Exp. Cell Res. 71:293–296, 1972.
37. Summers, M.D., and Smith, G.E. In: Texas Agricultural Experiment Station Bulletin No. 1555, 1987.
38. Bradley, M.K. In: Guide to Protein Purification (Deutschen, M.P., ed.). Volume 182. Academic Press, New York, pp. 112–132, 1990.
39. Cameron, I.R., Possee, R.D., and Bishop, D.H.L. Tends Biotechnol. 7:66–70, 1989.
40. Corsaro, B.G., and Fraser, M.J. J. Tissue Culture Methods 12:7–11, 1989.
41. Hink, W.F., Ralph, D.A., and Joplin, K.H. In: Biochemistry (Kerkut, G.A., and Gilbert, L.I., eds.). Pergamon Press, Oxford, England, pp. 547–570, 1985.

42. Maeda, S. In: Invertebrate Cell System Applications (Mitsuhashi, J., ed.). CRC Press, Boca Raton, FL, pp. 167–182, 1989.

43. Piwnica-Worms, H. In: Current Protocols in Molecular Biology (Ausubel, F.M., Brent, R., Kingston, R.E., Moore, D.D., Seidman, J.G., Smith, J.A., and Struhl, K. eds.). Wiley-Interscience, New York, pp. 16.8.1–16.11.7, 1990.

44. Cho, T., Shuler, M.L., and Granados, R.R. In: Advances in Cell Culture (Maramorosch, K., and Sato, G.H., eds.). Volume 7. Academic Press, New York, pp. 261–277, 1989.

45. Sohi, S.S. In: Invertebrate Systems In Vitro (Kurstak, E., Maramorosch, K., and Dubendorfer, A., eds.). Elsevier/North Holland Biomedical Press, Amsterdam, pp. 35–43, 1980.

46. Hsu, S.H., Li, S.Y., and Cross, J.H. J. Med. Entomol. 9:86, 1972.

47. Wilkie, G.E.I., Stockdale, M., and Firt, S.T. Dev. Biol. Stand. 46:29–37, 1980.

48. Ogonah, O., Shuler, M.L., and Granados, R.R. Biotechnol. Lett. 13:265–270, 1991.

49. Kuno, G. In Vitro 19:707, 1983.

50. Hink, W.F., Strauss, E.M., and Lynn, D.E. (Abstract). In Vitro 13:177, 1977.

51. Wang, P., Granados, R.R., and Shuler, M.L. J. Invertebr. Pathol. 59:46–53, 1992.

52. Léry, X. and Fédèire, G. J. Invertebr. Pathol. 55:342–349, 1990.

53. Brooks, M.A., Tsang, K.R., and Freeman, F.A. In: Invertebrate Systems in Vitro (Kurstak, E., Maramorosch, K., and Dubendorfer, A., eds.). Elsevier/North Holland Biomedical Press, Amsterdam, p. 67, 1980.

54. Pant, V., Mascarenhas, A.F., and Jagannathan, V. Indian J. Exp. Biol. 15:244, 1977.

55. Röder, A. Naturwissenschaften 69:92–93, 1982.

56. Gardiner, G.R., and Stockdale, H. J. Invertebr. Pathol. 25:363–370, 1975.

57. Weiss, S.A., Godwin, G.P., Whitford, W.G., Gorfien, S.F., and Dougherty, E.M. In: Proc. 10th Aust. Biotech. Conf. (Prince, I.G., ed.). pp. 67–71, 1992.

58. Maiorella, B., Inlow, D., Shouger, A., and Horano, I. Bio/technology 6:1406–1410, 1988.

59. Inlow, D., Shauger, A., and Maiorella, B. J. Tissue Culture Methods 12:13–16, 1989.

60. Hink, W.F. In Vitro Cell. Dev. Biol. 27:397–401, 1991.

61. Vaughn, J.L., and Louloudes, S.J. In: Metabolic Aspects of Lipid Nutrition in Insects (Mittler, T.E., and Dad, R.H., eds.). A Westview Science Study, Boulder, CO, pp. 223–244, 1983.

62. Goodwin, R.H., and Adams, J.R. In: Invertebrate Systems In Vitro (Kurstak, E., Maramorosch, K., and Dubendorfer, A., eds.). Elsevier/North Holland, Amsterdam, pp. 493–509, 1980.

63. Mitsuhashi, J., and Goodwin, R.H. In: Invertebrate Cell System Applications (Mitsuhashi, J., ed.). Volume I. CRC Press, Boca Raton, FL, pp. 31–43, 1989.

64. Vaughn, J.L., and Fan, F. In Vitro Cell. Dev. Biol. 25:143–145, 1989.

65. Schneider, I. J. Cell Biol. 42:603–606, 1969.

66. Pudney, M., and Varma, M.G.R. Exp. Parasitol. 29:7–12, 1971.

67. Dolphini, S. Chromosoma 33:196–208, 1971.

68. Sohi, S.S., and Ennis, T. J. Proc. Entomol. Soc. Ont. 112:45–48, 1981.

69. Crawford, A.M., Parslow, M., and Sheenan, C. N.Zl. J. Zool. 10:405–408, 1983.

70. Lynn, D.E., and Stoppleworth, A. In Vitro 20:365–368, 1984.

71. Nickols, W.W., Braat, C., and Boune, W. Immunology 55:61–69, 1971.

72. Schneider, I. In: Tissue Culture Methods and Applications (Kruse, P.F., and Patterson, M.K., eds.). Academic Press, New York, pp. 788–790, 1973.

73. Disney, J.E., and McCarthy, W.J. In Vitro Cell. Dev. Biol. 21:563–568, 1985.

74. Vaughn, J.L., and Weiss, S.A. In: Large-Scale Mammalian Cell Culture Technology (Lubiniecki, A.S., ed.). Marcel Dekker, New York, pp. 597–617, 1991.

75. Greene, A.E., and Charney. J. Immunol. 55:51–61, 1971.

76. Vaughn, J.L., Goodwin, R.H., Tompkins, G.J., and McCawley, P. In Vitro 13:213–217, 1977.
77. Aldridge, C.A., and Knudson, D.L. In Vitro 16:384–391, 1980.
78. Tabachnick, W.J., and Knudson, D.L. In Vitro 16:392–398, 1980.
79. Brown, S.E., and Knudson, D.L. In Vitro 18:347–350, 1982.
80. Harvey, G.T., and Sohi, S.S. Can. J. Zool. 63:2270–2276, 1985.
81. Rochford, R., Dougherty, E.M., and Lynn, D.E. In Vitro 20:823–825, 1984.
82. Lynn, D.E., Dougherty, E.M., McClintock, J.T., and Loeb, M. In: Invertebrate and Fish Tissue Culture (Kuroda, Y., Kurstak, E., and Maramorosch, K., eds.). Springer-Verlag, Berlin, pp. 239–242, 1988.
83. Maskos, C.B., and Miltenburger, H.G. Mitt. Dtsch. Ges. Allg. Angew. Ent. 4:44–47, 1983.
84. McIntosch, A.H., and Ignoffo, C.M. Appl. Ent. Zool. 18:262–269, 1983.
85. O'Brien, S.J., Shannon, J.E., and Gail, M.H. In Vitro 16:119–135, 1980.
86. Mitsuhashi, J. J. Appl. Ent. Zool. 25:535–537, 1990.
87. Poiley, J.A. In: Large-Scale Mammalian Cell Culture Technology (Lubiniecki, A.S., ed.). Marcell Dekker, New York, pp. 483–494, 1990.
88. Hirumi, H. In: Invertebrate Tissue Culture Research Applications (Maramorosch, K., ed.). Academic Press, New York, pp. 233–268, 1976.
89. Plus, N. In: Invertebrate Systems In Vitro (Kurstak, E., Maramorosch, K., and Dubendorfer, A., eds.). Elsevier/North Holland, Amsterdam, pp. 435–439, 1980.
90. Barile, M.F. In: Contamination in Tissue Culture (Fogh, J., ed.). Academic Press, New York, pp. 729–735, 1973.
91. Barile, M.F., and Kern, J. Proc. Soc. Exp. Biol. Med. 138:432–437, 1971.
92. Merril, C.R., Friedman, T.B., Attallah, A.F.M., Greier, M.R., Krell, K., and Yarkin, R. In Vitro 8:91–93, 1972.
93. Nuttall, P.A., Luther, P.D., and Scott, E.J. Nature 266:835–837, 1977.
94. Price, P.J., and Gregory, E.A. In Vitro 18:576–584, 1982.
95. Molander, C.W., Paley, A., Boone, C.W., Kniazeff, A.J., and Imagawa, D.T. In Vitro 4:148, 1968.
96. Molander, C.W., Kniazeff, A.J., Boone, C.W., Paley, A., and Imagawa, D.T. In Vitro 7:168–173, 1971.
97. Chu, F.C., Johnson, J.B., Orr, H.C., Probst, P.G., and Petricciani, J.C. In Vitro 9:31–34, 1973.
98. Kniazeff, A.J. In: Contamination in Tissue Culture (Fogh, J., ed.). Academic Press, New York, pp. 233–242, 1973.
99. Ludovici, P.P., and Holmgren, N.B. Methods Cell Biol. 6:143–208, 1973.
100. Vaughn, J.L. Dev. Biol. Stand. 76:319–324, 1992.
101. Steiner, T., and McGarrity, G. In Vitro 19:672–682, 1983.
102. Plus, N. In Vitro 14:1015–1021, 1978.
103. Plus, N., Gissman, L., Veyrunes, J.C., Pfister, H., and Gateff, E. Ann. Virol. (L'institut Pasteur) 132:261–270, 1981.
104. Koch, E.A., and King, R.C. J. Morphol. 119:283–304, 1966.
105. Maramorosch, K., Shikata, E., and Granados, R.R. Trans. NY Acad. Sci. 130:841–855, 1968.
106. Maramorosch, K., Granados, R.R., and Hirumi, H. Adv. Virus Res. 16:135–193, 1970.
107. Brooks, M.A. J. Invertebr. Pathol. 16:249–258, 1970.
108. Brinton, L.P., and Burgdorfer, W. J. Bacteriol. 105:1149–1159, 1971.
109. Davis, R.E., and Whitcomb, R.F. Annu. Rev. Phytopathol. 9:119–154, 1971.
110. Maillet, P.L. Bull. Biol. Fr. Belg. 105:95–111, 1971.
111. Schwemmler, W., Quiot, J.M., and Amagier, A. Ann. Soc. Entomol. Fr. [N.S.] 7:423–438, 1971.

112. Granados, R.R., Nguyen, T., and Cato, B. Intervirology 10:309–317, 1978.
113. Weiss, S.A., Vaughn, J.L. In: The Biology of Baculoviruses (Granados, R.R., and Federici, B.A., eds.). Vol. 2. CRC, Press, Boca Raton, FL, pp. 65–67, 1986.
114. Black, L.M. Adv. Virus Res. 25:191–271, 1979.
115. Knudson, D.L., and Buckley, S.M. Methods Virol. 6:323–391, 1977.
116. Granados, R.R. Adv. Virus Res. 20:189–236, 1976.
117. Granados, R.R. In: Pathogenesis of Invertebrate Microbial Diseases (Davidson, E., ed.). Allanhel Osmun, Inc. pp. 101–126. 1981.
118. Belloncik, S. Adv. Virus Res. 37:173–209, 1989.
119. Francki, R.I.B., Fauguet, C.M., Knudson, D.L., and Brown, F. (eds.) In: Classification and Nomenclature of Viruses. 5th Rept. of ICTV. Arch. Virol.(Suppl. 2):117–123. 1991.
120. Van der Beck, C.P. Ph.D. Thesis. H. Veenmanand Zoneu B.V. Wageningen, The Netherlands, pp. 1–74, 1980.
121. Capinera, J.L., and Kanost, M.R. J. Econ. Entomol. 72:570–572, 1979.
122. Hinck, W.F. Nature. 220:466–467, 1970.
123. Luckow, V.A. In: Recombinant DNA Technology and Applications (Prokop, A., Bajpar, R.K., and Ho, C.S., eds.). McGraw Hill, pp. 97–152, 1991.
124. King, L.A., Mann, S.G., Lawrie, A.M., and Mulshaw, S.H. Virus Res. 19:93–104, 1991.
125. Wickham, T.J., Davis, T.R., Granados, R.R., Shuler, M.L., and Wood, H.A. Biotechnol. Prog. 8:391–396.
126. Davis, T.R., Wickham, T.J., McKenna, K.A., Granados, R.R., Shuler, M.L., and Wood, H.A. In Vitro 1993 (in press).
127. Wickham, T.J., and Nemerow, G.R. Biotechnol. Prog. 9:25–30, 1993.
128. McIntosh, A-H. J. Invertebr. Pathol. 57:441–442, 1991.
129. Granados, R.R., and Hashimoto, Y. In: Invertebrate Cell System Applications (Mitsuhashi, J. ed.). Vol. 2. CRC Press, Boca Raton, FL, pp. 3–13, 1989.
130. Granados, R.R., Lawler, K.A., and Burand, J.P. Intervirology 16:72–79, 1981.
131. Goodwin, R.H., Vaughn, J.L., Adams, J.R., and Louloudes, S.J. Misc. Pub. Ent. Soc. Am. 9:66–72, 1973.
132. Corsaro, B., and Fraser, M.J. In Vitro Cell. Dev. Biol. 23:855–862, 1987.
133. Rice, W.C., McIntosh, A.H., and Ignoffo, C.M. In Vitro Cell. Dev. Biol. 25:201–204, 1989.
134. Inoue, H., and Mitsuhashi, J. J. Seric. Sci. Jpn. 53:108–113, 1984.
135. Raghow, R., and Grace, T.D.C. J. Ultrastruct. Res. 47:384–399, 1974.
136. Himeno, M., Yasuda, S., Kohsaka, T., and Onodera, K.J. Invertebr. Pathol. 11:516–519, 1968.
137. Kawarabata, T., and Matsumoto, K. Appl. Ent. Zool. 8:227–233, 1973.
138. Granados, R.R., Dwyer, K.G., and Derksen, A.C.G. In: Biotechnol. Invertebr. Pathol. Cell Culture (Maramorosch, K., ed.). Academic Press, NY, pp. 167–181, 1987.
139. Winstanley, D., and Crook, N.E. In: Proc. Vth Int. Colloq. Invertebr. Cath. Microbial Control, Adelaide, Australia, August, pp. 431–433, 1990.
140. Crawford, A.M., and Granados, R.R. In: Proc. Int. Colloq. Invertebr. Pathol. 3rd. pp 154–159, 1982.
141. Quiot, J.M., Monsarrat, P., Meynadier, G., Croizier, G., and Vago, C.C. R. Acad. Sci. 276:3229–3231, 1973.
142. Crawford, A.M., and Sheehan, C. J. Gen. Virol. 66:529–539, 1985.
143. Burand, J.P., Stiles, B., and Wood, H.A. J. Virol. 46:137–142, 1983.
144. Volkman, L.E., and Goldsmith, P.A. Virology 143:185–195, 1985.
145. Granados, R.R. Biotechnol. Bioeng. 22:1377–1405, 1980.

146. Volkman, L.E., Summers, M.D., and Hsieh, C.H. J. Virol. 19:820–832, 1977.
147. Blissard, G.W., and Rohrmann, G.F. Annu. Rev. Entomol. 35:127–155, 1990.
148. Carbonell, L.F., Klowden, M.J., and Miller, L.K. J. Virol. 56:153–160, 1985.
149. Carpenter, W.M., and Bilimoria, S.L. Virology 130:227–231, 1983.
150. Kondo, A., and Maeda, S. J. Virol. 65:3625–3632, 1991.
151. Crawford, A.M., and Sheehan, C. Arch. Virol. 78:65–79, 1983.
152. Burand, J.P., and Wood, H.A. J. Gen. Virol. 67:167–173, 1986.
153. Potter, K.N., Faulkner, P., and MacKinnon, E.A. J. Virol. 18:1040–1050, 1976.
154. Fraser, M.J., Smith, G.E., and Summers, M.D. J. Virol. 47:287–300, 1983.
155. Burand, J.P., Wood, H.A., and Summers, M.D. J. Gen. Virol. 64:391–398, 1983.
156. Wickham, T.J., Davis, T., Granados, R.R., Hammer, D.A., Shuler, M.L., and Wood, H.A. Biotech. Lett. 13:483–488, 1991.
157. Kool, M., Voncken, J.W., Van Lier, F.L.J., Tramper, J., and Vlak, J.M. Virology 183:739–746, 1991.
158. Wyatt, S.S. J. Gen. Physiol. 39:841–857, 1956.
159. Grace, T.D.C. Nature 195:788–789, 1962.
160. Martignoni, M.E., and Scallion, R.J. Biol. Bull. 121:507–520, 1961.
161. Mitsuhashi, J., and Maramorosch, K. Contrib. Boyce Thompson Inst. 22:435–460, 1964.
162. Yunker, C.E., Vaughn, J.L., and Cory, J. Science 155:1565–1556, 1967.
163. Kuno, G. Ph.D. Thesis, Ohio State Univ., Columbus, 1970.
164. Kuno, G., Hink, W.F., and Briggs, J.D. J. Insect. Physiol. 17:1865, 1971.
165. Mitsuhashi, J. In: Development of Productive Activity of Organisms. Res. Rep. Educ. Comm., Ministry of Education, Japan, p. 364, 1980.
166. Peters, D., and Black, L.M. Tagungsber. Dtsch. Akad. Landwirtschaftswiss. Berlin 115:129, 1971.
167. Landureau, J.C., and Steinbach, M.Z. Naturoforsch. 25:231, 1970.
168. Weiss, S.A., Smith, G.C., Kalter, S.S., Vaughn, J.L., and Dougherty, E. Intervirology 15:213–222, 1981.
169. Carstens, E.B., Tijia, S.T., and Doerfler, W. Virology 99:386–398, 1979.
170. Mitsuhashi, J. In: Invertebrate Cell System Applications (Mitsuhashi, J. ed.). Vol. 1. CRC Press, Boca Raton, FL, pp. 3–20, 1989.
171. Broussard, D.R., and Summers, M.D. J. Invertebr. Pathol. 54:144–150, 1989.
172. Quelle, F.W., Caslake, L.F., Burkert, R.E., and Wojchowski, D.M. Blood 74:652–657, 1989.
173. Landureau, J.C., and Jolles, P. Exp. Cell Res. 54:391–398, 1969.
174. Kitamura, S., Imai, T., and Grace, T.D.C. J. Med. Entomol. 10:488, 1973.
175. Goodwin, R.H. In Vitro 12:303, 1976.
176. Vail, P.V., Jay, D.L., and Romine, C.L. J. Insect Physiol. 28:263–267, 1976.
177. Becker, J., and Landureau, J.C. In Vitro 17:471–479, 1981.
178. Brooks, M.A., and Tsang, K.R. In Vitro 16:222, 1980.
179. Goodwin, R.H., and Adams, J.R. In Vitro 14:351, 1978.
180. Mitsuhashi, J. Appl. Entomol. Zool. 17:575, 1982.
181. Mitsuhashi, J. In: Eigth International Conference on Invertebrate and Fish Tissue Culture (Fraser, M.J. ed.). Tissue Culture Assoc., Columbia, MD, pp. 83–89, 1991.
182. Vaughn, J.L., and Fan, F. In: Eighth International Conference on Invertebrate and Fish Tissue Culture (Fraser, M.J. ed.). Tissue Culture Assoc., Columbia, MD, pp. 111–116, 1991.
183. Goodwin, R.H. In Vitro Cell. Dev. Biol. 27:470–478, 1991.

184. Lynn, D.E., and Hung, A.C.F. In Vitro 22:440–448, 1986.

185. Mitsuhashi, J. J. Tissue Culture Method. 12:21–22, 1989.

186. O'Reilly, D.R., Miller, L.K., and Luckow, V.A. In: Baculovirus Expression Vectors: A Laboratory Manual. W.H. Freeman & Co., New York, pp. 109–122, 1992.

187. Knudson, D.L., and Tinsley, T.W. J. Virol. 14:934, 1974.

188. Dougherty, E.M., Weiner, R.M., Vaughn, J.L., and Reichelderfer, C.F. Appl. Environ. Microbiol. 41:1166, 1981.

189. McCarthy, W.J., Lambiase, J.T., and Henchal, L.S. J. Invertebr. Pathol. 36:48, 1980.

190. Knudson, D.L. In: Safety Aspects of Baculoviruses as Biological Insecticides (Miltenburger, H.G., ed.). Bundesministerum Fur Forschung Und Technologie, Bonn, p. 241, 1980.

191. Roberts, P.L., and Wecker, E. In: Safety Aspects of Baculoviruses as Biological Insecticides (Miltenburger, H.G. ed.). Bundesministerum Fur Forschung Und Technologie, Bonn, p. 135, 1980.

192. Wood, H.A., Johnston, L.B., and Burand, J.P. Virology 119:245, 1982.

193. Volkman, L.E., and Summers, M.D. J. Virol. 16:1630, 1975.

194. Watanabe, H. Appl. Ent. Zool. 22:397–398, 1987.

195. Mitsuhashi, J. In: Invertebrate Tissue Culture (Maramorosch, K., ed.). Academic Press, New York, pp. 13–38, 1976.

196. Sohi, S.S., and Smith, C. Can. J. Zool. 48:427–442, 1970.

197. Goodwin, R.H., and Adams, J.R. In Vitro 16:222, 1980.

198. Kuno, G. In Vitro 19:707, 1983.

199. Weiss, S.A., Whitford, W.G., and Godwin, G. In: Eighth International Conference on Invertebrate and Fish Tissue Culture (Fraser, M.J. ed.). Tissue Culture Assoc., Columbia, MD, pp. 153–159, 1991.

200. Belisle, B., Knoch, C., Celeri, C., Montgomery, T., Gong, T., Dougherty, E., Lynn, D., Shapiro, M., and Linduska, J. In: Eighth International Conference on Invertebrate and Fish Tissue Culture (Fraser, M.J. ed.). Tissue Culture Assoc., Columbia, MD, pp. 117–126, 1991.

201. Godwin, G., Gorfien, S., Tilkins, M.L., and Weiss, S.A. In: Eighth International Conference on Invertebrate and Fish Tissue Culture (Fraser, M.J. ed.). Tissue Culture Assoc., Columbia, MD, pp. 102–110. 1991.

202. Weiss, S.A., Godwin, G.P., and Whitford, W.G. In: Proceedings of the Tenth Australian Biotechnology Conference (Prince, I.G. ed.). pp. 67–71, 1992.

3

Comparison of Mammalian and Insect Cell Cultures

Michael L. Shuler

School of Chemical Engineering, Cornell University, Ithaca, New York 14853

INTRODUCTION

Many readers are probably already familiar with the culture techniques used with mammalian cells. Generally these techniques can be applied to insect cells with minor adaptations. However, the blind application of mammalian cell culture techniques can be unproductive. A clear understanding of the differences between insect cell and mammalian cell culture systems is crucial. In Chapter 1 some of the advantages and disadvantages of the insect cell–baculovirus system compared with mammalian cell expression systems were summarized. Details of culture conditions, infection, and protein production and processing follow in subsequent chapters. In this chapter I focus on comparing some of the intrinsic differences in culturing insect cells versus mammalian cells.

The basic cell biology of insect cells under noninfected conditions has not been well examined except for *Drosophilia*. This derth of knowledge about insect cell biology, especially regarding fundamental processes such as post-translational processing of proteins, makes it difficult to generalize with respect to the insect cell and to compare it to much better studied mammalian cell systems. Excellent reviews of culture methods for mammalian [1] and insect [2,3] cell systems are available. The purpose of this chapter is to provide an overview of intrinsic differences rather than a detailed comparison of techniques.

INTRINSIC DIFFERENCES

Immortality and Anchorage Dependence

Insect cell lines obtained from ovaries or embryos appear to be naturally continuous cell lines (i.e., capable of indefinite replication). Furthermore, some cell lines

Baculovirus Expression Systems and Biopesticides, pages 41–49

established from more differentiated tissue, such as the midgut, and fat body appear to be capable of indefinite replication. Sources and methods to establish these cell lines were discussed in Chapter 2. Mammalian cells obtained from normal tissue typically are mortal (i.e., capable of a limited number of cell divisions before senesence). For use in large-scale systems this limitation on cell division is an important constraint. Consequently most large-scale mammalian culture systems use "transformed" cells. These "transformed" cells become immortal. However, since all cancer cells are transformed cells and the process of transformation is not completely understood, there is considerable concern about the safety of using transformed cells. Consequently the ability to obtain continuous cell lines from insects without transformation is highly significant.

Although the insect cell lines are continuous, their growth characteristics are subject to alteration upon sustained subculture (especially in serum-free medium). They can be preserved successfully by freezing using cryopreservation techniques rather analogous to those used for mammalian cells [3].

Except for cells related to the circulatory system, most normal mammalian cells are attachment or anchorage dependent. However, transformation of a mammalian cell often leads to anchorage-independent growth and the use of suspension culture techniques. Although commercial-scale cultivation systems exist for both anchorage-dependent and -independent cell lines, suspension culture systems are normally considered to be easier to implement.

Most insect cell lines can be adapted to grow either attached or in suspension. An attached cell line can be adapted to suspension culture and then quickly readapted to attached cell growth. Shuler et al. [4] describe an experiment in which 16 passages were necessary to adapt TN368 (a *Trichoplusia ni* cell line) to suspension culture. During the adaption period cell morphology, as well as the maximum achieved cell density, changed. Goosen [5] reports needing six passages to adapt *Spodoptera frugiperda* (SF) cells to suspension conditions and a lack of success in adapting *T. ni* cells to suspension cultures. Cellular changes upon adaptation are common but can vary in amount from one laboratory to another, making reproduction of results from one laboratory to another problematic.

Whether attached or suspension culture growth alters potential productivity has not been completely explored. It appears from the available data that protein production under optimal conditions is not usually altered by the choice of growth mode, although this statement is not a general conclusion. For example, Bilimoria and Carpenter [6] selected for variants of *T. ni* that were more adhesive than the parent strain. The attached variant showed greater infectivity when challenged with the baculovirus than the parental strain, indicating possible differences in protein production.

One other important characteristic of transformed mammalian cells is the loss of "contact inhibition" upon transformation. *Contact inhibition* refers to the inhibition of cell division when cells are in contact with each other. Insect cells show inhibition [7] of both cellular and viral DNA when attached cells establish cell-to-cell contact. Because of this inhibition of DNA synthesis, productive infection by baculovirus requires infection prior to confluency—usually before cells are 70%

confluent. Clumping of cells and partial loss of "contact" inhibition can occur if cultures are stressed by suboptimal conditions.

Thus insect cell cultures have characteristics somewhat intermediate between normal and transformed mammalian cell lines. Insect cells are continuous cell lines subject to contact inhibition and are capable of being cultured in either anchorage-dependent or -independent modes.

Growth Characteristics

Although cellular chemistry is very similar between insect and mammalian cells, specific differences have significant impact on how insect cells are cultured and the choice of media. For example, optimum temperatures for cultivation vary: from 25°C to 30°C for most insect lines versus 37°C for mammalian cell lines. Optimal pH is 6.2–6.6 with insect cell lines versus 7.0–7.5 for mammalian cell lines. The use of CO_2 incubators is common with mammalian cell lines but unnecessary with insect cell lines.

Insect cells can be cultured in T-flasks, and on microcarriers using equipment designed for mammalian cells. With mammalian cells, cell removal from monolayers established in T-flasks normally involves removal of "old" medium that may contain trypsin inhibitors, washing with PBS (phosphate-buffered saline), then adding an enzyme or cell dissociation medium (e.g., trypsin sometimes in conjunction with EDTA), and then followed by physical dislodgement. For insect cell cultures trypsin is not used; physical dislodgement by sharply tapping the flask on the corner of a bench or hand is typically used. Less frequently a rubber policeman is used to dislodge cells. A nonenzymatic cell dissolution solution for insect cells is commercially available.

Mammalian cells grow best at an osmolarity of 280–320 mOsm/kg, with a value of 290 mOsm/kg being typical. Insect cells prefer a slightly high osmolarity, typically values in the range of 360–400 mOsm/kg.

Most insect cell lines have a diameter of 12–20 μm, which is slightly larger than most mammalian cell lines. The growth rates (doubling times of 12–48 hours) are similar. Both insect and mammalian cells are quite sensitive to shear [8], although insect cells appear to be more sensitive to laminar shear stresses (see Chapter 9). Oxygen requirements are similar to those of murine hybridomas [5] but vary depending on culture phase. [4,5] Oxygen requirements increase dramatically (at most twofold) during viral replication. Oxygen effects are discussed in more detail in Chapter 8.

The typical media used for insect cell culture are quite complex in comparison to those developed for many mammalian cells. However, insect cells have been cultured in a normal mammalian cell medium [10]. Table 1 compares the composition of Grace's medium for insect cell culture with an Eagle's Minimum Essential Medium for mammalian cells. Note that the Grace's medium contains many TCA cycle intermediates and a high level of sucrose. Although sucrose is added primarily to control osmotic pressure, some cell lines (e.g., *Bombyx mori*) are capable of hydrolyzing sucrose into glucose and fructose and then consuming both monosac-

TABLE 1. Comparison of the Compositions of Grace's Insect Cell Culture Medium [5] and Eagle's Minimal Essential Medium [11]

Component	Grace's (mg/liter)	Eagle's (mg/liter)
Inorganic salts		
$CaCl_2$ (anhyd.)	750	200
KCl	4,100	400
$MgCl_2 \cdot 6H_2O$	2,280	—
$MgSO_4 \cdot 7H_2O$	2,780	200
$NaHCO_3$	350	2,200
$NaH_2PO_4 \cdot H_2O$	1,013	140
NaCl	—	6,800
$Fe(NO_3)_3 \cdot 9H_2O$	—	0.10
Other components		
α-Ketoglutaric acid	370	—
Fructose	400	—
Fumaric acid	55	—
D-Glucose	700	1,000
Malic acid	670	—
Succinic acid	60	—
Sucrose	26,680	—
Phenol red	—	10
Amino acids		
β-Alanine	200	—
L-Alanine	225	—
L-Arginine HCL	700	126
L-Asparagine	350	—
L-Aspartic acid	350	—
L-Cystine	22	24
L-Glutamic acid	600	—
L-Glutamine	600	292
Glycine	650	—
L-Histidine	2,500	42
L-Isoleucine	50	52
L-Leucine	75	52
L-Lysine HCL	625	73
L-Methionine	50	15
L-Phenylalanine	150	33
L-Proline	350	—
DL-Serine	1,100	—
L-Threonine	175	48
L-Tryptophan	100	10
L-Tyrosine	50	36
L-Valine	100	47
Vitamins		
Biotin	0.01	—
D-Ca pantothenate	0.02	1.0

TABLE 1. (*Continued*)

Component	Grace's (mg/liter)	Eagle's (mg/liter)
Choline chloride	0.20	1.0
Folid acid	0.02	1.0
i-Inositol	0.02	2.0
Niacin	0.02	—
Nicotinamide	—	1.0
Para-aminobenzoic acid	0.02	—
Pyridoxine HCL	0.02	1.0
Riboflavin	0.02	0.1
Thiamine HCl	0.02	1.0
Fetal bovine serum	5%–10%	5%–10%

charides. [12] Because of sucrose degradation, glucose can accumulate in some portions of the culture cycle.

Also from Table 1 it is clear that a much greater number of amino acids are used in insect cell culture than mammalian cultures and that many amino acids (e.g., histidine) are used at much higher levels. Interestingly, glutamine, the dominant amino acid supplied in mammalian cell media, is not dominant in insect cell media. However, glutamine is usually exhausted in most batch growth experiments [12,13] and may be the growth-limiting nutrient for many cell lines. Vaughn and Weiss [13] have reported on the usage of amino acids by SF9 cells grown in 250 ml shaker culture. Alanine, a major metabolic product of glutamine metabolism in mammalian cells, accumulates four- to fivefold in 6–7 days, while glutamine is totally consumed. These data indicate that alanine is the major byproduct of glutamine metabolism. Approximately 70% of the initial cysteine as well as 50% of the leucine were consumed. A little more than 40% of serine and tyrosine were consumed. Less than 40% of the other amino acids was removed from solution by the cells.

In mammalian cells ammonia toxicity is a major limitation on cell culture; ammonia is primarily generated from glutamine metabolism. Insect cell cultures produce uric acid, and ammonia added to the medium will be consumed by the cells. [13] Since ammonia toxicity is difficult to circumvent in cell culture, the ability of insect cells to convert ammonia to uric acid is a distinct technological advantage.

The other major inhibitory metabolic byproduct in mammalian cell culture is lactate formation, resulting primarily from glucose catabolism. The inhibitory effects of lactate are primarily due to disturbances in intracellular pH. With insect cell cultures lactate is formed, primarily as the cells enter the stationary phase. [4,12,13] In most media glucose and glutamine are exhausted simultaneously. Although lactate is inhibitory to these cells, lactate formation appears to be less of a problem than with mammalian cells.

In summary, insect cell cultures can be cultured in devices analogous to those used for mammalian cell culture provided that the subtle differences in nutritional and physical requirements are recognized.

Protein Synthesis and Processing

Three areas of difference in protein synthesis and processing between the insect cell–baculovirus system and most mammalian cell expression systems are 1) the lytic nature of the insect cell system, 2) limitations on post-translational processing machinery, and 3) sensitivity to high cell density. Chapters 4, 7, 8, 10, and 11 deal with the specifics of these issues.

Baculovirus systems are invariably lytic. Lysis presents potential problems with respect to modification of previously secreted proteins and release of contaminating material that can complicate downstream recovery. Secreted proteins can be processed by insect cells, so lysis is not a process requirement. The use of protease inhibitors can postpone lysis (T. Wickham, personal communication) for 24 hours or more, but lysis still occurs.

The lytic nature of the system makes it difficult to apply many of the novel reactor techniques that have proved effective for sustained protein production in nongrowing or slowly replicating mammalian cell cultures. [14] For example, the hollow fiber reactor systems used with hybridomas or the continuous culture system using a fluidized bed of collagen beads with CHO cells would be difficult to implement. Thus, the insect cell–baculovirus system is probably restricted to batch and multistage continuous systems [15].

An alternative to the lytic system would be the approach by Jarvis et al. [16] with which they obtained continuous expression and efficient processing of foreign gene products in stably transformed cells using an immediate early gene promoter (IE1). Although stable production was obtained, the level of β-galactosidase made decreased by more than 100-fold from the lytic system using the polyhedrin promoter. However, with tissue plasminogen activator (TPA), the transformed cells secrete TPA more quickly, with production levels slightly less than 50% of TPA production from the traditional baculovirus expression system. The use of the IE1 promoter would probably not be competitive with a good CHO expression system due to low expression level.

An intriguing idea might be to use persistently infected cells (see Chapter 2), but it is unclear how this strategy could be used for stable, high-level protein production. Another alternative is an insect cell line that does not use the baculovirus. Culp et al. [17] have described a stable, continuous culture *Drosophilia* expression system using a *Drosophila* metallothionein promoter with multicopy insertion. This system has been used for the production of the HIV envelope gp120 protein.

Another effect of importance is the reduction in specific productivity (i.e., amount of protein produced per unit time per cell) at high cell densities. In attached cell cultures the cell density effect has been demonstrated with respect to protein production [18] and with polyhedra [7,19,20]. Although definitive studies with suspension cultures have not been published, a cell density effect almost certainly exists. Caron et al. [21] provide data indicating that above approximately 2×10^6

cells/ml (SF9 cells) specific production of heterologous protein was greatly inhibited. Lindsay and Betenbaugh [22] found no cell density effect below 10^6 cells/ml. In studies in our own laboratory we observe a cell density effect with β-galactosidase at more than 3×10^6 cells/ml. The cell density effect can be shifted to approximately 6×10^6 ml if high oxygen levels are maintained and the medium is supplemented with additional glutamine.

Insect cells can be cultured to high cell densities (1 to 2×10^7 cells/ml), but the utilization of high-density systems is limited if specific productivity decreases. In contrast to insect cell systems, high cell densities can be achieved in mammalian cell reactors without dramatic decreases in specific productivity [14]. Until the cell density effect is better understood and a method to circumvent it found, the use of many of the novel animal cell reactors will be inappropriate for insect cell cultures.

The third factor of concern is limitations on post-translational processing. The two primary concerns have been a very large reduction in expression levels for glycosylated, secreted proteins and in differences in glycosylation patterns compared with mammalian cells.

A reduction of 100-fold in expression levels between a cytoplasmic protein such as β-galactosidase and a secreted, glycosylated protein such as TPA is commonly observed. Potentially intrinsic rate-limiting steps exist in the post-translational pathways in insect cells. However, as discussed in Chapter 7, it is possible to select for cell lines with greatly increased (about 10-fold) expression levels of secreted, glycosylated proteins over the commonly used SF9 cell line. One can also envision a program to engineer the insect cell metabolically to improve the rate of protein processing. Such a strategy involves identifying the rate-limiting step(s) in the secretion and post-translational pathways and cloning the appropriate gene encoding for proteins that will increase processing at that point in the pathway [23]. In mammalian cells there appears to be a better match of rate of synthesis and processing a protein than in insect cells, where high rates of synthesis can possibly "overwhelm" the processing machinery.

The other primary difference is that the post-translational processing by insect cells does not necessarily yield a protein with glycosylation patterns (and other forms of processing) identical to the natural protein or the protein produced in mammalian cell culture. Insect cells frequently substitute truncated high-mannose structures at sites where mammalian cells would add complex type structures leading to a conclusion that "insect cells appear to synthesize only high-mannose type N-linked oligosaccharides and truncated derivatives thereof" [24]. Davidson and Castellino [25,26] have shown that cultured insect cells possess the machinery needed to assemble N-linked complex-type oligosaccharides, but the expression of that potential varies with the time postinfection. It is unknown whether cultured cells of different insects will have different abilities to conduct post-translational processing, since most studies done thus far have been with the SF21 and SF9 cell lines, but intriguing differences between cell lines do exist (see Chapter 7). The details of protein processing are discussed in Chapter 4. When comparing insect cell and mammalian expression systems, the reader must be aware of intrinsic differences in protein processing. Further differences in response to high cell density and the lytic nature of the insect cell system must be considered in selecting appropriate culture conditions.

CONCLUSIONS

Intrinsic differences in biochemistry between insect and mammalian cells require different media, temperature, pH, and gas phase composition in culturing these cells. The insect cell–baculovirus expression system cannot be well utilized in some of the bioreactor systems developed for mammalian cell cultures.

REFERENCES

1. Freshney, R.I. Culture of Animal Cells: A Manual of Basic Technique. 2nd Ed. Alan R. Liss, Inc., New York, 1987.
2. O'Reilly, D.R., Miller, L.K., and Luckow, V.A. Baculovirus Expression Vctors: A Laboratory Manual, W.H. Freeman and Company, New York, 1992.
3. Summers, M.D., and Smith, G.E. A Manual of Methods for Baculovirus Vectors and Insect Cell Culture Procedures. Texas A&M. Experimental Station Bulletin No. 1555.
4. Shuler, M.L., Cho, T., Wickham, T., Ogonah, O., Kool, M., Hammer, D.A., Granados, R.R., and Wood H.A. Ann. N.Y. Acad. Sci. 589:399–422, 1990.
5. Goosen, M.F.A. Can. J. Chem. Eng. 69:450–456, 1991.
6. Billmoria, S.L., and Carpenter, W.M. In Vitro 19:870–874, 1983.
7. Wood, H.A., Johnston, L.B., and Burand, J.P. Virology 119:245–254, 1982.
8. Tramper, J., and Vlak, J.M. Ann. N.Y. Acad. Sci. 469:279–288, 1986.
9. Streett, D.A., and Hink, W.F. J. Invertebr. Pathol. 32:112–113, 1978.
10. McIntosh, A.H., Maramorosch, K., and Reehtoris, C. In Vitro 8:375–378, 1973.
11. GIBCO BRL Catalog and Reference. Life Technologies, Inc., Gaithersburg, MD, pp. 582–591, 1990.
12. Stavroulakis, D.A., Kalogerakis, N., Behie, L.A., and Iatrou, K. Can. J. Chem. Eng. 69:457–464, 1991.
13. Vaughn, J.L., and Weiss, S.A. BioPharm. 4:16–19, 1991.
14. Lydersen, B.K. (ed). Large Scale Cell Culture Technology, Hanser, New York, 1987.
15. vanLier, F.L.J., van den End, E.J., deGooijer, C.D., Vlak, J.M., and Tramper, J. Appl. Microbiol. Biotechnol. 33:43–47, 1990.
16. Jarvis, D.L., Fleming, J-A.G.W., Kovacs, G.R., Summers, MD., and Guarino, L.A. Bio/Technology 8:950–955, 1990.
17. Culp, J.S., Johansen, H., Hellmig, B., Beck, J., Matthews, T.J., Delers, A., and Rosenberg, M. Bio/Technology 9:173–177, 1991.
18. Wickham, T.J., Davis, T., Granados, R.R., Shuler, M.L., and Wood, H.A. Biotechnol. Prog. 8:391–396, 1992.
19. Hink, W.F., Strauss, E.M., and Ramoska, W.A. J. Invertebr. Pathol. 30:185–191, 1977.
20. Stockdale, H., and Gardiner, G.R. J. Invertebr. Pathol. 30:330–336, 1977.
21. Caron, A.W., Archambault, J., and Massie, B. Biotechnol. Bioeng. 36:1133–1140, 1990.

22. Lindsay, D.A., and Betenbaugh, M. Biotechnol. Bioeng. 39:614–618, 1992.
23. Lee, E.U., Roth, J., and Paulsson, J.C. J. Biol. Chem. 264:13848–13855, 1989.
24. Thomsen, D.R., Post, L.E., Elhammer, A.P. J. Cell. Biochem. 43:67–79, 1990.
25. Davidson, D.J., and Castellino, F.J. Biochemistry 30:6167–6174, 1991.
26. Davidson, D.J., and Castellino, F.J. Biochemistry 30:6689–6696, 1991.

4

Protein Production and Processing From Baculovirus Expression Vectors

Verne A. Luckow

Molecular and Cellular Biology, Monsanto/Searle, Chesterfield, Missouri 63198

INTRODUCTION

Baculovirus vectors have become widely used to direct the expression of foreign genes. Foreign genes placed under the transcriptional control of the strong polyhedrin promoter of the *Autographa californica* nuclear polyhedrosis virus (AcMNPV) or *Bombyx mori* nuclear polyhedrosis virus (BmSNPV) are often abundantly expressed during the late stages of infection in cultured insect cells or in insect larvae. The recombinant proteins are in many cases soluble and antigenically, immunogenically, and functionally similar to their authentic counterparts [1–7].

This review summarizes recent developments concerning the expression of foreign genes in insect cells with baculovirus vectors. This chapter focuses on the development of improved baculovirus transfer vectors and the post-translational processing of recombinant proteins expressed in baculovirus-infected insect cells. Novel applications of this technology will also be described. Baculovirus molecular biology and genetics, small- and large-scale insect cell culture, and the use of wild-type and genetically engineered baculoviruses as viral pesticides are summarized in companion chapters in this volume.

The early development of the baculovirus expression vector system has been summarized in many reviews [1–4,8–16]. Specific methods for identification and propagation of wild-type and recombinant baculoviruses in cultured insect cells have also been previously described [4,10,17–23]. Detailed protocols, covering all aspects of insect cell culture and the construction of recombinant baculoviruses, can be found in recently published laboratory manuals [6,7,24]. The biology and molecular genetics of baculoviruses, and other insect viruses, are described in several

Baculovirus Expression Systems and Biopesticides, pages 51–90

general reviews and books devoted to baculoviruses [25–31]. These sources, and the other chapters within this volume, reference most of the background literature described below.

A BRIEF BIOLOGY OF BACULOVIRUSES

Baculoviruses are a large group of DNA viruses capable of infecting over 500 species of insects. They usually have a very restricted host range, limited to specific insect species, and do not infect vertebrates or plants. The rod-shaped virions have an envelope and contain a circular double-stranded DNA genome ranging in size from 80 to 150 kb in length. Three subgenera of baculoviruses are recognized. Nuclear polyhedrosis viruses (NPVs) produce large polyhedral virus particles late in infection that serve to protect the virus from inactivation by environmental factors, such as dessication or ultraviolet light, during transmission from insect to insect. These particles, or occlusion bodies of nuclear polyhedrosis viruses, are also called *polyhedra*. Granulosis viruses produce oval occlusion bodies that contain a single nucleocapsid. Nonoccluded viruses are not packaged into an occluded form at any stage in their life cycle.

Two baculoviruses, the *Autographa californica* nuclear polyhedrosis virus (AcMNPV) and the *Bombyx mori* nuclear polyhedrosis virus (BmSNPV), have been studied extensively at the genetic and molecular levels and used to develop expression vectors. AcMNPV, originally isolated from an alfalfa looper, can infect over 30 species of insects. Cell lines are not available for all of these species, and AcMNPV is most commonly propagated in the laboratory in cultured cells derived from the fall armyworm *Spodoptera frugiperda* or from the cabbage looper *Trichoplusia ni*. BmSNPV, on the other hand, has a very strict host range limited to the silkworm *Bombyx mori*.

The replication cycle of AcMNPV and other baculoviruses is very complex. Caterpillars feeding on infected plants ingest the occluded viral particles, which dissolve in the alkaline juices of the midgut to release viruses that enter cells by endocytosis. Upon entry into the nucleus of a susceptible cell, the virus is uncoated and specific viral genes are transcribed in a tightly coordinated cascade. Four phases of viral gene expression are generally recognized. Immediate early (α) genes are expressed using functions supplied by the host cell. Expression of early (β) genes are controlled by the products of the immediate early genes. Several of the products of the immediate early genes are transactivators involved in regulation of transcription of specific β genes. The products of some of the β genes are known to be involved in virus-specific transcription of late genes. One β gene encodes a viral DNA polymerase. Late (γ) genes are expressed following DNA replication (about 5–7 hours post infection [hpi]) and are involved in structure and assembly of nucleocapsids, transport and budding of virions from the cell surface, and degradation of the host cell genome. Budded virions (BV) are released between 10 and 20 hpi into the hemolymph of the insect and spread the infection to other tissues. Occluded viral particles (OV) are detected in the nucleus by 18 hpi and accumulate for 3–5 days until the infected cells lyse. Polyhedrin, a 29 kd structural protein found in the crystalline matrix of the occlusion body, is abundantly produced in the late stages of infection. The polyhedrin (*polh*) gene is one of several members of a fourth, very late (δ) class of genes that are highly transcribed. The 10 kd product of the *p10* gene, which forms fibrous arrays in the nucleus of infected cells, is also abundantly produced very late in infection, but its role in the larval host is unclear.

Viruses containing deletions of the polyhedrin gene are viable and stably maintained in cultured cells, indicating that this gene is not essential for replication or production of the budded form of the virus [32]. This observation led to the development of methods for the construction of recombinant baculoviruses that contain foreign genes under the control of the polyhedrin promoter [33–35].

CONSTRUCTION OF RECOMBINANT BACULOVIRUSES

The large size of the typical baculovirus genome precludes many of the strategies used to facilitate construction of genetically engineered bacterial, plant, and animal viruses, whose genomes are often much smaller and amenable to manipulation in one step by cut-and-paste methods with standard recombinant DNA techniques. Recombinant baculoviruses, however, are traditionally constructed in two steps. First, a foreign gene is cloned into a plasmid transfer vector downstream from a baculovirus promoter, flanked by baculovirus DNA derived from a nonessential locus, usually the polyhedrin gene. Baculovirus transfer vector DNA and genomic parental baculovirus DNA are then mixed and introduced into cultured insect cells where they recombine at a low frequency to produce a virus containing an integrated copy of the foreign gene. Recombinant viruses are then identified by a variety of methods and purified free of contaminating parental virus, typically by plaque purification. Usually three or more rounds of plaque purification are necessary to purify a recombinant virus. Since transfections, plaque assays, and the preparation of pure virus stocks typically take 5–7 days each, the entire procedure to generate a recombinant virus using traditional methods can take from 4 to 6 weeks.

Transfection

A number of techniques have been used to transfect baculovirus DNAs into insect cells. Protocols based on the coprecipitation of DNA with calcium phosphate [21] have been widely used, because the reagents are inexpensive and simple to prepare and the results are generally reproducible. A variety of cationic lipids, such as Lipofectin (GIBCO/BRL), have also been used as reagents for transfecting baculovirus DNAs into insect cells [36]. The formation of the liposomes with Lipofectin is inhibited by serum, so this step of the process must be carried out in serum-free media. This method is reported to be 10-fold more efficient than the calcium phosphate transfection method. It is not clear, however, whether this improves the frequency of recombinant viruses among wild-type parental viruses in a transfection mixture, but only ensures that more cells are infected earlier with baculovirus DNAs. Other reagents such as DEAE-Dextran [6] and protamine [37] also have been used to replace calcium phosphate, but these techniques are not widely used. Electroporation has been used to introduce AcMNPV DNA into *Spodoptera frugiperda* cells [38], but the equipment is expensive, and the method does not appear to be much more efficient, or simpler, than methods using calcium phosphate or Lipofectin reagent.

Identifying and Confirming the Structure of Recombinant Viruses

Recombinant viruses containing the foreign gene inserted into the polyhedrin locus can be identified by an altered plaque morphology characterized by the absence of occluded virus in the nucleus of infected cells. Plaque assays are used to visualize and purify the recombinant virus away from the wild-type parental virus. Dilutions of virus-containing media are typically incubated with uninfected cells attached to a tissue culture dish for an hour and replaced with media containing agarose that forms a semisolid overlay. After incubation for 5–7 days at 27°C in a humidified incubator, the plates are examined for plaques under a stereo dissecting microscope. Plaques are distinguished from the background of uninfected cells by a number of criteria. First, cells are more sparsely distributed in a plaque because once a cell is infected with a virus, it does not divide. Adjacent cells are soon infected with budded virus released from the initial infected cell. Uninfected cells, however, continue to divide and nearly reach confluence over the course of the plaque assay. Second, infected cells swell to a diameter that is about 50% larger than uninfected cells. Third, those cells originally in the center of the plaque occasionally lyse, accentuating the sparse appearance of the plaque. When an agarose plate is illuminated from the side or at a small angle with a strong light source, wild-type plaques have a slightly golden glow due to refraction of light through the crystalline-like polyhedra present in the nucleus of infected cells. Recombinant viruses with insertions into the polyhedrin locus lack polyhedra and have occlusion-minus plaques that appear gray among the mostly colorless background of uninfected cells. Occlusion-minus plaques, particularly those in a population containing 99% wild-type plaques, are not easily distinguished by the untrained eye.

Seeding density dramatically influences the appearance of plaques in a plaque assay. Monolayers that are seeded too heavily will give rise to plaques that are very small. Sparsely seeded monolayers, however, give rise to large plaques with poorly defined boundaries. The best seeding density for a given cell type may need to be determined empirically in pilot experiments using a range of different cell densities. Doubling times of a given cell line in different media can also vary significantly. Cells transferred to a new type of media, particularly serum-free media, will grow slower until they have adapted to the new media. Different cell lines can also vary dramatically in virus susceptibility, budded virus production, and occluded virus production [39], so it is wise to carry out plaque assays in a consistent fashion using one type of cell line propagated in one kind of media in order to minimize those variables that affect virus production.

A number of stains can be used to enhance the visualization of plaques against the background of uninfected cells in a plaque assay. The stains are usually added as dilute (0.1%–0.5% w/v) solutions that diffuse into the agarose overlay over several hours or added to a second agarose overlay. Neutral red and trypan blue have been used for this purpose [40,41], but the contrast enhancement is not particularly dramatic. Recently, Shanafelt [42] described the use of the yellow dye MTT (3-[4,5-dimethylthiazol-2-yl]-2,5-diphenyltetrazolium bromide) to increase the contrast between the dead cells in a plaque and the uninfected background cells. The MTT is metabolized to form a purple precipitate within the cytoplasm of living cells. Plaques in this case are faint yellow against a deep purple background, which first appears by 30 minutes and is fully apparent by 2 hours after addition of the MTT solution. MTT, trypan blue, and neutral red are vital stains and do not help to distinguish wild-type virus plaques from recombinant virus plaques dramatically.

The construction of recombinant baculoviruses by standard transfection and plaque assay methods can take as long as 4–6 weeks, and many methods to speed up the identification and purification of recombinant viruses have been tried in recent years. These methods include plaque lifts [21], serial limiting dilutions of virus [43], and cell affinity techniques [44,45]. Each of these methods requires confirmation of the recombination event by visual screening of plaque morphology (described above), DNA dot blot hybridization [46], immunoblotting [47], or amplification of specific segments of the baculovirus genome by polymerase chain reaction (PCR) techniques [48–50]. PCR is a particularly powerful tool to characterize the genetic composition of recombinant baculoviruses and to monitor the ratio of wild-type and recombinant viruses during purification by sequential plaque assays or serial dilution methods. Viral DNA prepared from infected cells is used as a template in a PCR reaction containing universal primers that flank the cloning site in the baculovirus transfer vector. In most cases, the primers will amplify a short DNA segment from the wild-type or parental baculovirus DNA used in the original transfection. A longer segment containing sequences from the foreign gene flanked by baculovirus DNA is amplified from the recombinant baculoviruses. Partially purified viral stocks will give rise to both PCR products while completely purified recombinant virus stocks will give rise to only the larger PCR product. PCR can also be used to facilitate the engineering of foreign genes for insertion into baculovirus transfer vectors [51].

BACULOVIRUS TRANSFER VECTORS

A wide variety of plasmids have been developed for use in constructing recombinant viruses that express fused or nonfused foreign proteins under the control of a baculovirus promoter. The earliest vectors were constructed by inserting the AcMNPV *Eco*RI-I fragment containing the polyhedrin gene into a high copy number bacterial plasmid such as pUC8 [33]. Portions of the coding sequence for the polyhedrin gene were removed by nuclease digestion and one or more unique restriction sites introduced downstream from the polyhedrin promoter. These plasmids are called *baculovirus transfer vectors* or *allelic transplacement vectors*. Foreign genes, usually cDNAs lacking introns, must then be inserted into the unique cloning site(s) in the correct orientation, and the resulting plasmid DNA is introduced into insect cells with genomic (parental) baculovirus DNA. Most transfer vectors confer resistance to ampicillin in *E. coli* and are easily isolated and characterized.

Although transfer vectors designed to generate recombinant viruses that express a single gene under the control of the polyhedrin promoter are most commonly used, many new transfer vectors and strategies for constructing recombinant viruses expressing one or more foreign genes have been developed within the past several years. These can be grouped into three major categories: 1) vectors for expressing a single foreign gene; 2) coexpression vectors that are used to express a single foreign gene and contain a screenable genetic marker, such as the *Escherichia coli lacZ* gene; and 3) dual-expression vectors, which are designed to accept and express two foreign genes. Each major category can be further subdivided on the basis of the baculovirus promoter used to drive the expression of the foreign gene and the presence or absence of an ATG start codon that can be used to direct expression of fusion proteins. The expression cassette can also be targeted to different nonessential regions of the baculovirus genome, depending on the origin of the flanking baculovirus DNA. Vectors that target insertions to a region other than the polyhedrin locus, such as the *p10* gene, or carry an intact polyhedrin gene, can be used to generate occlusion-positive recombinant viruses that may be useful for oral ingestion by insect larvae. Key features of each type of plasmid are summarized in Table 1 and described briefly in the following sections. Since transfer vectors are easily modified for specialized purposes, only those that have been or are likely to be widely used will be described. The types of parent viruses that are available to generate recombinant viruses are discussed in subsequent sections.

Polyhedrin-Based, Single Gene Transfer Vectors

The transfer vector pAc373 [1,52] was originally widely used to generate recombinant viruses for the production of nonfused foreign proteins. A unique *Bam*HI site was inserted between a position eight bases upstream from the polyhedrin ATG start codon (where the ATG is at positions +1, +2, and +3) and a natural *Bam*HI site within the polyhedrin gene at position +177. Several similar vectors, including pEV51 [53] and the pAcRP series [54,55], which have cloning sites upstream from the polyhedrin start codon, were also used, until it was realized that steady-state mRNA levels driven by the polyhedrin promoter were dramatically reduced if portions of the untranslated polyhedrin leader sequence were deleted [46,56–59].

The transfer vector pAcYM1 [57] has all of the upstream untranslated sequences of the polyhedrin gene, including the A of the ATG start codon followed by a *Bam*HI linker (CGGATCCG). The plasmids pAcCL29-1 [60] and pBacPAK1 [61] are derivatives of pAcYM1, which contains an M13 origin of replication to facilitate production of single-stranded DNA (ssDNA) in *E. coli*. The ssDNA can then be used as a template for DNA sequencing and for site-directed mutagenesis. Plasmids pBacPAK8 and pBacPAK9 are derivatives of pBacPAK1 that have 18 unique cloning sites downstream from the polyhedrin promoter. *A Bam*HI site is located

TABLE 1. Baculovirus Transfer Vectors

Vector	Promoter	Cloning sites	Locus	Notes	Reference (Source)
Polyhedrin locus, polyhedrin promoter, single gene, no ATG supplied					
p36C	P_{polh}	*Bam*HI	*polh*		[63]
pAc373	P_{polh}	*Bam*HI	*polh*		[1,52]
pAcCL29-1	P_{polh}	*Sst*I, *Kpn*I, *Sma*I, *Bam*HI, *Xba*I, *Sal*I, *Pst*I	*polh*	M13 ori	[60]
pAcYM1	P_{polh}	*Bam*HI	*polh*		(Invitrogen)
pBac3	P_{polh}	*Nhe*I, *Bam*HI, *Xho*I, *Sac*I, *Bgl*II, *Pst*I, *Pvu*II, *Kpn*I, *Nco*I, *Eco*RI, *Hin*dIII			
pBacPAK1	P_{polh}	*Bam*HI	*polh*	M13 ori	[61,81] (Clontech)
pBacPAK8	P_{polh}	*Bam*HI, *Sse*83871, *Pst*I, *Stu*I, *Xho*I, *Bst*BI, *Xba*I, *Bgl*II, *Asp*718I, *Kpn*I, *Ecl*136II, *Sac*I, *Eco*RI, *Xma*I, *Sma*I, *Not*I, *Eag*I, *Pac*I	*polh*	M13 ori	(Clontech)
pBacPAK9	P_{polh}	*Bam*HI, *Xma*I, *Sma*I, *Eco*RI, *Ecl*136II, *Sac*I, *Asp*718I, *Kpn*I, *Bgl*II, *Xba*I, *Bst*BI, *Xho*I, *Stu*I, *Sse*8387I, *Pst*I, *Not*I, *Eag*I, *Pac*I	*polh*	M13 ori	(Clontech)
pEV55	P_{polh}	*Bgl*II, *Xho*I, *Eco*RI, *Xbq*I, (*Cla*I)[a], *Kpn*I	*polh*		[53]
pEVmod	P_{polh}	*Bgl*II, *Xho*I, *Eco*RI, *Xba*I, (*Cla*I), *Kpn*I	*polh*		[62]
pEVmXIV	P_{XIV}	*Bgl*II, *Xho*I, *Eco*RI, *Xba*I, (*Cla*I), *Kpn*I	*polh*		[62]
pVL941	P_{polh}	*Bam*HI	*polh*		[56]
pVL1392	P_{polh}	*Bgl*II, *Pst*I, *Eco*RI, *Eag*I, *Not*I, *Xba*I, (*Kpn*I), *Xma*I/*Sma*I, *Bam*HI	*polh*		[2,6]
pVL1393	P_{polh}	*Bam*HI, *Xma*I/*Sma*I, (*Kpn*I), *Xba*I, *Eco*RI, *Not*I, *Eag*I, *Pst*I, *Bgl*II	*polh*		[2,6]

(continued)

TABLE 1. Baculovirus Transfer Vectors *(Continued)*

Vector	Promoter	Cloning sites	Locus	Notes	Reference (Source)
Polyhedrin locus, late promoter, single gene, no ATG supplied					
pc/pS1	$P_{cap/polh}$	*Bgl*II, (*Bsu*36I), *Sma*I, *Sst*II, (*Bsu*36I), *Xba*I, (*Eco*RI), *Xho*I, (*Bam*HI), *Pst*I, *Kpn*I	*polh*		Sihler, O'Reilly, and Miller cited in ref. 6
pAcMP1	P_{cor}	(*Dra*I), *Bam*HI	*polh*		[66]
Polyhedrin locus, polyhedrin promoter, coexpression of single gene and *lac*Z marker gene, no ATG supplied					
pAcDZ1	P_{polh}	*Bam*HI	*polh*	*lac*Z	[72]
pJV*Nhe*I	P_{polh}	*Nhe*I	*polh*	*lac*Z, fl ori	[70]
pBlueBac (pJVETL)	P_{polh}	*Nhe*I	*polh*	*lac*Z, fl ori	[71] (Invitrogen)
pBlueBac2 (pJVETL2)	P_{polh}	*Nhe*I, *Bam*HI	*polh*	*lac*Z	[71] (Invitrogen)
pBlueBac3	P_{polh}	*Bam*HI, *Bgl*II, *Pst*I, *Nco*I, *Hin*dIII	*polh*	*lac*Z	[71] (Invitrogen)
pBlueBac3;2	P_{polh}	*Nhe*I, *Bam*HI, *Xho*I, *Bgl*II, *Pst*I, *Kpn*I, *Nco*I, *Hin*dIII	*polh*	*lac*Z	(Invitrogen)
Polyhedrin locus, coexpression of single gene and polyhedrin, no ATG supplied					
pSynXIV VI$^+$	$P_{SynXIV^{(-)}}$	*Eco*RI, *Pst*I, (*Bam*HI), *Spe*I, *Xba*I, *Not*I, *Bst*XI, *Sst*II, *Sst*I	*polh*		[62]
pSynXIV VI$^+$ X3	$P_{SynXIV^{(-)}}$	*Eco*RI, *Xho*I, *Pst*I, *Bgl*II, *Sal*I, *Sma*I, poly(A), *Sst*I	*polh*		[62]
pAcUW2B	$P_{p10^{(-)}}$	*Bgl*II, (*Eco*RI), (*Pvu*II)	*polh*		[74]
pAcUW21	$P_{p10^{(-)}}$	*Bgl*II, *Eco*RI	*polh*		(Pharmingen)
Polyhedrin locus, polyhedrin promoter, single gene, ATG supplied					
pAcC4	P_{polh}	*Nco*I, *Sma*I, *Kpn*I, *Pst*I, *Bgl*II, *Xba*I, *Eco*RI, *Bam*HI	*polh*		O'Rourke and Kawasaki, cited in ref. 6

pAcC5	P_{polh}	*Nco*I, *Bam*HI, *Eco*RI, *Xba*I, *Bgl*II, *Pst*I, *Kpn*I, *Sma*I	*polh*		O'Rourke and Kawasaki, cited in ref. 6
pAcCL29-8	P_{polh}	*Pst*I, *Sal*I, *Xba*I, *Bam*HI, *Sma*I, *Kpn*I, *Sst*I	*polh*	M13 ori	[60]
Polyhedrin locus, polyhedrin promoter, single gene, ATG fused to polyhedrin coding sequences					
pAc360	P_{polh}	*Bam*HI	*polh*		[33]
pAc700, pAc701, pAc702	P_{polh}	*Bam*HI, *Kpn*I	*polh*		[1]
Polyhedrin locus, polyhedrin promoter, single gene, ATG fused to metal-binding domain/enterokinase cleavage site					
pBlueBacHIS −A, −B, −C	P_{polh}	*Bam*HI, *Bgl*II, *Pst*I, *Nco*I, *Hin*dIII	*polh*		(Invitrogen)
pBlueBacHIS −A;2, −B;2, −C;2	P_{polh}	*Nhe*I, *Bam*HI, *Xho*I, *Bgl*II, *Pst*I, *Pvu*II, *Kpn*I, *Nco*I, *Eco*RI, *Hin*dIII	*polh*		(Invitrogen)
Polyhedrin locus, polyhedrin promoter, single gene, ATG fused to honey bee mellitin signal peptide					
pVT-Bac	P_{polh}	*Bam*HI, *Sma*I, *Pst*I, *Sst*I, *Not*I, *Nhe*I, *Eco*RI, *Kpn*I	*polh*	fl ori	[68]
Polyhedrin locus, coexpression of heterologous gene and polyhedrin, ATG supplied					
pSynXIV VI$^+$ X3/2, X3/3, X3/4	$P_{SynXIV^{(-)}}$	*Eco*RI, *Pst*I, *Bgl*II, *Sal*I, *Sma*I, poly(A), *Sst*I	*polh*		[62]
Polyhedrin locus, dual expression of two heterologous genes					
p2XIV VI$^-$	$P_{XIV^{\#1\,(-)}}$	*Eco*RI, *Pst*I, *Bam*HI, *Spe*I, (*Xba*I), *Not*I, *Bst*XI, *Sst*II, *Sst*I	*polh*		[62]
	$P_{XIV^{\#2\,(+)}}$	*Bgl*II, *Xho*I, (*Xba*I), (*Cla*I), *Kpn*I			
p2Bac;2	$P_{p10}{}^{(-)}$	*Bsi*WI, *Spl*I, *Not*I, *Eag*I, *Srf*I, *Xma*I, *Sma*I, *Asc*I, *Bss*HI, *Afl*II, *Stu*I, *Sse*8387I, *Apa*I, *Sac*II, *Bgl*II, *Xba*I	*polh*		(Invitrogen)

(*continued*)

TABLE 1. Baculorvirus Transfer Vectors *(Continued)*

Vector	Promoter	Cloning sites	Locus	Notes	Reference (Source)
	P_{polh} (+)	*Nhe*I, *Bam*HI, *Xho*I, *Spe*I, *Pst*I, *Pvu*II, *Kpn*I, *Nco*I, *Eco*RI, *Hin*dIII			
pAcUW3	P_{p10} (−)	*Bgl*II, (*Eco*RI), (*Pvu*II)	*polh*		[75]
	P_{polh} (+)	*Bam*HI			
pAcUW31	P_{p10} (−)	*Bgl*II	*polh*		(Lopez-Farber and Possee, Clontech)
	P_{polh} (+)	*Bam*HI			
pAcUW51	P_{p10} (−)	*Bgl*II, *Eco*RI	*polh*		(Pharmingen)
	P_{polh} (+)	*Bam*HI			
p10 locus, *p10* promoter, single gene, no ATG supplied					
pAcAs2	P_{p10}	*Bam*HI, *Sma*I, (*Hin*dIII)	*p10*		[76]
pAcUW1	P_{p10}	*Bgl*II	*p10*		[74]
pEP252	P_{p10}	*Bam*HI	*p10*		Kuzio and Faulkner, cited in ref. 6
p10 locus, *p10* promoter, coexpression of heterologus gene and *lac*Z, no ATG supplied					
pAcAS3	P_{p10}	*Bam*HI	*p10*	*lac*Z	[76]
p10 locus, dual expression of two heterologus genes, no ATG supplied					
pAcUW41	P_{p10} (+)	*Bgl*II	*p10*		(Pharmingen)
	P_{polh} (+)	*Bam*HI			

[a]Restriction sites in the polylinker rgion downstream from a promoter that are not unique are shown in parentheses.

closest to the promoter and a *Pac*I site is located at the end of the cloning region in both plasmids. The two plasmids differ from each other only in the orientation of the 16 other unique restriction sites between the *Bam*HI and *Pac*I sites. The *Pac*I site provides translational stop codons in all three reading frames for expressing truncated proteins. Plasmid pBac3 (Invitrogen) is a derivative of pBlueBac3 (see below) and contains 11 unique cloning sites in a polylinker region downstream from the polyhedrin promoter.

Plasmids pEVmod [62] and pEVmXIV [62] are derivatives of pEV55 [53] that contain the same short polylinker located between positions +1 and +631 of the polyhedrin gene. Inconvenient restriction sites in the flanking regions of the polyhedrin gene in pEV55 were deleted to generate pEVmod. pEVmXIV is a derivative of pEVmod that contains a modified polyhedrin promoter, P_{XIV}, that is 50% stronger than the wild-type polyhedrin promoter P_{polh} [58]. P_{XIV} is a linker-scan mutant with a *Hin*dIII linker (CCAAGCTTGG) located between positions −72 and −61, just upstream from the transcriptional start site within the polyhedrin promoter at position −50.

Luckow and Summers [56] used a different strategy to design transfer vectors that would achieve high-level expression of nonfused foreign genes. This approach was based on the observation that the levels of accumulation for β-galactosidase or chloramphenicol acetyltransferase (CAT) proteins fused to the N terminus of polyhedrin were four- to fivefold greater than when they were expressed as nonfused proteins using pAc373. In pVL941, the polyhedrin start codon of the fusion vector pAc311 [33] was changed from ATG to ATT by site-directed mutagenesis, which introduced an *Ssp*I site (AATATT) from positions −3 to +3. This did not affect the level of steady-state mRNA directed by the polyhedrin promoter. Foreign genes are inserted into a unique *Bam*HI site located in the former polyhedrin coding sequences after position +35. A similar vector, p36C, was derived from the fusion vector pAc360 [33] by changing the polyhedrin ATG to ATC [63]. Plasmids pVL1392 and pVL1393 are derivatives of pVL941 that differ only in the orientation of a polylinker that contains eight unique restriction sites [described in ref. 61].

Translation of the hybrid mRNAs in viruses constructed from pVL941, pVL1392, or pVL1393 should begin at the first ATG encountered in the foreign gene inserted into the cloning site downstream from the polyhedrin promoter. In recent studies, however, translation was also observed to initiate at the ATT at a low level, to give two proteins if the ATT was in the same reading frame as the ATG of the cloned foreign gene [64]. The major species was the unfused foreign protein, and the minor, unexpected, species was a polyhedrin fusion protein. The second species was not produced if the ATG of the cloned gene is out-of-frame with the ATT at +1. Undesired fusion species might also be produced using other vectors that have codons such as ACG or ATT that can function as a translational start site upstream from and in-frame with the foreign ATG. A minor species believed to be a polyhedrin/CAT fusion protein appeared to be produced by AcYM1-CAT [56], and the levels of nonfused CAT produced by AcYM1-CAT and VL941-CAT were identical. In this study, the CAT gene in AcYM1-CAT was inserted in-frame with an ACG codon at position +1, but was out-of-frame with respect to the ATT at +1 in VL941-CAT.

When pVL941 was constructed, translation initiation at ACG codons had been reported in mammalian systems [56], but not at AUU triplets. In retrospect, these observations are not so surprising, if the bases immediately preceding the polyhedrin AUG constitute an optimal ribosomal binding site. Recently, systematic studies have tested the efficiency of translation initiation at non-AUG codons in plant protoplasts by monitoring the expression of CAT from mRNAs that differ by one residue from the AUG start codon [65]. CUG showed the most activity (30% of the AUG activity), followed by GUG and ACG (both 15%), AUA and AUU (both 5%), and UUG (3%). No significant activity ($< 0.1\%$) could be detected if the codons were AAG or AGG. Although these results are consistent with the production of minor fusion proteins by the recombinant baculoviruses described above, similar systematic studies of translation initiation will need to be carried out in insect cells.

Expression of heterologous genes inserted into pAcMP1 [66] or pc/pS1 [Sihler, O'Reilly, and Miller, cited in ref. 6] are expressed under the control of promoters that become active earlier in infection than the polyhedrin promoter. In pAcMP1, transcription is under the control of the promoter for the basic core protein of AcMNPV. Transcription from P_{cor} is active between 8 and 24 hpi, with peak synthesis between 12 and 15 hpi. Recombinant viruses containing the *E. coli lazZ* gene under the control of P_{cor} accumulate β-galactosidase at moderate levels compared with those vectors that contain the *lacZ* gene under the control of the polyhedrin or *p10* promoter. A hybrid capsid/polyhedrin promoter, $P_{cap/polh}$ [67], is used to drive expression of genes inserted into pc/pS1. The capsid promoter is fused to the *Hin*dIII site of the modified polyhedrin promoter P_{XIV}. Transcription from the distal capsid promoter initiates in the late stage of infection, and transcription from the stronger proximal polyhedrin promoter initiates at the very late stage of infection. Expression of heterologous genes at earlier times in infected cells may lead to more uniform processing or post-translational modifications than at the very late stages of infection where the capacity of the cell to carry out normal metabolic processes may be diminished, particularly for proteins that are abundantly expressed under the control of the *p10* or polyhedrin promoter.

Several transfer vectors have been constructed that supply an ATG start codon upstream from a cloning site for insertion of foreign genes. In pAcC4 and pAcC5 [O'Rourke and Kawasaki, cited in ref 6] two nucleotides at positions −2 and −1 were altered so that the polyhedrin ATG start codon embedded is within an *Nco*I site (CCATGG) followed by a polylinker region. The fusion vectors pAc311 and pAc360 [33] contain a *Bam*HI site located within the polyhedrin coding sequences after positions +35 and +34, respectively. Genes inserted in frame with the polyhedrin ATG are expressed as fusions to the N terminus of polyhedrin. These types of fusion proteins often accumulate to higher levels than the nonfused protein, apparently as the result of stabilization by the polyhedrin sequences. Plasmids pAc700, pAc701, and pAc702 [1] are derived from pAc373 and contain *Bam*HI and *Kpn*I sites in all three reading frames behind the polyhedrin ATG start codon. Plasmid pAcCL29-8 [60], which was derived from pAcYM1, has an ATG start codon embedded within a synthetic sequence upstream from the multiple cloning sites.

Tessier et al. [68] described construction of pVT-Bac, which contains a honeybee melittin signal peptide to facilitate the secretion of heterologous proteins. A synthetic polylinker containing eight unique restriction sites is positioned downstream from the sequence encoding the signal peptide. The secretion of the precursor of papain (propapain) was increased five- to eightfold using this signal peptide compared with the native plant signal peptide.

Polyhedrin-Based Coexpression Vectors

Pennock et al. [34] first demonstrated that the *E. coli lac*Z gene could be fused in-frame to various portions of the AcMNPV polyhedrin gene that are expressed as active β-galactosidase fusion proteins. Recombinant viruses form blue occlusion-minus plaques when the agarose overlay in plaque assays contains a chromogenic substrate for β-galactosidase, such as X-gal. These results suggested that a broadly applicable simple visual screening system could be developed for many eukaryotic viruses. Chakrabarti et al. [69] subsequently described the construction of plasmid vectors that direct the insertion of a foreign gene together with the *lac*Z gene into the thymidine kinase locus of the vaccinia virus genome. These plasmids, termed coexpression vectors, greatly facilitated the visual screening for recombinant vaccinia viruses.

Vialard et al. [70] constructed a baculovirus coexpression vector, pJV*Nhe*I, that contains a DNA cassette consisting of the *p10* promoter (P_{p10}), the *lac*Z gene, and an SV40 poly(A) signal inserted into the *Eco*RV site upstream from the polyhedrin promoter. Foreign genes are inserted into an *NheI* site located downstream from the polyhedrin promoter. Transcription by the inserted *p10* promoter is directed in the opposite (−) orientation from the polyhedrin promoter. Recombinant viruses generated from pJV*Nhe*I expressed β-galactosidase at moderate levels, and blue plaques could be detected after 3 days compared with the 5–7 days typically required for optimal visualization of colorless occlusion-minus plaques. Although this expedites the identification and purification of recombinant viruses, a significant proportion of the blue plaques are often occlusion-positive when wild-type baculovirus DNA is used as the parent virus in the transfection. In most cases, viruses with a blue occlusion-plus phenotype are the result of single crossovers between the transfer vector and the parent baculovirus genome. These viruses are usually unstable and can be resolved by a second crossover at the polyhedrin locus to generate a wild-type or a pure recombinant virus. A small proportion of these viruses may have coexpression vector inserted randomly into other regions of the viral genome. The large size of the *lac*Z gene also limits the number of convenient restriction sites in the transfer vector and increases the chance of nonblue recombinant viruses. A series of improved coexpression vectors have since been introduced that are smaller and contain more convenient cloning sites downstream from the polyhedrin promoter. These include pBlueBac, pBlueBac2, and pBlueBac3 [71]. A weak early promoter (P_{ETL}) was substituted for the *p10* promoter to drive transcription of the *lac*Z gene, which is adequate to generate blue plaques and not interfere with the expression from the polyhedrin promoter late in infection. A more recent version of pBlueBac3, designated here as pBlueBac3;2 (Invitrogen), has a more versatile polylinker region and changes in the polyhedrin leader sequence that should improve translation.

Zuidema et al. [72] described construction of a different coexpression vector, pAcDZ1, which contains the promoter from the *Drosophila* heat shock protein 70 (P_{hsp70}) driving transcription of the *lac*Z gene. An SV40 poly(A) signal is inserted downstream from the polyhedrin promoter at the 3′ end of the *lac*Z gene. Transcription from P_{hsp70} is directed in the opposite orientation from that directed by the polyhedrin promoter. This differs from the pBlueBac series of vectors that have the *lac*Z transcriptional cassette inserted upstream from the polyhedrin promoter. Smaller derivatives of pAcDZ1, such as pAcDZ6, may be more convenient for routine cloning of foreign genes downstream from the polyhedrin promoter. Similar coexpression vectors have also described that contain the *lac*Z gene under the control of the promoter for the Rous sarcoma virus long terminal repeat [73].

The pBlueBacHIS A, B, and C series of vectors (Invitrogen) were designed to facilitate expression of fusion proteins that can be purified in one step over metal affinity columns. The vectors contain a natural polyhedrin leader sequence and an ATG start codon followed by a sequence that encodes six histidine residues and an enterokinase cleavage site. Foreign genes are inserted into a polylinker region positioned downstream from this sequence. The three vectors differ only in the reading frame of the start of polylinker region with respect to the ATG start codon. The fusion proteins are purified by binding to nickel-charged sepharose resin and eluted under denaturing conditions (pH 4.0 and containing urea) or under native conditions (pH 6.0). The eluted protein can then be treated with enterokinase to cleave off the metal binding domain present on the N terminus of the fusion protein. These vectors also contain the *E. coli lac*Z gene under the control of P_{ETL} to facilitate screening for recombinant viruses in the presence of X-gal or Bluo-gal. A newer set of vectors, designated here as pBlueBacHIS-A;2, -B;2, and C;2 (Invitrogen), have a more versatile polylinker region and changes in the polyhedrin leader sequence that should improve translation.

A number of plasmids have been designed that permit the expression of a single foreign gene in an occlusion-positive recombinant virus. The occluded form of these viruses can be added to the diet of insect larvae that become infected and express the desired foreign protein in a wide variety of tissues. The plasmids pSynXIV VI$^+$ and pSynXIV VI$^+$ X3 contain the hybrid promoter P_{synXIV} inserted upstream and in the opposite direction from the intact polyhedrin gene [62]. The hybrid promoter contains a synthetic promoter P_{syn} and the linker-modified polyhedrin promoter P_{XIV} arranged in tandem. The two plasmids differ only in the sequence of the polylinker region downstream from the hybrid promoter. The plasmids pAcUW2B [74] and a smaller derivative, pAcUW21 (Pharmingen), are similar, but have a *p10* promoter and an SV40 poly(A) region inserted upstream and in the opposite orientation from the intact polyhedrin gene. An occlusion-minus parent virus should be used with these plasmids to generate the recombinant virus (see section on parent viruses, below).

Plasmids pSynXIV VI$^+$X3/2, X3/3, X3/4 are derived from pSynXIV VI$^+$ X3 [62], differing only in the polylinker region that contains an ATG upstream from the cloning sites. The cloning sites are positioned in all three reading frames downstream from the ATG start codon. Heterologous genes lacking an ATG can easily be inserted into one of the three vectors that is used to generate an occlusion-positive rcombinant virus.

Polyhedrin-Based Dual-Expression Vectors

Dual-expression vectors permit the simultaneous expression of two foreign genes in a single recombinant virus. In pAcUW3 [75], one gene is expressed under the control of the polyhedrin promoter, while the other is expressed under the control of the *p10* promoter. This plasmid is a derivative of pAcUW2B [74], which has the *p10* promoter and SV(40) poly(A) sequences inserted upstream and directed in the opposite orientation from the polyhedrin promoter. The polyhedrin gene of pAcUW2B was replaced with the polyhedrin promoter and cloning site of pAcYM1 [57]. Plasmids pAcUW31 (M. Lopez-Ferber and Possee, Clontech) and pAcUW51 (Pharmingen) were derived from pAcUW3 by deleting redundant sequences and adding an M13 origin. Plasmid p2XIV VI$^-$ [62] contains two P_{XIV} promoters arranged in opposite orientations at the polyhedrin locus. Eight unique restriction sites are present in the polylinker linked to the inverted P_{XIV} promoter, while three unique sites are present in the other polylinker. Plasmid p2Bac (Invitrogen) has the polyhedrin promoter and *p10* promoter arranged in a back-to-back orientation at the polyhedrin locus. A segment containing the bovine growth hormone poly(A) transcriptional termination signals is inserted downstream from the *p10* promoter. The polylinker located immediately downstream from the *p10* promoter has at least 19 unique restriction sites, and the other polylinker downstream from the polyhedrin promoter has at least 7 unique restriction sites. The plasmid p2Blue (Invitrogen) is a derivative of p2Bac that contains the P_{ETL}-*lac*Z-SV40 poly(A) cassette from the pBlueBac vectors to facilitate identification of recombinant viruses.

p10-Based Single Gene Transfer Vectors

Transfer vectors that direct the insertion of foreign genes into the *p10* locus have also been developed. Plasmid pAcAS2 [76] contains a segment of the AcMNPV genome that includes the *p10* promoter and flanking regions inserted into the pUC19 cloning vector. Portions of the *p10* coding sequences between +2 and +282 were removed and replaced by a short segment with a unique *Bam*HI and *Sma*I site. Plasmids pAcUW1 (74) and pEP252 [J. Kuzio and P. Faulkner, cited in ref. 6] are very similar to pAcAS2 except that they have much shorter 5′ flanking regions upstream from the *p10* promoter. pAcUW1 has a unique *Bgl*II site and pEP252 has a unique *Bam*HI site after position +1. Recombinant viruses are typically generated with these plasmids by using parent viruses, such as vAcAS3, AcUW1-*lac*Z, or Ac228z (see below), which contain the *lac*Z gene inserted into the *p10* locus. Recombinant viruses are identified as white occlusion-positive plaques in a background of blue occlusion-positive plaques. If AcUW1-PH is used as a parent virus, recombinant viruses have white occlusion-minus plaques in a background of white occlusion-positive plaques. In this parent virus, an intact polyhedrin gene is located at the *p10* locus, but not at the polyhedrin locus.

p10-Based Coexpression Vectors

Plasmid pAcAS3 [76] is a derivative of pAcAS2 [76] that contains a P_{hsp70}-*lac*Z-SV40 poly(A) cassette inserted downstream and in the opposite orientation from the *p10* promoter. Recombinant viruses are identified as blue occlusion-positive plaques in a background of colorless occlusion-positive plaques if wild-type AcMNPV is used as a parental virus.

p10-Based Dual-Expression Vectors

Only one dual-expression vector has been described that targets genes for insertion at the *p10* locus. Plasmid pAcUW41 (Pharmingen) is a derivative of pAcUW1 [74] and contains a copy of the polyhedrin promoter inserted downstream from the *p10* promoter in the same orientation. Transcription from the *p10* promoter is terminated within an SV40 poly(A) region that separates the two promoters. Transcription from the polyhedrin promoter terminates in the *p10* poly(A) signals. Recombinant viruses can be identified as white occlusion-positive plaques in a background of blue occlusion-positive plaques if AcUW1-*lac*Z or a similar virus is used as a parent virus.

PARENT VIRUSES

A variety of AcMNPV derivatives can be used as parent viruses for the generation of recombinant baculoviruses. These are briefly summarized in Table 2, which lists their plaque phenotype, ancestry, and notes about their construction. The plaque phenotype of recombinant viruses generated from different combinations of parent viruses and transfer vectors is shown in Table 3.

AcRP6-SC [77] is an occlusion-minus virus that contains a unique *Bsu*36I site inserted downstream from the polyhedrin promoter. Linearized (*Bsu*36I-digested) DNA is 15- to 150-fold less infectious than circular AcRP6-SC or wild-type AcMNPV. When linearized AcRP6-SC DNA was used as the parent virus in transfections with an appropriate transfer vector, the proportion of recombinant viruses after one plaque assay approached 30%. Two other AcMNPV derivatives have also been constructed to contain unique restriction sites, but retain the polyhedrin gene. Sewall and Srivastava [78] inserted an *Sse*8387I site upstream from the polyhedrin gene (in a virus designated here as AcMNPV-Inv), and Hartig and Cardon [79]

TABLE 2. AcMNPV-Based Parent Viruses for Use in Constructing Recombinant Baculoviruses

Virus name	Promoter	Marker	Notes	Locus	Plaque phenotype	AcNPV variant	Reference
AcRP6-SC			*Bsu*36I site in polylinker replaces polyhedrin	*polh*	occ$^-$, white	C-6	[77]
vAcInv	P_{polh}	*polh*	Sse8387I site upstream of polyhedrin	*polh*	occ$^+$, white	E2	[78]
AcV-EPA	P_{polh}	*polh*	*Bsu*36I site downstream of polyhedrin ATG (+845)	*polh*	occ$^+$, white	E2	[79]
AcMNPV-*lac*Z	P_{polh}	*lac*Z		*polh*	occ$^-$, blue	C-6	[80]
BacPAK6, BaculoGold	P_{polh}	*lac*Z	3 *Bsu*36I sites, one in ORF603, one in *lac*Z, one in ORF1629	*polh*	occ$^-$, blue	C-6	[61, 81]
vSynVI-gal	P_{Syn}	*lac*Z		*polh*	occ$^-$, blue	L-1	[62]
AcAs3	P_{hsp70}	*lac*Z		*p10*	occ$^+$, blue	E2	[76]
AcUW1-*lac*Z	P_{p10}	*lac*Z		*p10*	occ$^+$, blue	C-6	[74]
AcUW1-PH	P_{p10}	*polh*		*p10*	occ$^+$, white	C-6	[74]
Ac228z	P_{p10}	*lac*Z	Out-of-frame *lac*Z at *p10* locus	*p10*	occ$^+$, blue	HR3	Kuzio and Faulkner, cited in ref. 6
vDA26	P_{DA26}	*lac*Z		*DA26*	occ$^+$, blue	L-1	[82]

TABLE 3. Plaque Phenotype With Different Types of Transfer Vectors and Parent Viruses

	Virus				
Transfer vector	wt ACMNPV, vAcInv, vAcEPA (o^+, z^-)	vBacPAK6, vSynVI-gal, AcNPV-*lacZ* (o^-, z^+)	RP6-SC (o^-, z^-)	Ac228z AcUW1-*lacZ* AcAS3 (o^+, z^+)	vDA26Z (o^+, z^+)
None	occ^+, white	occ^-, blue	occ^-, white	occ^+, blue	occ^+, blue
Polyhedrin locus, single or dual gene (pVL1393, p2bac) (o^-,z^-)	occ^-, white	occ^-, white	ND	occ^-, blue	occ^-, blue
Polyhedrin locus, single or dual gene, coexpress *lacZ* (pBlueBac2, p2blue) (o^-, z^+)	occ^-, blue	ND	occ^-, blue	occ^-, blue	occ^-, blue
Polyhedrin locus, single gene, coexpress polyhedrin (pSynXIV VI+) (o^+, z^-)	ND	occ^+, white	occ^+, white	ND	ND
p10 locus, single or dual gene (pAcAS2, pEP252, pAcUW41) (z^-)	ND	ND	ND	occ^+, white	ND
p10 locus, single gene, coexpress *lacZ* (pAcAS3) (z^+)	occ^+, blue	ND	occ^-, blue	ND	ND

Abbreviations: ND, recombinant virus phenotype is indistinguishable from the parent virus; occ^-, occlusion (polyhedrin) minus; occ^+, occlusion (polyhedrin) positive; blue, expression of β-galactosidase by *lacZ* detectable in agarose overlays containing X-gal or Bluo-gal; white, no expression of β-galactosidase.

inserted a *Bsu*36I site downstream from the polyhedrin gene (at position +845) in AcV-EPA. A high frequency of the recombinant viruses generated from linearized AcMNPV-Inv or AcV-EPA will have an occlusion-minus phenotype in a background of occlusion-positive parental viruses (Table 3).

Several recombinant viruses that contain the *E. coli lac*Z gene inserted into the polyhedrin locus can be used to screen for recombinant viruses that form white plaques in a background of blue occlusion-minus plaques. AcMNPV-*lac*Z [80] and Ac360-βgal [21] contain the *lac*Z gene under the control of the polyhedrin promoter. vSyn VI$^-$gal [62] contains the *lac*Z gene under the control of a synthetic promoter (P_{syn}) that is inserted as a cassette into the polyhedrin locus but in the opposite (−) orientation from normal polyhedrin transcription. A single *Bsu*36I site is present in the *lac*Z gene, so it is possible to linearize the genomic DNA of these viruses to enhance the recovery of recombinant progeny viruses. All three can be used as parental viruses with most polyhedrin-based transfer vectors, except for the *lac*Z coexpression vectors (Table 3). White occlusion-minus plaques are generated by vectors such as pVL1393 or p2Bac, and white occlusion-positive plaques are generated by vectors such as pSynSIV VI$^+$ that contain an intact polyhedrin gene. The plaque phenotype of a *lac*Z coexpression vector, such as pBlueBac2, is indistinguishable from the parent virus in this case.

A novel AcMNPV derivative has recently been described that can be linearized with *Bsu*36I and used to generate recombinant viruses at a high frequency [61,81]. This virus (marketed as BacPAK6 by Clontech and as BaculoGold by Pharmingen) contains three *Bsu*36I sites near the polyhedrin locus. One site is in a *lac*Z gene under the control of the polyhedrin promoter. A second site is located in a nonessential gene (ORF603) located upstream from the polyhedrin promoter, and a third site is located in an essential gene (ORF1629) located immediately downstream from the polyhedrin coding sequences. BacPAK6 DNA digested with *Bsu*36I yields two small fragments, one containing portions of ORF603 and *lac*Z and the other containing the remaining portion of *lac*Z and part of ORF1629. If the remainder of the genome recircularizes, it is missing part of an essential gene and cannot produce viable viruses. When linearized BacPAK6 DNA is transfected into insect cells along with a transfer vector, viable progeny are produced that have the ORF1629 gene restored by homologous recombination. About 95% (range 86%–99%) of the progeny are the desired recombinant viruses in a background of blue occlusion-minus viruses. The remainder may result from loss of only the *Bsu*36I fragment containing portions of ORF603 and the *lac*Z genes to yield white occlusion-minus viruses. Theoretically, this background could be decreased if the *Bsu*36I digestions were complete and the small fragments were removed from the large fragment by some method of size selection. Alternatively, unique sites not recognized by *Bsu*36I, such as Sse8387I or other infrequent cutters, might be inserted into the genome so that the sticky ends of the large fragment or the large fragment containing an intact ORF1629 (as the result of incomplete digestion or re-ligation) are incompatible and cannot promote the recircularization of a viable genome.

Several viruses that contain the *lac*Z gene inserted into the *p10* locus can be used as parent viruses for the construction of recombinant viruses. AcAs3 [76],

AcUW1-*lac*Z [74], and Ac228z [Kuzio and Faulkner, cited in ref. 6] all have a blue occlusion-positive plaque phenotype. In AcAS3, the *lac*Z gene is under the control of the *Drosophila* hsp 70 promoter, and in AcUW1-*lac*Z and Ac228z it is under the control of the *p10* promoter. The *lac*Z gene in Ac228z is inserted out-of-frame from the normal ATG start codon, so this virus produces plaques that are faint blue compared to in-frame insertions that produce dark blue plaques in the presence of typical amounts of X-gal. When transfer vectors such as pAcAS2 or pEP252 are used with these parent viruses, to direct the insertion of genes at the *p10* locus, the recombinant progeny produce white occlusion-positive plaques (Table 3). Blue occlusion-minus recombinant viruses are generated when transfer vectors that direct the insertion of genes into the polyhedrin locus are used instead.

The virus vDA26Z [82] contains the *lac*Z gene inserted into the coding sequences of the nonessential *da26* gene, which is transcribed early in infection. The virus produces light blue occlusion-positive plaques due to the weak activity of the *da26* promoter. When polyhedrin-based transfer vectors such as pVL1393 or pBlueBac are used with vDA26 as a parent virus, the recombinant progeny produce blue occlusion-minus plaques (Table 3).

OTHER STRATEGIES TO GENERATE RECOMBINANT BACULOVIRUSES

In Vitro Site-Specific Recombination

Peakman et al. [83] recently developed an efficient in vitro method for the construction of recombinant baculoviruses using the Cre-*lox* system of bacteriophage P1. Oligonucleotides containing recognition sequences (*lox*) for the Cre enzyme were inserted into the AcMNPV genome to generate a parent virus (vAclox) and into a baculovirus transfer vector containing the *lac*Z gene (ploxZ). When these DNAs are incubated in the presence of Cre enzyme, they are able to recombine at the inserted *lox* sites. High-recombination frequencies (5×10^7 recombinant viruses/μg starting plasmid DNA) make this an attractive system for future development of expression cloning methods, even though no more than 50% of the viral progeny are recombinants. Careful optimization of the reaction conditions are required, however, to maximize single insertions of the transfer vector into the parent virus and minimize multiple insertion events or the reverse process. Multiple plaque assays are required to purify the recombinant viruses.

Yeast-Based Baculovirus Shuttle Vectors

Patel et al. [84] recently described the construction of a baculovirus shuttle vector (YCbv::SUP4-o) that can propagate in the yeast *Saccharomyces cervisiae*. Recombinant baculoviruses can be generated by homologous recombination between the shuttle vector and a modified baculovirus transfer that contains a segment of yeast DNA. The shuttle vector contains yeast ARS and CEN sequences, which ensure stable replication in yeast, and two selectable marker genes (URA3 and SUP4-o) inserted as a cassette into the AcMNPV genome downstream from the polyhedrin promoter (P_{polh}) in the order P_{polh}, SUP4-o, ARS, URA3, and CEN. The SUP4-o marker expresses an ochre-suppressing allele of a $tRNA^{tyr}$ gene, and the URA3 marker permits growth in a ura3 mutant host strain. Two transfer vectors were also constructed, pAcY1 and pAcY2, which contain a number of unique cloning sites flanked on the 5′ end by baculovirus sequences and on the 3′ end by the yeast ARS sequence. The cDNAs for two members of a family of transcription factors, CREB1 and CREB2, were then inserted into pAcY1 to generate pAcCREB1 and pAc-CREB2. The yeast strain y657 (mat α hist3-11,15 trp1-1 ade2-1 leu2-3,112 ura3-52 can1-100 his4::HIS3) harboring the shuttle vector was transformed with the transfer vector derivatives and large pink colonies purified and scored for growth on plates containing canavanine, a toxic arginine analog. Colonies containing recombinant shuttle vectors (which lack the SUP4-o gene) are pink due to suppression of the ochre mutation in the host ADE2 gene, which causes accumulation of an intermediate product of the adenine pathway. The colonies harboring the recombinants are also canavanine resistant because the loss of the SUP4-o gene prevents expression of arginine permease, which is the product of the CAN1 gene. Colonies containing the parent shuttle vector alone are white (Ade^+) and canavanine sensitive. It was not possible to select directly for canavanine-resistant colonies due to the low frequency of transformation and a high reversion rate to canavanine resistance. Large colonies were chosen, since the presence of the SUP4-o results in slower yeast growth. The recombinant shuttle vector DNA was then partially purified from total yeast DNA over sucrose gradients and used to transfect insect cells. The recombinant shuttle vectors were able to propagate in insect cells and express the CREB1 or CREB2 genes under the control of the polyhedrin promoter. Expression of CREB1 and CREB2 by recombinant viruses constructed using homologous recombination in insect cells was not reported, so it is not possible to tell from these results if viruses constructed by this method offer any advantage as far as protein expression is concerned. One great advantage of this method is that the time-consuming steps of plaque purification are eliminated, since all of the viral DNA isolated from yeast contains the foreign gene inserted into the baculovirus genome and there is no background of contaminating parent virus. Using this method, stocks of recombinant viruses can be generated within 10–12 days. Its two main drawbacks, however, are the low transformation efficiency of yeast and the tedious purification of recombinant shuttle vector DNA from total yeast DNA over sucrose gradients.

E. coli–Based Baculovirus Shuttle Vectors (Bacmids)

Recently Luckow et al. [85] described a novel strategy to generate recombinant baculoviruses rapidly and efficiently by site-specific transposition of a DNA cassette into a baculovirus shuttle vector (bacmid) propagated in *E. coli.* Two baculovirus shuttle vectors were constructed, bMON14271 and bMON14272, that contain a mini-F replicon, a kanamycin resistance marker, and a segment of DNA that encodes the *lac*Zα peptide from a pUC-based cloning vector. The mini-F replicon contains all the genes required for replication and stable segregation of low-copy-number plasmids. A short segment of DNA containing the attachment site for the bacterial transposon Tn7 (*att*Tn*7*) is also inserted into the N terminus of the *lac*Zα gene so that the reading frame of the α-peptide is not disrupted. The two shuttle vectors differ only in the orientation of the mini-F-Kan-*lac*Zα-mini-*att*Tn7 cassette inserted into the polyhedrin locus of the baculovirus genome. When shuttle vector DNA isolated from insect cells is transformed into the *E. coli* strain DH10B, it can propagate as a large plasmid that confers resistance to kanamycin and can complement a *lac*Z deletion mutation located on the chromosome. Colonies containing bMON14271 or bMON14272 are blue (Lac$^+$) on plates containing X-gal and IPTG. Shuttle vector DNA isolated from *E. coli* is infectious when transfected into insect cells.

Unlike most transposable elements, Tn*7* inserts at a high frequency into the single *att*Tn*7* site located on the *E. coli* chromosome and into DNA segments carrying *att*Tn*7* on a target plasmid. A plasmid containing a deletion derivative of Tn*7* was modified to contain an expression cassette consisting of a gentamicin resistance gene, the polyhedrin promoter, a foreign gene, and SV40 poly(A) transcription termination sequences between the left and right arms of Tn7. The mini-Tn*7* on the donor plasmid can transpose efficiently to the *att*Tn7 on the chromosome or the mini*att*Tn*7* present in the bacmid when Tn*7* transposition functions are provided in trans by a helper plasmid. Insertions of the mini-Tn*7* donor element into the mini-*att*Tn*7* in the bacmid can easily be distinguished from insertions into the *att*Tn*7* in the chromosome by screening for white colonies in a background of blue colonies on agar plates containing X-gal and IPTG. Typically 10%–25% of the total colonies are white, and all contain a composite bacmid with a single insertion of the mini-Tn*7* into the baculovirus genome. When composite bacmid DNAs are isolated from *E. coli* and transfected into insect cells, the foreign gene is expressed at levels that are similar to those observed by viruses constructed by traditional methods.

Generating recombinant baculoviruses by site-specfic recombination in *E. coli* has a number of advantages over homologous recombination in insect cells or in yeast. Since the composite viral DNA isolated from bacteria is already pure, plaque assays are eliminated, and it is possible to obtain pure stocks of recombinant viruses within 7–10 days. The donor plasmids are also small and easily manipulated to facilitate the cloning of foreign genes or other genetic elements within the mini-Tn*7* element. Composite bacmid DNAs can be easily purified from *E. coli* and introduced into insect cells by standard methods. This approach not only greatly facilitates the rapid and simultaneous generation of recombinant baculoviruses compared with traditional methods, but also permits the development of strategies for protein engineering and expression cloning.

PROCESSING OF FOREIGN PROTEINS EXPRESSED IN BACULOVIRUS-INFECTED INSECT CELLS

The use of baculovirus vectors to produce a wide variety of foreign proteins in insect cells has dramatically increased in recent years. In this section, the post-translational processing and modifications observed on proteins expressed in baculovirus-infected insect cells will be briefly summarized, and key examples from recent literature will be cited to illustrate these events. Several review articles and manuals, which list many of the genes expressed in insect cells and the biological properties of the recombinant proteins, should be consulted for additional examples and discussion of earlier literature [1–7]. A survey of the types of post-translational chemical modifications observed in proteins from all sources is described in a recent review [86].

Glycosylation

Many newly synthesized membrane or secretory proteins pass through the endoplasmic reticulum (ER) where they are modified by addition of a high-mannose oligosaccharide core to an asparagine residue in an appropriate sequence context (Asn-X-Ser/Thr, where X is not proline) [reviewed in ref. 87]. The core residues are systematically modified by addition of other sugars by glycosyltransferases and trimming of terminal residues by glycosidases as the protein is transferred to the Golgi complex (GC). Complex oligosaccharides, which are found on many mammalian glycoproteins, contain sialic acid, galactose, and glucosamine residues. *O*-linked glycosylation involves the addition of ologosaccharides to serine and threonine residues. This process is initiated by the addition of *N*-acetylgalactosamine to the protein. No consensus sequence has been determined for *O*-linked glycosylation.

Glycosylation can affect the folding, stability, and solubility of a protein by stabilizing its conformation or protecting it from proteases [88,89]. These factors will influence recovery and purification of a glycoprotein. The antigenicity of a protein may also be influenced by the nature of the oligosaccharides attached to it, and the heterogeneity of these structures can, in some cases, modulate the biological activity, in vivo clearance rate, and immunogenicity of a protein [88,89]. Unfortunately, there are no hard and fast rules to predict what effect the presence or absence of certain oligosaccharide structures will have on these biological properties for any given protein. Differences in the microheterogeneity of oligosaccharide structures are often observed for mammalian glycoproteins expressed in different mammalian cell lines or by individual cell lines under different culture conditions, which may or may not reflect the structure or heterogeneity of the protein in its "native" environment [90,91]. The relative importance of these issues seems to vary significantly among researchers, differing primarily in whether their focus is on events mediated by the protein at the cellular level or on structure–function relationships at the molecular level. Knowledge about the exact nature of the oligosaccharide structures present on a protein expressed in a particular type of cell may be more valuable for a protein intended to be used as a therapeutic biopharmaceutical or for a vaccine compared with a protein used as a reagent in diagnostic assays or as a target in a structure and activity screening program.

There are now hundreds of examples describing the expression of proteins modified by *N*-glycosylation in baculovirus-infected insect cells. In many cases, glycosylation was examined only indirectly by treatment with tunicamycin, by incorporation of radiolabeled sugars, or by treatment with endogylcosidases. The proteins are usually separated on SDS-PAGE gels and detected by staining, immunoblotting, lectin blotting, or autoradiography if they were metabolically labeled with radioactive amino acids or sugars. In the majority of examples, some differences in the sizes and structures of the attached oligosaccharides are observed compared with structures observed in mammalian or plant cells. Usually the insect cell–derived glycoproteins run faster on SDS-PAGE gels than the "native" protein, implying that the oligosaccharide structures are smaller than those produced in the natural host cell. When *N*-glycosylation is inhibited by tunicamycin or when all of the sugars are removed by endoglycosidase digestion using *N*-glycanase, the insect cell–derived proteins are the same size as the native proteins that have undergone the same treatment. This is usually the same size as the recombinant protein recovered from *E. coli,* which cannot *N*-glycosylate proteins. The significance of these differences for a given protein is not always clear, for many baculovirus-expressed glycoproteins retain full biological activity in in vitro assays. Studies directly comparing the in vivo properties, such as targeting, clearance rate, and immunogenicity of baculovirus-derived HIV gp160, for example, are very complex [92–94].

Kuroda et al. [95] directly determined the chemical structures of the oligosaccharide side chains attached to the hemagglutinin of fowl plague (influenza) virus in AcMNPV-infected *Spodoptera frugiperda* cells. The majority were oligomannosidic side chains consisting of $Man_{5-9}GlcNac_2$, and the truncated oligosaccharide cores $Man_3GlcNac_2$ and $Man_3[Fuc]GlcNac_2$. Since complex oligosaccharides occupy most of the glycosylation sites of hemagglutinin isolated from vertebrate cells, these studies demonstrated that baculovirus-infected insect cells have the capacity to trim the oligomannosidic structures to trimannosyl cores and to modify these by the addition of fucose.

The structures of *O*-linked oligosaccharides present on two proteins expressed in AcMNPV-infected SF9 cells have been reported. The major *O*-linked structure present on pseudorabies virus glycoprotein gp50 was the monosaccharide GalNac [96]. This protein lacks the recognition sequences for *N*-linked glycosylation. Smaller amounts of Gal-β-1–3GalNac were also observed on gp50 (12%) and on a truncated version of the protein gp50T (26%). A secreted chimeric glycoprotein composed of the extracellular domains of respiratory syncytial virus F and G proteins contains both *N*- and *O*-linked oligosaccharide structures [97]. The majority were the *O*-linked structures Gal-β-1–3GalNac (66%) and GalNac (17%), and the remainder was the single *N*-linked structure $Man_3[Fuc]GlcNac_2$.

Svoboda et al. [98] determined the structure of mouse interleukin-3 expressed in BmSNPV-infected *Bombyx mori* larvae by mass spectrometry. Two *N*-glycosylation sites are present and modified to contain structures that consist of mannose, fucose, and glucosamine. Incomplete processing of the *N*-linked oligosaccharides was suggested by the presence of glucose residues. Galactosamine was also present, indicating the presence of *O*-glycosylated residues.

Complex glycosylation has been observed on human plasminogen (r-HPg) expressed in several baculovirus-infected insect cell lines, and the nature of the side chains have been characterized directly by chemical methods [99–102]. Heterogeneity was observed in the oligosaccharide structures attached to the sole Asn that resides in a recognition sequence for *N*-glycosylation at position 289. In their first study, the majority of the side chains consisted of high-mannose structures ($Man_9GlcNac_2$) and truncated high-mannose species ($Man_{3-5}GlcNac_2$). The remaining 40% consisted of bisialo-biantennary complex carbohydrate (Sia-Gal-GlcNac-Man)$_2$-Man-$GlcNac_2$) structures. When r-HPg was expressed in *Mammestra brassicae* (IZD-MBO503) cells and purified at 48 hpi, 63% of the oligosaccharides were of the complex type, and the remainder were various high-mannose species. Bisialyo-biantennary (28%), fucosylated bisialo-biantennary (25%), asialo-biantennary (7%), and fucosylated asialo-biantennary (3%) oligosaccharides constituted the majority of complex sugars that were identified. Subsequent studies have demonstrated that the ratio of complex to high-mannose types of structures depended on the time the r-HPg was isolated from the infected cells. Early in infection (0–20 hpi) 96% of the species were the high-mannose type ($Mann_{3-9}GlcNac_2$) compared with late in infection (60–96 hpi) when biantennary, trantennary, and tetraantenarary complex classes with various amounts of outer arm completion was observed. From 20 to 60 hpi, 77% of the structures were of the complex type, and the remainder had various high-mannose structures. These results suggested that *S. frugiperda* cells do contain the appropriate glycosyltransferases required for the assembly of complex oligosaccharides and that these enzymes become induced or unmasked in baculovirus-infected cells. Recent studies have shown that extracts from uninfected *S. frugiperda* cells contain an enzyme that will trim $Man_9GlcNac_2$ to $Man_6GlcNac_2$. Extracts isolated from cells infected with wild-type AcMNPV or from a recombinant virus expressing r-HPg, however, contain enzymes that led to the rapid trimming of $Man_6GlcNac_2$ to $Man_5GlcNac_2$ and $Man_3GlcNac_2$. These activities increased in extracts prepared later in infection.

In a recent study designed to compare the level of expression for several proteins in 23 insect cell lines infected with three recombinant AcMNPV derivatives, Hink et al. [103] observed variations in the size of pseudorabies gp50T separated by SDS-PAGE that suggest that different cell lines may glycosylate to different extents. It is also interesting to note that there was no single best combination of cell line and media that could be used to express all three foreign proteins. The best four cell lines for expression of gp50T were different from the best five for expression of r-HPg and the best three for expression of *E. coli* β-galactosidase. The five best cell lines for r-HPg, for example, produced more enzyme than Sf9 cells, which are widely used to propagate wild-type and recombinant baculoviruses. Most cell lines propagated in serum-free media produced higher yields of recombinant protein than the same cells propagated in serum-supplemented media.

Fatty Acid Acylation

Baculovirus-infected insect cells are capable of modifying specific proteins by fatty acid acylation. Modifications by myristoylation, palmitylation, or isoprenylation are usually demonstrated by the incorporation of their tritium-labeled acid precursors. Since myristic acid can be metabolized to palmitic acid, it is important to determine the structures of the incorporated fatty acids chemically.

The gag precursor proteins of human, simian, and feline immunodeficiency viruses (HIV-1, HIV-2, SIV, FIV) are modified by myristoylation at the gly2 residue at the N terminus [104–110]. Myristoylation appears to be necessary for targeting of the protein to the cell membrane and for incorporation into virus-like structures that bud from the plasma membrane and are released into the culture medium. Unmyristoylated forms are also produced, but these tend to accumulate inside the cells. HIV-1 negative factor (Nef) protein [111] and foot-and-mouth disease virus capsid precursor proteins also are myristoylated [112] in AcMNPV-infected cells.

A variety of membrane-associated proteins involved in signal transduction, such as the ras-related family of proteins that bind GTP, are isoprenylated and palmitylated. These include Ha-ras [113], Kirsten-ras [4B] p21 [114], rab1b, rab3a, and rab6 [115,116], *Xenopus* xlcaax-1 [117], Rap1A (Krev-1) [118], and mammalian G-protein α, β, and γ subunits [119–121]. The Rap1A protein, for example, was synthesized as a 23 kd cytosolic precursor that binds to membranes and was converted into a 22 kd form that incorporated label derived from radiolabeled mevalonate and contained a COOH-terminal methyl group. Chemical analysis indicated that the Rap1A was modified by a C20 (geranylgeranyl) isoprenoid. In contrast, H-ras was modified by a C15 (farnesyl) group.

α-Amidation

Several proteins are amidated at their C terminus. This process involves recognition and cleavage to leave a C-terminal glycine residue that is modified by amidation. When the frog enzyme peptidylglycine α-amidating monooxygenase was purified from baculovirus-infected insect cells, it converted a model peptide to a C-terminal α-hydroxylglycine–extended peptide, X-Gly(OH), instead of the expected amidated product X-NH_2 [122]. A second, distinct insect enzyme, present in the media of infected cells, could complete the reaction to yield the amidated product. Biologically active and amidated cecropin A could be recovered from the hemolymph of *T. ni* larvae infected with a recombinant AcMNPV derivative containing the cDNA for preprocecropin A [123,124]. Recombinant sarcotoxin IA expressed by a BmSNPV vector in Bm-N cells contained glycine at the C terminus and not an amidated arginine as found in the authentic enzyme purified from fleshflies [125]. Gastrin-releasing peptide precursor was not amidated and consequently was inactive, even though it was properly cleaved in insect cells [126].

N-Terminal Acetylation

The N-terminal methionines of many intracellular mammalian proteins are removed, and the second amino acid is modified to contain an acetyl group [127–129]. This process is strongly influenced by the sequence of the first several amino acids. Methionine removal and subsequent N-terminal modification have not been examined systematically, however, in insect cells. Human muscle aldose reductase expressed in insect cells could not be sequenced using Edman degradation chemistry unless the enzyme was treated with acylamino-acid–releasing enzyme [130]. The blocked N-terminal amino acid was found to be acetylalanine. Three forms of human leukotriene A_4 hydrolase (hLTA$_4$H) are found in the media of AcMNPV-infected cells [131], one of which could not be sequenced. The blocked and unblocked forms were present in about equal proportions, and the biological activities of all three forms were indistinguishable. Electrospray ionization mass spectroscopy indicated that the largest two forms differed by a molecular mass of 42, indicating that the blocking group is an acetyl group. In this case, the acetyl group is present on the N-terminal methionine residue. The cDNA for this gene was originally engineered to contain an alanine codon inserted between the N-terminal methionine codon and proline codon for expression in *E. coli,* which is very efficient at cleaving between MetAla and not very efficient at cleaving MetPro [127–129]. Half of the unblocked forms of hLTA$_4$H isolated from the media of infected cells began with the sequence MetAlaPro, and the other half began with AlaPro. This microheterogeneity at the N terminus may be a reflection of the inability of the cells to remove efficiently the methionine late in infection due to the overexpression of the hLTA$_4$H, coupled with a diminished supply of the appropriate processing enzyme that results from the virus-induced shutoff of host proteins. A cDNA lacking the alanine codon was not constructed, so it is not known if the N-terminal amino acid of the native MetPro form of hLTA$_4$H would be modified in insect cells.

Phosphorylation

Phosphorylation of serine, threonine, or tyrosine residues by endogenous protein kinases and the removal of phosphate residues by phosphatases are widely recognized as regulatory mechanisms that modulate the targeting and activities of a protein. Many phosphorylated proteins have been expressed in insect cells, and in most cases the specific amino acids that are modified in insect cells are the same as those observed in the native protein. Some exceptions exist, however, primarily for proteins that are abundantly expressed very late in infection and are underphosporylated compared with the native protein [132]. Altered phosphorylation patterns are also observed when the same proteins are overexpressed in mammalian cells [132]. Phosphorylation of proteins expressed in insect cells is extensively reviewed by O'Reilly et al. [6].

One notable application, recently described by Parker et al. [133], illustrates the use of baculovirus-expressed proteins to understand the regulation of kinases involved in the induction of meiosis. *Schizosaccharomyces pombe* p34cdc2 is a histone H1 kinase and p107wee1 is a kinase that participates in the phosphorylation of p34cdc2. When p34cdc2 and p107wee1 were coexpressed in insect cells, a minor fraction of the p34cdc2 was phosphorylated on tryosine. When clam cyclins A and B, which are involved in the transition into meiosis I and II in the absence of protein synthesis, were coexpressed with p34cdc2 and p107wee1, p34cdc2 formed a complex with both cyclins. The p34cdc2 was extensively phosphorylated and exhibited histone H1 kinase activity. When p107wee1 is incubated with p34cdc2 in vitro, pp34cdc2 is not phosphorylated and the histone H1 kinase activity of p34cdc2 is inhibited. Histone H1 activity of pp34cdc2 is restored only if the both cyclins are present. These and other results indicate that p107wee1 is a dual-specificity protein kinase that functions as a mitotic inhibitor by directly phosphorylating p34cdc2 and that the preferred substrate for phosphorylation is the p34cdc2/cyclin complex.

Subcellular Localization

Generally, proteins expressed in baculovirus-infected cells are directed to the appropriate cellular compartments for processing and targeted to their natural locations in the nucleus, cytoplasm, organelles, plasma membrane, or extracellular space. Some notable exceptions are the secretion of human leukotriene A_4 hydrolase [131], human acid β-galactosidase [134], and human aldose reductase [135]. Large amounts of LTA_4 hydrolase accumulate in the media even though it lacks a signal peptide and is normally thought to be cytoplasmic in mammalian cells. This unexpected observation facilitated the purification of this enzyme since it accounted for nearly 90% of the protein present in the media of cells propagated in a serum-free media. Similarly, large amounts of human aldose reductase can be recovered from the media of infected cells. The acid β-galactosidase is glycosylated, and large amounts are found in the media or diffusely distributed in the cytoplasm and near the nuclear membrane, but not translocated to lysosomes, its normal cellular compartment in human cells.

Human mineralocorticoid receptor (hMR) [136,137] and rat glucocorticoid receptor (rGR) [138] are functional receptors for specific steroid hormones that have been expressed in insect cells. The human MR binds aldosterone, cortisol, cortexolone, and progesterone. The hMR is synthesized and normally resides in the cytoplasm until exposed to a glucocorticoid agonist, which causes it to translocate to the nucleus. Upon exposure to a glucocorticoid agonist, the hMR translocates to the nucleus from the cytoplasm and is able to act as transcriptional enhancer by binding specific DNA sequences. The sedimentation, elecrophoretic mobility, and immunological properties of the recombinant hMR are also indistinguishable from the antive receptor. The properties of the rat glucocorticoid receptor are similar to that reported for hMR, and the recombinant and native proteins are indistinguishable. Between 1 and 3×10^6 rat receptor molecules are expressed per infected ell, which is 15- to 45-fold greater than that normally expressed in a hepatocyte. The unactivated rcombinant rGR could be purified in a simple three-step procedure.

Strong evidence that the product of the cystic fibrosis gene (CFTR) is a cyclic AMP–stimulated chloride anion channel protein was demonstrated by expressing CFTR in baculovirus-infected SF9 cells [139–141]. Membrane localization, epitope mapping, and anion flux, monitored by radioiodide efflux and patch clamping, mimicked in many ways the properties of CFTR expressed in epithelial cells.

Proteolytic Processing

With few exceptions, insect cells are able to recognize and properly cleave signal peptides that direct a protein into the endoplasmic reticulum for further processing and subsequent targeting to cellular membranes or secretion into the extracellular space. In most studies, proper cleavage is inferred from the proper location of the product and its size compared with the native protein. Direct sequence analysis of the N terminus is used to confirm the proper cleavage and has been reported for several proteins. One exception that should be noted, however, is the protective antigen (PA) of *Bacillus anthracis* [142,143]. The baculovirus-expressed product was slightly larger (86 kd) than the PA produced in *B. anthracis* (83.5 kd), and trypsin mapping indicated that the signal sequence remained on the insect cell–derived PA. The recombinant PA was immunogenic, however, and protected guinea pigs and mice against a lethal *B. anthracis* spore challenge.

A number of baculovirus vectors have been constructed that contain synthetic signal peptides downstream from the polyhedrin promoter to facilitate the secretion of proteins. Tessier et al. [68] constructed pVTBac, which contains a sequence encoding the honey bee mellitin signal peptide inserted downstream from the polyhedrin promoter and upstream from a polylinker region to facilitate the fusion of inserted genes. Fivefold more prepropapain was secreted using the mellitin signal peptide compared with the native plant gene. Daugherty et al. [51] used PCR to insert five fragments encoding signal peptides for human immunoglobulin-κ (Ig-κ), *Drosophila* 68C glue, antistasin, bovine growth hormone (bGH), and human apolipoprotein (Apo E) upstream from the gene for eichistatin. Transient assays indicated that each differed in relative efficiency of secretion of eichistatin into the media of infected cells. Benatti et al. [144] used the leader peptide of the vesicular stomatitis virus G-protein to direct the secretion of variants of hirudin (HV1 and HV2) into the media. Amino acid sequencing of HV1 indicated that the heterologous leader peptide was correctly removed. An artifical signal sequence was used to direct the secretion of a fusion between the antibacterial peptide cecropin A and the antibody-binding part of protein A from *Staphylococcus aureus* [124]. Cecropin A could be released from the fusion protein by treatment with cyanogen bromide. Synthetic signal peptides were also used to direct the secretion of ricin B-chain and chimeric plasminogen activators in baculovirus-infected insect cells [145,146]. Soluble human 37 kd CD23 was also expressed and secreted into the culture medium using the interleukin-2 leader sequence [147].

The processing of polyproteins in baculovirus-infected insect cells has been extensively discussed in previous reviews [2,6,7,11,12]. Most of the examples involve viral proteins that are synthesized as long precursor proteins that are processed by proteolytic cleavage at specific sites in the molecule. Examples that demonstrate accurate or inaccurate and efficient or inefficient cleavage can all be found. Sindbis virus and rubella virus polyproteins encoding capsid and envelope proteins [148,149], for example, are processed to release the individual capsid (C) and envelope (E1 and E2) proteins that are indistinguishable from the same proteins expressed in BHK cells. The HIV-1 envelope glycoprotein gp160, however, is inefficently cleaved into gp120 and gp41 [150,151]. Overton et al. [106] have demonstrated that, when recombinant baculoviruses expressing HIV protease and p55gag are coinfected into insect cells, authentic processing of the p55gag precursor takes place to generate the products p17, which is myristoylated at its N terminus, and p24, which is phosphorylated. When p55gag is expressed alone, the uncleaved protein forms immature retroviral core-like particles within the cytoplasm. Several notable examples from recent literature are the self-processing of the Kex2 endoprotease of *Saccharomyces cervisiae* [152,153], the processing of human Alzheimer amyloid precursor protein [154–160], and the lack of C-terminal processing of mouse β-nerve growth factor [161].

Tertiary and Quarternary Structure Formation

There are now over 100 examples demonstrating the proper folding, disulfide bond formation, or hetero- and homooligomeric assembly of recombinant proteins in baculovirus-infected insect cells. Many of these involve the processing and assembly of various structural proteins (capsid, core, envelope, spike) from a wide variety of viruses. An extensive list of structural and nonstructural viral proteins, organized taxonomically, can be found in a recent manual [6]. Proteins from the following virus families have been expressed in insect cells: adeno-, arena-, bunya-, caulimo-, como-, corona-, flavi-, hepadna-, herpes-, nepo-, orthomyxo-, papova-, paramyxo-, parvo-, picorno-, polydna-, reo-, retro-, rhabdo-, and togaviruses.

Two approaches can be used to study the interactions between two or more recombinant proteins. The insect cells can be multiply-infected with several recombinant viruses at a high MOI or they can be infected with one or more viruses constructed from dual- and single-expression vectors. Dual-expression vectors have the advantage that both genes will be expressed and form complexes in a single cell regardless of the input MOI, but suffer if the two genes are not expressed in equal amounts. This can be controlled to some extent, if mutiple infection is used, by varying the input MOI for each recombinant virus. This approach requires high titer stocks of virus to ensure that the MOI is above 5 and all cells are infected simultaneously.

One of the most striking examples of complex formation was described by Stow [162], who constructed seven recombinant baculoviruses each of which contained herpes simplex virus type 1 (HSV-1) genes required for viral origin-dependent DNA synthesis. All seven viruses were used to superinfect *S. frugiperda* cells previously transfected with a plasmid containing a functional HSV-1 origin of replication. The plasmid DNA was amplified only when HSV-1 origin was not mutated and all seven HSV-1 replication proteins were present. Extension of the system to test the structure and function relationships of the individual replication proteins will be a powerful tool in revealing the molecular details of HSV DNA replication.

Three of the essential HSV-1 DNA replication genes encode a helicase-primase that consists of the products of the UL5, UL8, and UL52 genes [163–167]. The active three-protein complex could only be purified from triply infected insect cells and could not be reconstituted from equimolar mixtures of cells infected with single recombinant viruses expressing UL5, UL8, or UL52 or from the individual purified proteins. Both UL5 and UL52 could form a subcomplex that has DNA-dependent ATPase activity, DNA helicase activity, and the ability to prime DNA synthesis on a poly(dT) template. The UL8 protein appears to associate more loosely with the

UL5/UL52 subcomplex and plays a role in stabilizing the association of newly synthesized oligoribonucleotide primers and the template DNA.

There are many examples of complex virus assembly in baculovirus-infected cells, some of which are described below. Noninfectious, double-shelled, virus-like particles were expressed when insect cells were coinfected with two recombinant baculovirus dual-expression vectors, one expressing the two outer capsid proteins VP2 and VP5 of bluetongue virus (BTV) and one expressing the two major core proteins VP3 and VP7 [168]. The particles had the same size and appearance as authentic BTV virons and were immunogenic, giving rise to antibodies that would recognize authentic BTV proteins. The antibodies could also neutralize BTV in vitro. When polyomavirus major capsid protein VP1 was expressed in insect cells, it could form uniform capsid-like particles in the nucleus [169]. Cytoplasmic VP1 could also form particles, but only after the intracellular calcium concentration was raised by treatment with a calcium ionophore. Empty parovirus B19 capsids could be recovered from insect cells expressing VP2 or both VP1 and VP2 [170]. Double-shelled rotavirus-like particles were produced in cells coinfected with baculoviruses expressing VP6 and VP7 [171]. The two major capsid proteins L1 and L2 of human papillomavirus (HPV) type 16 are able to form a complex in the absence of HPV DNA [172].

The abundant production of these and other proteins in baculovirus-infected cells has greatly facilitated the testing and development of subunit vaccines for the prevention or therapy against viral or parasitic diseases. Individual purified proteins or mixtures of several proteins that form virus-like particles are often able to provide protection to viral challenge in animals. One benefit of this technology is the absence of safety issues that arise when viral proteins are purified from vertebrate cells infected with potentially hazardous viruses that are normally propagated in specialized containment facilities. In any case, the insect cell–derived recombinant proteins are often ideally suited for use in diagnostic kits. These and other important issues concerning the preparation of vaccines are explored in a recent review [16].

Several other examples illustrate the diversity of protein complex formation in baculovirus-infected insect cells. These include the assembly of immunoglobulin heavy and light chains to form functional antibodies [173–180], the formation of disulfide-linked homodimers of interleukin-5 [181,182], homodimers of human nerve growth factor [183], and dimers of interferon-γ receptor [184].

FUTURE PROSPECTS

The widespread use of baculovirus vectors to produce recombinant proteins has increased dramatically in recent years, and this trend is likely to continue. Several factors contribute to the popularity of this eukaryotic expression system. In most cases, the recombinant proteins are processed, modified, and targeted to their appropriate cellular location where they have the expected biological properties. The high level expression of many recombinant proteins under the control of the polyhedrin promoter occurs at a time late in infection when background cellular synthesis is diminished. Although insect cells provide a eukaryotic environment for expression, they are in many ways sufficiently different from mammalian cells so that the endogenous background of cellular proteins does not interfere with the detection and purification of recombinant proteins. This feature can be particularly important if the desired protein is a member of a family of closely related proteins that are relatively abundant in the mammalian cell lines commonly used for expression studies. Exceptions to these general statements are sometimes more interesting than observations that confirm previous knowledge, because they reveal limitations of the insect cell expression system and provide opportunities for the development of novel applications that cannot be obtained in other expression systems. The inability of AcMNPV-infected *S. frugiperda* cells to keep up with certain kinds of post-translational modifications very late in infection, for example, can be exploited to produce proteins used as targets for purified modification enzymes from other sources, thereby permitting the identification of factors that affect the antigenic, immunogenic, and biological properties of the target protein. The comparisons of post-translational modifications present on proteins expressed in insect, mammalian, plant, and bacterial expression systems have led to a greater understanding of these processes, not only in insect cells but in these other host organisms as well.

The rapid development of methods to facilitate the construction of recombinant baculoviruses will continue. Improved baculovirus transfer vectors containing convenient cloning sites for the insertion of one or more foreign genes have been developed, and many are now commercially available. More convenient parent viruses, which have an easily screenable plaque phenotype and can be linearized to improve the frequency of recombinant viral progeny, will also be improved and used more widely in the coming years. Many nonessential regions of baculovirus genomes will be identified as soon as the sequences of the AcMNPV and BmSNPV genomes become available and will lead to the development of many new vectors that may have specialized applications, particularly in the development of improved viral pesticides. Previously unidentified promoters, which are constitutive or are active at one or more stages of infection, should prove useful in new vectors designed to express proteins at different times and at different levels. Promoters from other organisms might also be used in the presence of their transcriptional regulators to build vectors that have expression controlled in an inducible fashion. Inducible promoters would permit the development of exquisite systems to separate and determine the roles of *cis*- and *trans*-acting factors affecting the regulation of viral gene expression, much in the same way the genetic elements of the *E. coli lac* operon and bacteriophage T7 have been used to dissect the regulation of vaccinia virus gene expression [185–187]. Further development of yeast- and *E. coli*–based shuttle vectors will continue and lead to very rapid and efficient methods for the construction of recombinant baculoviruses. These vectors may be particularly valuable for protein engineering projects, which require the expression of many protein variants, or for expression cloning of previously uncharacterized genes. Shuttle vectors also have the potential to facilitate greatly the identification and genetic dissection of essential viral genes.

An important area for future development, particularly for biotechnology and pharmaceutical companies interested in commercial applications of this system, is the optimization of conditions for expression in insect cells propagated in large-scale bioreactors. Airlift bioreactors and traditional stirred-tank vessels are now commonly tested and used, but since large capital investments are required to purchase all the necessary equipment, very little research into novel bioreactor designs that could boost cellular productivity has been reported in recent years. Other chapters in this volume, focusing on the engineering aspects of scale-up, describe potential areas for improvement in great detail.

The development of new insect cell culture media and the isolation of cell lines with improved growth properties are perhaps more important than the size or shape of a culture vessel and the methods used to maintain dissolved oxygen. New cell lines, which outperform others in the expression of particular kinds of proteins, will be valuable to companies optimizing commercial processes that lead to vaccines and other protein-based biopharmaceuticals. Prolific cell lines that do not grow well in suspension cultures are less valuable for use on a commercial scale, however, than less prolific cell lines that are well-adapted for growth in large-scale bioreactors. Not all cell lines continuously grown as attached cells will adapt successfully to growth in suspension without changing some of their properties. Since any recombinant baculovirus can be used to infect a susceptible cell line, uninfected cells can be continuously propagated on a moderate scale for use as inoculum for large-scale batch production runs. This can be seen as one advantage over continuous expression in stably transfected mammalian or insect cells that must be gradually scaled-up and customized for particular cell lines. Insect cells are also particularly suited for the expression of cytotoxic or nonsecreted proteins, which are not often expressed well or easily purified from cells used in continuous expression processes. New types of insect cell culture media optimized for particular cell lines will also be a key factor in acheiving the goals of many commercial processes. Serum-free media have several advantages over serum-supplemented media. These include lower cost, consistent performance lot-to-lot, and very low protein content, which reduces downstream processing costs for secreted proteins. Optimized formulations, available in research- and in GMP-quality grades, that lead to high yields and perform at all scales of production are all critical factors for success.

Expression of foreign genes by baculovirus vectors is an *enabling* technology that permits the production of proteins used as reagents for structure–function studies, as components in diagnostic kits, as subunit vaccines, or as therapeutic biopharmaceuticals that cannot often be achieved for technical or economic reasons with other expression systems. The use and development of this technology will continue to grow and be a major influence in shaping the future of basic biology and industrial biotechnology.

ACKNOWLEDGMENTS

I thank Ellen Gonis Luckow and Martin L. Bryant for reviewing this manuscript.

REFERENCES

1. Luckow, V.A., and Summers, M.D. Bio/Technology 6:47–55, 1988.
2. Luckow, V.A. In: Recombinant DNA Technology and Applications (Prokop, A., Bajpai, R.K., and Ho, C., eds.). McGraw-Hill, New York, pp. 97–152, 1991.
3. Miller, L.K. Annu. Rev. Microbiol. 42:177–199, 1988.
4. Maeda, S. Annu. Rev. Entomol. 34:351–372, 1989.
5. Murhammer, D.W. Appl. Biochem. Biotechnol. 31:283–310, 1991.
6. O'Reilly, D.R., Miller, L.K., and Luckow, V.A. Baculovirus Expression Vectors: A Laboratory Manual. W.H. Freeman and Company, New York, 1992.
7. King, L.A., and Possee, R.D. The Baculovirus Expression System: A Laboratory Guide. Chapman & Hall, London, 1992.
8. Atkinson, A.E., Weitzman, M.D., Obosi, L., Beadle, D.J., and King, L.A. Pestic. Sci. 28:215–224, 1990.
9. Bishop, D.H.L. Curr. Biol. 1:62–67, 1990.
10. Cameron, I.R., Possee, R.D., and Bishop, D.H.L. Trends Biotechnol. 7:66–70, 1989.
11. Fraser, M.J. In Vitro Cell. Dev. Biol. 25(3 part I):225–235, 1989.
12. Miller, L.K. In: Biotechnology in Invertebrate Pathology and Cell Culture (Maramorosch, K., eds.). Academic Press, San Diego, pp. 295–304, 1987.
13. Miller, L.K. Bioessays 11:91–5, 1989.
14. Summers, M.D. In: Current Communications in Molecular Biology: Viral Vectors (Gluzman, Y., and Hughes, S.H., eds.). Cold Spring Harbor, New York, pp. 91–97, 1988.
15. Summers, M.D. In: Concepts In Viral Pathogenesis (Notkins, A.L., and Oldstone, M.B.A., eds.). Springer-Verlag, New York, pp. 77–86, 1989.
16. Vlak, J.M., and Keus, R.J.A. Adv. Biotechnol. Processes 14:91–128, 1990.
17. Bradley, M.K. Methods Enzymol. 182:112–132, 1990.
18. Webb, N.R., and Summers, M.D. Technique 2:173–188, 1990.
19. Weiss, S.A., and Vaughn, J.L. In: The Biology of Baculoviruses (Granados, R.R., and Federici, B.A., eds.). Vol. 2. Practical Applications for Insect Control. CRC Press, Boca Raton, FL, pp. 63–87, 1986.
20. Page, M.J., and Murphy, V.F. Methods Mol. Biol. 5:573–588, 1990.
21. Summers, M.D., and Smith, G.E. Texas Agric. Exp. Station Bull. 1555:1–57, 1987.
22. Piwnica-Worms, H. In: Current Protocols In Molecular Biology (Ausubel, F.M., Brent, R., Kingston, R.E., Moore, D.D., Seidman, J.G., Smith, J.A., and Struhl, K., eds.). Wiley-Interscience, New York, pp. 16.8.1–16.11.7, 1990.
23. Maeda, S. In: Invertebrate Cell System Applications, (Mitsuhashi, J., eds.). Vol. 1. CRC Press, Boca Raton, FL, pp. 167–182, 1989.
24. Richardson, C., eds. Baculovirus Expression Protocols. Humana Press, Clifton, NJ, 1994.
25. Bilimoria, S.L. In Viruses of Invertebrates (Kurstak, E., eds.). Marcel Dekker, New York, pp. 1–72, 1991.
26. Blissard, G.W., and Rohrmann, G.F. Annu. Rev. Entomol. 35:127–55, 1990.
27. Doerfler, W. Curr. Top. Microbiol. Immunol. 131:51–68, 1986.
28. Friesen, P.D., and Miller, L.K. Curr. Top. Microbiol. Immunol. 131:31–50, 1986.
29. Granados, R.R., and Federici, B.A., eds. The Biology of Baculoviruses. Vol. 1. Biological Properties and Molecular Biology. CRC Press, Boca Raton, FL 1986.
30. Granados, R.R., and Federici, B.A., eds. The Biology of Baculoviruses. Vol. 2. Practical Applications for Insect Control. CRC Press, Boca Raton, FL, 1986.

31. Kurstak, E., eds. Viruses of Invertebates. Marcel Dekker, New York, 1991.
32. Smith, G.E., Fraser, M.J., and Summers, M.D. J. Virol. 46:584–593, 1983.
33. Smith, G.E., Summers, M.D., and Fraser, M.J. Mol. Cell. Biol. 3:2156–2165, 1983.
34. Pennock, G.D., Shoemaker, C., and Miller, L.K. Mol. Cell. Biol. 4:399–406, 1984.
35. Maeda, S., Kawai, T., Obinata, M., Fujiwara, H., Horiuchi, T., Saeki, Y., Sato, Y., and Furusawa M. Nature 315:592–594, 1985.
36. Hartig, P.C., Cardon, M.C., and Kawanishi, C.Y. Biotechniques 11:312–313, 1991.
37. Wienhues, U., Hosokawa, K., Hoveler, A., Siegmann, B., and Doerfler, W. DNA 6:81–9, 1987.
38. Mann, S.G., and King, L.A. J. Gen. Virol. 70:3501–3505, 1989.
39. Lenz, C.J., McIntosh, A.H., Mazzacano, C., and Monderloh, U. J. Invertebr. Pathol. 57:227–233, 1991.
40. Brown, M., and Faulkner, P. Appl. Environ. Microbiol. 36:31–35, 1978.
41. Estes, M.K., Crawford, S.E., Penaranda, M.E., Petrie, B.L., Burns, J.W., Chan, W.K., Ericson, B., Smith, G.E., and Summers, M.D. J. Virol. 61:1488–1494, 1987.
42. Shanafelt, A.B. Biotechniques 11:330 [erratum, Biotechniques 11:473,] 1991.
43. Fung, M.C., Chiu, K.Y.M., Weber, T., Chang, T.W., and Chang, N.T. J. Virol. Methods 19:33–42, 1988.
44. Ghiasi, H., Nesburn, A.B., Kaiwar, R., and Wechsler, S.L. Arch. Virol. 121:163–178, 1991.
45. Farmer, J.L., Hampton, R.G., and Boots, E. J. Virol. Methods 26:279–290, 1989.
46. Luckow, V.A., and Summers, M.D. Virology 167:56–71, 1988.
47. Capone, J. Gene Anal. Technol. 6:62–66, 1989.
48. Sisk, W.P., Bradley, J.D., Seivert, L.L., Vargas, R.A., and Horlick, R.A. Biotechniques 13:186, 1992.
49. Malitschek, B., and Schartl, M. Biotechniques 11:177–178, 1991.
50. Webb, A.C., Bradley, M.K., Phelan, S.A., Wu, J.Q., and Gehrke, L. Biotechniques 11:512–519, 1991.
51. Daugherty, B.L., Zavodny, S.M., Lenny, A.B., Jacobson, M.A., Ellis, R.W., Law, S.W., and Mark, G.E. DNA Cell. Biol:453–459, 1990.
52. Smith, G.E., Ju, G., Ericson, B.L., Moschera, J., Lahm, H., Chizzonite, R., and Summers, M.D. Proc. Natl. Acad. Sci. USA 82:8404–8408, 1985.
53. Miller, D.W., Safer, P., and Miller, L.K. In: Genetic Engineering (Setlow, J.K., and Hollaender, A., eds.). Vol. 8. Plenum, New York, pp. 277–298, 1986.
54. Matsuura, Y., Possee, R.D., and Bishop, D.H.L. J. Gen. Virol. 67:1515–1530, 1986.
55. Possee, R.D. Virus Res. 5:43–60, 1986.
56. Luckow, V.A., and Summers, M.D. Virology 170:31–39, 1989.
57. Matsuura, Y., Possee, R.D., Overton, H.A., and Bishop, D.H. J. Gen. Virol. 68:1233–1250, 1987.
58. Ooi, B.G., Rankin, C., and Miller, L.K. J. Mol. Biol. 210:721–736, 1989.
59. Rankin, C., Ooi, B.G., and Miller, L.K. Gene 70:1–39–49, 1988.
60. Livingstone, C., and Jones, I. Nucleic Acids Res. 17:0305–1048, 1989.
61. Kitts, P.A. CLONTECHniques VII: 1–3, 1992.
62. Wang, X.Z., Ooi, B.G., and Miller, L.K. Gene:131–137, 1991.
63. Page, M.J. Nucleic Acids Res. 17:0305–1048, 1989.
64. Beames, B., Braunagel, S., Summers, M.D., and Lanford, R.E. Biotechniques 11:378–383, 1991.
65. Gordon, K., Futterer, J., and Hohn, T. Plant J. 2:809–813, 1992.
66. Hill-Perkins, M.S., and Possee, R.D. J. Gen. Virol. 71:971–976, 1990.

67. Thiem, S.M., and Miller, L.K. Gene 91:87–94, 1990.
68. Tessier, D.C., Thomas, D.Y., Khouri, H.E., Laliberte, F., and Vernet, T. Gene 98:177–83, 1991.
69. Chakrabarti, S., Brechling, K., and Moss, B. Mol. Cell. Biol. 5:3403–3409, 1985.
70. Vialard, J., Lalumière, M., Vernet, T., Briedis, D., Alkhatib, G., Henning, D., Levin, D., and Richardson, C. J. Virol. 64:37–50, 1990.
71. Richardson, C., Lalumiere, M., Banville, M., and Vialard, J. In: Baculovirus Expression Protocols, Methods in Molecular Biology (Richardson, C., eds.). Humana Press, Clifton, NJ, 1994.
72. Zuidema, D., Schouten, A., Usmany, M., Maule, A.J., Belsham, G.J., Roosien, J., Klinge-Roode, E.C., van Lent, J.W.M., and Vlak, J.M. J. Gen. Virol. 71:2201–2209, 1990.
73. Zhao, L.J., Irie, K., Trirawatanapong, T., Nakano, R., Nakashima, A., Morimatsu, M., and Padmanabhan, R. Gene 100:147–154, 1991.
74. Weyer, U., Knight, S., and Possee, R.D. J. Gen. Virol. 71:1525–1534, 1990.
75. Weyer, U., and Possee, R.D. J. Gen. Virol. 72:2967–2974, 1991.
76. Vlak, J.M., Schouten, A., Usmany, M. Belsham, G.J., Klinge-Roode, C.E., Maule, A.J. van Lent, J.W. and Zuidema, D. Virology 179:312–320, 1990.
77. Kitts, P.A., Ayres, M.D., and Possee, R.D. Nucleic Acids Res. 18:5667–5672, 1990.
78. Sewall, A., and Srivastava, N. Digest 4:1, 1991.
79. Hartig, P.C., and Cardon, M.C. J. Virol. Methods 38:61–70, 1992.
80. Possee, R.D., and Howard, S.C. Nucleic Acids Res. 15:10233–10248, 1987.
81. Kitts, P.A., and Possee, R.D. Biotechniques 14:810–817, 1993.
82. O'Reilly, D.R., Passarelli, A.L., Goldman, I.F., and Miller, L.K. J. Gen. Virol. 71:1029–1037, 1990.
83. Peakman, T.C., Harris, R.A., and Gewert, D.R. Nucleic Acids Res. 10:495–500, 1992.
84. Patel, G., Nasmyth, K., and Jones, N. Nucleic Acids Res. 20:97–104, 1992.
85. Luckow, V.A., Lee, S.C., Barry, G.F., and Olins, P.O. J. Virol. 67:4566–4579, 1993.
86. Han, K.K., and Martinage, A. Int. J. Biochem. 24:19–28, 1992.
87. Kornfeld, R., and Kornfeld, S. Annu. Rev. Biochem. 54:631–664, 1985.
88. Parekh, R.B., Dwek, R.A., Edge, C.J., and Rademacher, T.W. Trends Biotechnol. 7:117–122, 1989.
89. Paulson, J.C. Trends Biochem. Sci. 14:272–276, 1989.
90. Goochee, C.F., and Monica, T. Bio/Technology 8:421–427, 1990.
91. Goochee, C.F., Gramer, M.J., Andersen, D.C., Bahr, J.B., and Rasmussen, J.R. Bio/Technology 9:1347–1355, 1991.
92. Viscidi, R., Ellerbeck, E., Garrison, L., Midthun, K., Clements, M.L., Clayman, B., Fernie, B., and Smith, G. AIDS. Res. Hum. Retroviruses 6:1251–1256, 1990.
93. Dolin, R., Graham, B.S., Greenberg, S.B., Tacket, C.O., Belshe, R.B., Midthun, K., Clements, M.L., Gorse, G.J., Horgan, B.W., Atmar, R.L., Karzon, D.T., Bonnez, W., Fernie, B.F., Montefiori, D.C., Stablein, D.M., Smith, G.E., and Koff, W.C. Ann. Intern. Med. 114:119–127, 1991.
94. Redfield, R.R., Birx, D.L., Ketter, N., Tramont, E., Polonis, V., Davis, C., Brundage, J.F., Smith, G., Johnson, S., Fowler, A., Wierzba, T., Shafferman, A., Volvovitz, F., Oster, C., Burke, D.S., and Military Medical Consortium for Applied Retroviral Research. N. Engl. J. Med. 324:1677–1684, 1991.
95. Kuroda, K., Geyer, H., Geyer, R., Doerfler, W., and Klenk, H.-D. Virology 174:418–429, 1990.
96. Thomsen, D.R., Post, L.E., and Elhammer, A.P. J. Cell. Biochem. 43:67–79, 1990.

97. Wathen, M.W., Aeed, P.A., and Elhammer, A.P. Biochemistry 30:2863–2868, 1991.
98. Svoboda, M., Przybylski, M., Schreurs, J., Miyajima, A., Hogeland, K., and Deinzer, M. J. Chromatogr. 562:403–419, 1991.
99. Davidson, D.J., Fraser, M.J., and Castellino, F.J. Biochemistry 29:5584–5590, 1990.
100. Davidson, D.J., and Castellino, F.J. Biochemistry 30:6689–6696, 1991.
101. Davidson, D.J., and Castellino, F.J. Biochemistry 30:6165–6174, 1991.
102. Davidson, D.J., Bretthauer, R.K., and Castellino, F.J. Biochemistry 30:9811–9815, 1991.
103. Hink, W.F., Thomsen, D.R., Davidson, D.J., Meyer, A.L., and Castellino, F.J. Biotechnol. Prog. 7:9–14, 1991.
104. Delchambre, M., Gheysen, D., Thines, D., Thiriart, C., Jacobs, E., Verdin, E., Horth, M., Burny, A., and Bex, F. EMBO J. 8:2653–2660, 1989.
105. Gheysen, D., Jacobs, E., de Foresta, F., Thiriart, C., Francotte, M., Thines, D., and De Wilde, M. Cell 59:103–112, 1989.
106. Overton, H.A., Fujii, Y., Price, I.R., and Jones, I.M. Virology 170:107–116, 1989.
107. Royer, M., Cerutti, M., Gay, B., Hong, S.S., Devauchelle, G., and Boulanger, P. Virology 184:417–422, 1991.
108. Royer, M., Hong, S.S., Gay, B., Cerutti, M., and Boulanger, P. J. Virol. 66:3230–3235, 1992.
109. Morikawa, S., Booth, T.F., and Bishop, D.H. Virology 183:288–297, 1991.
110. Luo, L., Li, Y., and Kang, C.Y. Virology 179:874–880, 1990.
111. Matsuura, Y., Maekawa, M., Hattori, S., Ikegami, N., Hayashi, A., Yamazaki, S., Morita, C., and Takebe, Y. Virology 184:580–586, 1991.
112. Belsham, G.J., Abrams, C.C., King, A.M., Roosien, J., and Vlak, J.M. J. Gen. Virol. 72:747–751, 1991.
113. Lowe, P.N., Sydenham, M., and Page, M.J. Oncogene 5:1045–1048, 1990.
114. Lowe, P.N., Page, M.J., Bradley, S., Rhodes, S., Sydenham, M., Paterson, H., and Skinner, R.H. J. Biol. Chem. 266:1672–1678, 1991.
115. Khosravi Far, R., Lutz, R.J., Cox, A.D., Conroy, L., Bourne, J.R., Sinensky, M., Balch, W.E., Buss, J.E., and Der, C.J. Proc. Natl. Acad. Sci. USA 88:6264–6268, 1991.
116. Yang, C., Mayau, V., Godeau, F., and Goud, B. Biochem. Biophys. Res. Commun. 182:1499–1505, 1992.
117. Kloc, M., Reddy, B., Crawford, S., and Etkin, L.D. J. Biol. Chem. 266:8206–8212, 1991.
118. Buss, J.E., Quilliam, L.A., Kato, K., Casey, P.J., Solski, P.A., Wong, G., Clark, R., McCormick, F., Bokoch, G.M., and Der, C.J. Mol. Cell. Biol. 11:1523–1530, 1991.
119. Labrecque, J., Caron, M., Torossian, K., Plamondon, J., and Dennis, M. FEBS Lett. 304:157–162, 1992.
120. Graber, S.G., Figler, R.A., and Garrison, J.C. J. Biol. Chem. 267:1271–1278, 1992.
121. Graber, S.G., Figler, R.A., Kalman Maltese, V.K., Robishaw, J.D., and Garrison, J.C. J. Biol. Chem. 267:13123–13126, 1992.
122. Suzuki, K., Shimoi, H., Iwasaki, Y., Kawahara, T., Matsuura, Y., and Nishikawa, Y. EMBO J. 9:4259–4265, 1990.
123. Hellers, M., Gunne, H., and Steiner, H. Eur. J. Biochem. 199:435–439, 1991.
124. Andersons, D., Engstrom, A., Josephson, S., Hansson, L., and Steiner, H. Biochem. J. 280:219–224, 1991.
125. Yamada, K., Nakajima, Y., and Natori, S. Biochem. J. 272:633–636, 1990.
126. Lebacq-Verheyden, A.M., Kasprzyk, P.G., Raum, M.G., Van Wyke Coelingh, K., Lebacq, J.A., and Battey, J.F. Mol. Cell. Biol. 8:3129–3135, 1988.

127. Ben-Bassat, A., and Bauer, K. Nature 326:316, 1987.

128. Flinta, C., Persson, B., Jörnvall, H., and von Heijne, G. Eur. J. Biochem. 154:193–196, 1986.

129. Sherman, F., Stewart, J.W., and Tsunasawa, S. BioEssays 3:27–31, 1985.

130. Nishimura, C., Yamaoka, T., Mizutani, M., Yamashita, K., Akera, T., and Tanimoto, T. Biochim. Biophys. Acta 1078:171–178, 1991.

131. Gierse, G.K., Luckow, V.A., Asconas, L.J., Duffin, K.L., Aykent, S., Bild, G.S., Rodi, C.P., Sullivan, P.M., Bourner, M.K., Kimack, N.M., and Krivi, G.G. Prot. Express. Purification 4:358–366, 1993.

132. Hoss, A., Moarefi, I., Scheidtmann, K.H., Cisek, L.J., Corden, J.L., Dornreiter, I., Arthur, A.K., and Fanning, E.J. Virol. 64:4799–4807, 1990.

133. Parker, L.L., Atherton-Fessler, S., Lee, M.S., Ogg, S., Falk, J.L., Swenson, K.I., and Piwnica-Worms, H. EMBO J. 10:1255–1263, 1991.

134. Itoh, K., Oshima, A., Sakuraba, H., and Suzuki, Y. Biochem. Biophys. Res. Commun. 167:746–753, 1990.

135. Nishimura, C., Matsuura, Y., Kokai, Y., Akera, T., Carper, D., Morjana, N., Lyons, C., and Flynn, T.G. J. Biol. Chem. 265:9788–9792, 1990.

136. Alnemri, E.S., Maksymowych, A.B., Robertson, N.M., and Litwack, G. J. Biol. Chem. 266:18072–18081, 1991.

137. Binart, N., Lombes, M., Rafestin-Oblin, M.E., and Baulieu, E.E. Proc. Natl. Acad. Sci. USA 88:10681–10685, 1991.

138. Alnemri, E.S., Maksymowych, A.B., Robertson, N.M., and Litwack, G. J. Biol. Chem. 266:3925–3936, 1991.

139. Kartner, N., Hanrahan, J.W., Jensen, T.J., Naismith, A.L., Sun, S.Z., Ackerley, C.A., Reyes, E.F., Tsui, L.C., Rommens, J.M., Bear, C.E., and Riordan, J.R. Cell 64:681–691, 1991.

140. Bear, C.E., Li, C.H., Kartner, N., Bridges, R.J., Jensen, T.J., Ramjeesingh, M., and Riordan, J.R. Cell 68:809–818, 1992.

141. Sarkadi, B., Bauzon, D., Huckle, W.R., Earp, H.S., Berry, A., Suchindran, H., Price, E.M., Olson, J.C., Boucher, R.C., and Scarborough, G.A. J. Biol. Chem. 267:2087–2095, 1992.

142. Iacono-Connors, L.C., Schmaljohn, C.S., and Dalrymple, J.M. Infect. Immun. 58:366–72, 1990.

143. Iacono-Connors, L.C., Welkos, S.L., Ivins, B.E., and Dalrymple, J.M. Infect. Immun. 59:1961–1965, 1991.

144. Benatti, L., Scacheri, E., Bishop, D.H., and Sarmientos, P. Gene 101:255–260, 1991.

145. Piatak, M., Lane, J.A., O'Rourke, E., Clark, R., Houston, L.L., and Apell, R. ICSU Short Rep. 8:62, 1988.

146. Devlin, J.J., Devlin, P.E., Clark, R., O'Rourke, E.C., Levenson, C., and Mark, D.F. Bio/Technology 7:286–292, 1989.

147. Graber, P., Jansen, K., Pochon, S., Shields, J., Aubonney, N., Turcatti, G., and Bonnefoy, J.Y. J. Immunol. Methods 149:215–226, 1992.

148. Oker-Blom, C., Pettersson, R.F., and Summers, M.D. Virology 172:82–91, 1989.

149. Oker-Blom, C., and Summers, M.D. J. Virol. 63:1256–1264, 1989.

150. Murphy, C.I., Lennick, M., Lehar, S.M., Beltz, G.A., and Young, E. Genet. Anal. Tech. Appl. 7:160–171, 1990.

151. Wells, D.E., and Compans, R.W. Virology 176:575–586, 1990.

152. Germain, D., Vernet, T., Boileau, G., and Thomas, D.Y. Eur. J. Biochem. 204:121–126, 1992.

153. Germain, D., Dumas, F., Vernet, T., Bourbonnais, Y., Thomas, D.Y., and Boileau, G. FEBS Lett. 299:283–286, 1992.

154. Bhasin, R., Van Nostrand, W.E., Saitoh, T., Donets, M.A., Barnes, E.A., Quitschke, W.W., and Goldgaber, D. Proc. Natl. Acad. Sci. USA 88:10307–10311, 1991.

155. Currie, J.R., Ramakrishna, N., Burrage, T.G., Hwang, M.C., Potempska, A., Miller, D.L., Mehta, P.D., Kim, K.S., and Wisniewski, H.M. J. Neurosci. Res. 30:687–698, 1991.

156. Gandy, S.E., Bhasin, R., Ramabhadran, T.V., Koo, E.H., Price, D.L., Goldgaber, D., and Greengard, P. J. Neurochem. 58:383–386, 1992.

157. Ghiso, J., Wisniewski, T., Vidal, R., Rostagno, A., and Frangione, B. Biochem. J. 282:517–522, 1992.

158. Knops, J., Johnson, W.K., Schenk, D.B., Sinha, S., Lieberburg, I., and McConlogue, L. J. Biol. Chem. 266:7285–7290, 1991.

159. Lowery, D.E., Pasternack, J.M., Gonzalez DeWhitt, P.A., Zurcher Neely, H., Tomich, C.C., Altman, R.A., Fairbanks, M.B., Heinrikson, R.L., Younkin, S.G., and Greenberg, B.D. J. Biol. Chem. 266:19842–19850, 1991.

160. Ramakrishna, N., Saikumar, P., Potempska, A., Wisniewski, H.M., and Miller, D.L. Biochem. Biophys. Res. Commun. 174:983–989, 1991.

161. Luo, Y., and Neet, K.E. J. Biol. Chem. 267:12275–12283, 1992.

162. Stow, N.D. J. Gen. Virol. 73:313–321, 1992.

163. Calder, J.M., and Stow, N.D. Nucleic Acids Res. 18:3573–3578, 1990.

164. Crute, J.J., Bruckner, R.C., Dodson, M.S., and Lehman, I.R. J. Biol. Chem. 266:21252–21256, 1991.

165. Dodson, M.S., Crute, J.J., Bruckner, R.C., and Lehman, I.R. J. Biol. Chem. 264:20835–20838 [erratum, J. Biol. Chem. 1990 265:4769, 1990], 1989.

166. Dodson, M.S., and Lehman, I.R. Proc. Natl. Acad. Sci. USA 88:1105–1109, 1991.

167. Sherman, G., Gottlieb, J., and Challberg, M.D. J. Virol. 66:4884–4892, 1992.

168. French, T.J., Marshall, J.J., and Roy, P. J. Virol. 64:5695–5700, 1990.

169. Montross, L., Watkins, S., Moreland, R.B., Mamon, H., Caspar, D.L., and Garcea, R.L. J. Virol. 65:4991–4918, 1991.

170. Brown, C.S., van Lent, J.W., Vlak, J.M., and Spaan, W.J. J. Virol. 65:2702–2706, 1991.

171. Sabara, M., Parker, M., Aha, P., Cosco, C., Gibbons, E., Parsons, S., and Babiuk, L.A. J. Virol. 65:6994–6997, 1991.

172. Xi, S.Z., and Banks, L.M. J. Gen. Virol. 72:2981–2988, 1991.

173. Hasemann, C.A., and Capra, J.D. Proc. Natl. Acad. Sci. USA 87:3942–3946, 1990.

174. Hasemann, C.A., and Capra, J.D. J. Biol. Chem. 266:7626–7632, 1991.

175. Hasemann, C.A., and Capra, J.D. J. Immunol. 147:3170–3179, 1991.

176. Nesbit, M., Fu, Z.F., McDonald Smith, J., Steplewski, Z., and Curtis, P.J. J. Immunol. Methods 151:201–208, 1992.

177. Reis, U., Blum, B., von Specht, B.U., Domdey, H., and Collins, J. Bio/Technology 10:910–9812, 1992.

178. zu Putlitz, J.Z., Kubasek, W.L., Duchêne, M., Marget, M., von Specht, B.U., and Domdey, H. Bio/Technology 8:651–654, 1990.

179. Stern, L.J., and Wiley, D.C. Cell 68:465–477, 1992.

180. Godeau, F., Casanova, J.L., Luescher, I.F., Fairchild, K.D., Delarbre, C., Saucier, C., Gachelin, G., and Kourilsky, P. Int. Immunol. 4:265–275, 1992.

181. Ingley, E., Cutler, R.L., Fung, M.C., Sanderson, C.J., and Young, I.G. Eur. J. Biochem. 196:623–629, 1991.

182. Tavernier, J., Devos, R., Van der Heyden, J., Hauquier, G., Bauden, R., Fache, I., Kawashima, E., Vandekerckhove, J., Contreras, R., and Fiers, W. DNA 8:491–501, 1989.

183. Barnett, J., Chow, J., Nguyen, B., Eggers, D., Osen, E., Jarnagin, K., Saldou, N., Straub, K., Gu, L., Erdos, L., Chaing, H.S., Fausnaugh, J., Townsend, R.R., Lile, J., Collins, F., and Chan, H. J. Neurochem. 57:1052–1061, 1991.

184. Fountoulakis, M., Schlaeger, E.J., Gentz, R., Juranville, J.F., Manneberg, M., Ozmen, L., and Garotta, G. Eur. J. Biochem. 198:441–450, 1991.

185. Alexander, W.A., Moss, B., and Fuerst, T.R. J. Virol. 66:2934–2942, 1992.

186. Zhang, Y.F., and Moss, B. Virology 187:643–653, 1992.

187. Zhang, Y., Keck, J.G., and Moss, B. J. Virol. 66:6470–6479, 1992.

5

Development and Testing of Genetically Improved Baculovirus Insecticides

H. Alan Wood

Boyce Thompson Institute for Plant Research, Cornell University, Ithaca, New York 14853

INTRODUCTION

Commercial and basic research interests in baculoviruses have recently focused on the use of baculoviruses as expression vector systems for the variety of products described in Chapter 4. Historically, however, baculovirology had its beginnings in the area of insect pathology and the use of baculoviruses as insecticides. The first written record of a baculovirus infection appeared in 1527 in a description of a jaundice disease of the silkworm *Bombyx mori* [for historical review, see ref. 1]. The first recorded attempt to use a baculovirus as an insecticide involved the nuclear polyhedrosis virus (NPV) of *Lymantria monacha,* the nun moth [2].

The interest in using baculoviruses as pesticides arose from observations involving the natural control of insect populations. In many instances, when pest populations occurred in high concentrations, the insect population would crash due to naturally occurring baculovirus epizootics. These natural epizootics often play an important role in the regulation of insect populations. In the development of viral pesticides, the strategy has been to attempt to create an epizootic prior to the time when pest population densities have reached sufficient levels to cause unacceptable economic damage.

After World War II, the development of the chemical pesticide industry was accompanied by an interest in developing baculoviruses and other microbial control agents as pesticides. Although baculoviruses are pervasive in the environment, the fact that they could replicate was cause for concern with their use as pesticides. Accordingly, extensive health and environmental safety testing was performed [for review, see ref. 3]. The data indicated that naturally occurring baculoviruses posed no foreseeable hazards to the environment or other arthropod hosts. The effective

Baculovirus Expression Systems and Biopesticides, pages 91–102

host ranges of these viral pesticides were generally limited to a small number of pests and did not include beneficial insects. Based on their potential as alternative pesticides and on their safety, the U.S. Environmental Protection Agency granted full pesticide registration to four baculoviruses, *Helicoverpa* (formerly *Heliothis*) *zea* (1975), *Orgyia pseudosugata* (1976), *Lymantria dispar* (1978), and *Neodiprion sertifer* (1983) NPVs.

From 1950 to 1980, numerous insect viruses were field tested to evaluate their pesticidal properties. Cunningham [4] documented the use of 24 different baculoviruses to control forest pests. These viruses were used to control 16 lepidopteran species and 8 hymenopteran species. Yearian and Young [5] similarly documented the use of 29 different baculoviruses to control vegetable, forage, and fruit crop pests.

However, concurrent with these viral insecticide studies, numerous synthetic chemical pesticides were developed that were broad-spectrum and inexpensive. Because of these developments and the higher cost/benefit ratio associated with production and use of viral pesticides, commercial interest in viral insecticides eroded.

Beginning in the late 1970s, there was a growing awareness of the real and potential problems that synthetic chemical pesticides posed to human health and the environment as a whole [6]. With this awareness came more rigorous registration requirements, which have lead to the registration of fewer and fewer new active ingredients and to a current average development cost per product of 50 million dollars [7]. This problem has been compounded by the fact that insect populations become resistant to many pesticides, resulting in abandonment of the product or the need for higher application rates.

Additionally, in 1988, Congress amended the Federal Insecticide, Fungicide and Rodenticide Act to strengthen and accelerate the Environmental Protection Agency's pesticide reregistration program. Because of the costs associated with this reregistration process, many pesticide companies have dropped the registration of products for small markets. This has caused a significant problem to small market growers. These factors, combined with the increasing difficulty and costs to develop and register new classes of chemical pesticides, have encouraged the search for alternative methods for insect control.

With the development of the baculovirus expression vector (BEV) system in the early 1980s [8,9] came a resurgence in baculovirus research. It was soon apparent that almost any foreign gene could be inserted and expressed in a BEV. Initial investigations centered on the production of pharmaceutical and basic research products and provided the impetus to develop large-scale reactors for insect cell cultures. The development of cell culture bioreactors for BEV also opened opportunities for the production biopesticides. Previously, baculovirus pesticide production had been limited to larval production systems. By the late 1980s, the BEV system was being used to develop new viral pesticides. These recombinant viruses were designed to overcome some of the obstacles that previously deterred the commercial development of baculovirus pesticides.

A major deterrent to the development of viral insecticides had been that it can take from 5 to 15 days postinfection to kill an insect pest. Based on this slow speed of action, baculoviruses, as well as many other microbial control agents, were considered to have poor commercial efficacy. An obvious solution to this problem, of course, was genetic engineering of baculoviruses. The idea was to add a new gene to the virus that would allow the virus to kill insects quickly or to prevent them from feeding after infection. As described in Chapter 4, the addition and expression of pesticidal genes to baculoviruses through the use of the BEV technology is rather simple and straightforward.

The major problem with development and commercialization of genetically enhanced baculoviruses is the production of an effective product that poses little or no risk to health or the environment. First, the release of any engineered organism into the environment raises questions of potential effects to human health and nontarget organisms. Second, there are concerns that engineered organisms will displace naturally occurring organisms from their niches, thereby causing ecological perturbations. Third, if an engineered organism exhibits unanticipated and deleterious environmental properties, it may be difficult to eliminate the organism from the environment. And last, but not least, the foreign genetic material may be transferred from the released organism to other organisms with unpredicted consequences. These issues must be satisfactorily addressed prior to the field release of genetically engineered organisms.

GENETIC ENHANCEMENT OF VIRAL PESTICIDES

The BEV system has been used to insert numerous foreign genes into the *Autographa californica* (Ac), *Helicoverpa zea* (Hz), *Lymantria dispar* (Ld), and *Bombyx mori* (Bm) nuclear polyhedrosis viruses (see Chapter 4), many of which have been evaluated with respect to improvement of their commercial pesticidal properties. In most instances, the polyhedrin gene of the recombinant virus was replaced with the foreign pesticidal gene. Expression of these foreign products was aimed at enhancing the speed with which infected larvae are killed or stop feeding.

Present techniques for assessing the pesticidal properties of engineered baculoviruses have been restricted to LD_{50} and ST_{50} assays [10]. The LD_{50} assay measures the (lethal) dosage at which 50% of the larvae become infected. Because LD_{50} values are based on the efficiency of infection (attachment, penetration, and uncoating), baculoviruses expressing foreign pesticidal genes should have LD_{50}s equivalent to the parental virus. The ST_{50} assays are used to determine the (survival) time after infection when 50% of the larvae have died. Accordingly, ST_{50} assays are currently the most meaningful assays for assessing the pesticidal properties of these viruses. Although time to death can be an important parameter, time to cessation of feeding is the critical commercial parameter of pesticides. Standardized assays designed to quantitate the time to cessation of feeding and/or to quantitate feeding damage need to be developed.

Buthus eupeus Insectotoxin-1

In 1988, Carbonell et al. [11] reported the first attempt to improve the pesticidal properties of a baculovirus through insertion and expression of a foreign gene. They inserted the scorpion, *Buthus eupeus,* insectotoxin-1 (BeIT) gene into the AcMNPV genome under the transcriptional control of the polyhedrin gene promoter. The BeIT is a 4 kd, insect-specific paralytic neurotoxin that was isolated from the venom of *B. eupeus* [12,13]. The BeIT gene was synthesized based on the amino acid sequence [13] using the codon preferences exhibited in the AcMNPV polyhedrin and p10 genes. The level of BeIT gene transcription was equivalent to polyhedrin gene transcription. However, only small amounts of BeIT protein were detected, and no BeIT biological activity was observed. The lack of biological activity was considered to result from low toxin concentrations (inefficient translation), instability of the protein, or an error in the amino acid sequence analysis.

Androctonus australis AsIT Toxin

The AsIT toxin gene from the North African scorpion *Androctonus australis* has been cloned into and expressed during the replication of AcMNPV [14,15] and the *B. mori* NPV [16]. Both recombinant AcMNPV isolates were constructed with the toxin gene under the control of p10 promoter and with a polyhedrin gene so that the progeny virions were occluded.

Stewart et al. [14] tested an AsIT protein that was fused with the secretory signal of the gp67 protein of AcMNPV. Using neonate larval assays, the AsIT–AcMNPV exhibited a 25% reduction in the ST_{50} as compared with wild-type virus infections at a dosage of 17 polyhedra per larva (lethal dosage values not given). Despite the small reduction in ST_{50}, feeding damage by third instar larvae infected with AsIT–AcMNPV was 50% less than that produced by those infected with the wild-type virus.

The AsIT–AcMNPV constructed by McCutchen et al. [15] had a silkworm bombyxin secretory signal. Using second instar *Heliothis virescens* larval assays, the AsIT–AcMNPV had a 30% reduction in ST_{50} as compared with the wild-type virus at a dosage of 250 polyhedra per larva (lethal dosage values not given). Although the LD_{50} value for the recombinant virus was lower than the wild-type virus, the values were not significantly different.

Maeda et al. [16] replaced the polyhedrin gene of BmNPV with the AsIT gene fused to the secretion signal sequence of the silkworm bombyxin gene. Therefore, unlike the AsIT–AcMNPV recombinants, the BmNPV recombinant lacked the polyhedrin gene and formed only nonoccluded progeny virions. Injection of 10^5 plaque-forming units of the AsIT–BmNPV into second instar larvae resulted in a cessation of feeding at 40 hours postinoculation and death at 60 hours. This time to death was approximately 40% faster than with wild-type virus infections.

Pyemotes tritici Neurotoxin Tox-34

The female mite *Pyemotes tritici* produces a TxP-I toxin that causes muscle contraction and paralysis of its insect prey. Tomalski and Miller [17] constructed an AcMNPV recombinant that produced the TxP-I toxin protein following transcription, translation, and processing of the Tox-34 gene product. The TxP-I protein had a 39 N-terminal residue that was considered to act as a secretory signal and was cleaved to produce the mature TxP-I protein. The nonoccluded TxP-I-AcMNPV and wild-type virus were injected into fifth instar *T. ni* larvae at a dosage that infected approximately 38% and 55% of the larvae, respectively. Although dosages differed and the ST_{50} was not determined, the data suggested that death or paralysis with TxP-I-AcMNPV infections occurred approximately 40% faster than with wild-type virus infections.

Subsequent studies were performed with a polyhedrin plus TxP-I–AcMNPV recombinant [18]. These studies suggested that the recombinant virus had a higher LC_{50} than the wild-type virus. Using first instar *T. ni* larval assays, the time to paralysis or death (no differentiation made) was 40% lower with the TxP-I–expressing versus the wild-type virus infections.

Bacillus thuringiensis Delta-Endotoxin

The *Bacillus thuringiensis* delta-endotoxin is a highly effective toxin for the control a large number of agronomic insect pests [19]. The protoxin form of the *B. thuringiensis* ssp. *kurstaki* (Bt) HD-73 delta-endotoxin gene was inserted into the AcMPNV under the control of the polyhedrin and the p10 gene promoters [20] producing polyhedrin-minus and polyhedrin-plus recombinants, respectively. Biologically active toxin was isolated from virus-infected cell cultures and *T. ni* larvae. The Bt toxin bioassays were complicated by the feeding deterrent properties of the toxin. Although virus bioassays with second instar *T. ni* larvae suggested that the LD_{50} was lower with the Bt-recombinant virus than with the wild-type virus, the authors felt that the differences were within the experimental error of the bioassay procedure. ST_{50} bioassays performed with the Bt-recombinant and wild-type viruses indicated no significant differences. These results are consistent with the fact that, following ingestion, the protoxin is cleaved in the midgut of susceptible insects, producing an active toxin that results in cessation of feeding and eventual death. Cleavage of the protoxin to the active form might not be expected to occur within a cell. In addition, the active toxin normally binds to a receptor on the exterior surface of a cell. If membrane toxin receptors were available within the cell, the insertion and expression of the active toxin sequences rather than the protoxin might effect faster death of infected larvae. The expression of the protoxin did not lead to a detectable enhancement of pesticidal properties. However, under field conditions both the progeny virus and Bt toxin produced in these larvae would contaminate plant tissues and might significantly increase the potential for secondary control.

Manduca sexta Diuretic Hormone

In 1989, Maeda [21] reported replacement of the *B. mori* nuclear polyhedrosis virus (BmMNPV) polyhedrin gene with the diuretic hormone (DH) gene from the tobacco hornworm *Manduca sexta*. The DH gene was synthesized based on the amino acid sequence of the amidated C-terminal tryptic hexapeptide hormone [22]. Since the effective dose of the nonamidate form of DH is 3 logs lower than the amidated form, a glycine residue was added to the C terminus. Insect diuretic and antidiuretic hormones are generally considered to regulate secretion and resorption, respectively, of water in response to environmental changes.

The DH-virus–infected larvae had reduced hemolymph volumes compared with control and wild-type-virus–infected larvae. The weights of infected larvae were reduced 10% compared with control larvae. Concomitant with these effects, the DH-virus infections resulted in a reduction in ST_{50} compared with the wild-type virus infections of fifth-instar larvae. Although LD_{50} values were not presented, the data suggest a 10%–20% reduction in the LD_{50}.

Heliothis virescens Juvenile Hormone Esterase

Hammock et al. [23] inserted the juvenile hormone esterase (JHE) gene from the tobacco budworm *Heliothis virescens* genome into the AcMNPV genome under the control of the polyhedrin gene promoter. The JH is part of the larva's endocrine system involved in metamorphosis. Early in the last larval instar the level of ecdysteroids increases, and, as the JHE levels increase, there is a decrease in JH. This is followed by cessation of feeding and metamorphosis to the pupal stage [24]. If JHE activity is inhibited, metamorphosis is delayed, resulting in giant larvae [25]. Accordingly, the expression of JHE during virus replication might be expected to cause cessation of feeding and premature pupation.

The inserted JHE gene sequence included a secretory leader sequence that accounted for secretion of 94%–98% of the total JHE activity during virus replication in tissue culture cells. The yields corresponded to 75 mg per liter of medium. The data suggest that recombinant JHE produced in SF21 cells was more highly glycosylated than JHE isolated from *H. virescens* hemolymph.

When first instar *T. ni* larvae were infected with the JHE virus, their feeding and growth were reduced compared with control and wild-type-virus–infected larvae. However, this effect was not observed following infection of later instar *T. ni* larvae, and no observations were reported on JHE-virus infections of *Heliothis* larvae.

Ecdysteroid UDP-Glucosyl Transferase

Ecdysteroids are a family of steroid hormones that are important components of the lepidopteran endocrine system. Ecdysteroid levels are elevated both prior to molting to successive larval instars and prior to metamorphosis from the larval to pupal stage. O'Reilly and Miller [26,27] discovered that the AcMNPV genome contained a gene coding for a 57 kd protein with ecdysteroid UDP-glucosyl transferase (EGT) activity. This enzyme can transfer a glucose molecule from UDP-glucose to ecdysteroid, thereby inactivating ecdysteroids.

O'Reilly and Miller [26] observed that most fourth instar larvae of the fall armyworm *Spodoptera frugiperda* infected with the wild-type AcMNPV did not molt to the fifth instar. However, all larvae molted during infections with an AcMNPV mutant lacking the EGT gene (EGT-minus). Two characteristics of impending molts, feeding cessation and wandering, were noted with EGT-minus–infected larvae, but no data were presented regarding any alteration in the pesticidal properties of the EGT-minus virus. It should be noted that the EGT-minus AcMNPV contained a β-galactosidase gene fused in frame to the N terminus of the EGT gene. Expression of β-galactosidase during AcMNPV infections of *T. ni* larvae has been shown to increase the ST_{50} by 35% when compared with wild-type-virus infections [28].

Eldridge et al. [29] constructed and evaluated the biological properties of an EGT-minus AcMNPV recombinant that expressed the JHE and did not express β-galactosidase. It was considered that the production of EGT could mask the potential effects of JHE. The hemolymph from infected fourth instar *T. ni* larvae contained 40-fold higher levels of JHE than control larvae; however, little or no differences were detected in time to death, developmental characteristics, or weight gain.

Based on lack of molting following infection and on Southern hybridization data, most baculovirus genomes probably contain an EGT gene. It is currently speculated that EGT-minus baculoviruses, by themselves or with foreign gene inserts, may be useful in the development of enhanced viral pesticides.

Viral Enhancing Factor

The viral enhancing factor (VEF) gene in the *T. ni* GV genome codes for a protein that increases the viral pesticidal activity [30]. The *T. ni* GV VEF has properties similar to the *Pseudaletia unipuncta* GV synergistic factor originally reported by Tanada [31,32]. The VEF is a component of the granules (comparable to NPV polyhedra). The biological activity of VEF appears to be restricted to the infection process. Following dissolution of granules and release of the virus particles, the VEF disrupts the peritrophic membrane, a barrier that must be traversed for virus particles to come in contact with susceptible midgut epithelial cells. The removal of this barrier significantly increases the efficiency of infection.

In dose–response assays with *T. ni* larvae inoculated with AcMNPV, TnSNPV, and *Anticarsia gemmatalis* MNPV, the addition of VEF resulted in 3- to 16-fold reductions in the LD_{50}s [33]. The addition of supplemental VEF in baculovirus pesticide products could significantly reduce application rates and would be particularly useful for the control of late instar larvae, which typically require very high application rates to obtain infection.

Other Pesticidal Genes

There are of course a large number of gene products that may prove useful in improving the pesticidal properties of baculoviruses. Besides toxic proteins, host hormones, hormone receptors, and metabolic enzymes are attractive candidates, especially if one takes into consideration the potential environmental consequences of a genetically modified viral pesticide. Keeley and Hayes [34] proposed the expression of "antipeptides" that would bind to neurohormones, thereby blocking their physiological functions. Unlike toxic protein genes isolated from predators, the use of host proteins should pose few if any toxicological problems.

ENGINEERING STRATEGIES

Expression of the above-mentioned foreign genes to improve the pesticidal properties has generally followed the strategy of the original baculovirus expression vector system [8,9] as described in Chapter 4. With the removal of the polyhedrin gene, only the nonoccluded form of the virus is produced. The nonoccluded virions are unstable, and infectivity is rapidly lost in the soil, on plant tissues, and in dead larval tissues [35,36]. In the absence of polyhedrin protein production, enveloped virions that normally would become occluded are still produced and are highly infectious per os, but they are too unstable for commercial applications [28].

The infectivity of baculovirus pesticides must be stabilized by occlusion in polyhedra or by some other means in order to be able to deliver an active pesticide to the field. Therefore polyhedrin-plus recombinant viruses expressing *Bacillus thuringiensis* delta-endotoxin [20] and *Androctonus australis* AsIT toxin [14,15] were constructed under the tanscriptional control of the p10 gene. Like the polyhedrin gene, the p10 gene is a nonessential gene that is under the control of a strong late promoter. In 1988, Vlak et al. [27] proposed replacement of the p10 gene with foreign gene inserts. Recombinant viruses lacking the p10 gene produce polyhedra lacking a polyhedral envelope. It should be noted that polyhedra plus recombinant pesticides are environmentally stable and will persist for years in an ecosystem. Accordingly, prior to their use as pesticides, they will require special evaluation for the potential problems that they might pose to the environment.

An alternative method to occlude and stabilize poly-minus baculoviruses is with the co-occlusion strategy. Host cells can be coinfected with both wild-type and recombinant, poly-minus virus isolates [36,38,39]. The wild-type virus produces polyhedrin protein that will occlude both the wild-type and poly-minus virus particles within polyhedra. The co-occlusion strategy provides a means of placing an engineered virus at a selective disadvantage in the environment. The continued propagation of a co-occluded, poly-minus virus in a virus population is controlled by the probability of coinfection of individual larvae and cells with both virus types as the virus is passed from insect to insect. As documented in larval assay procedures, the probability of the production of co-occluded, poly-minus virus under field conditions is extremely low [36,40].

Transcription of foreign gene inserts to date have generally been under the control of either the polyhedrin or the p10 gene promoters. Both are strong, late viral promoters. It has been considered that using early gene promoters would effect faster expression of foreign gene products. Early viral gene promoters, however, are not strong and result in quicker but weaker expression. The use of early promoters may be restricted to the expression of foreign gene proteins that are fast acting and have extremely high specific activities. However, Tomalski and Miller [18] showed that the expression of the *Pyemotes tritici* neurotoxin Tox-34 under the control of the polyhedrin was more effective than expression under control of the early ETL promoter. A possible alternate promoter would be a constitutively expressed insect gene promoter.

ENVIRONMENTAL ISSUES

To date most research toward improving the pesticidal properties of baculoviruses has centered on the insertion and expression of genes that will effect quick cessation of feeding and/or death. Based on progress to date, it is clear that many such recombinant viruses will soon be available. However, in addition to improving the commercial properties, the field release of genetically engineered baculoviruses or any other organism should satisfy a host of ecological and environmental issues [41]. The challenge in the future will be to construct these viruses in such a manner that the regulatory agencies, scientific community, and public are assured of safety both to human health and to the environment as a whole.

One of the positive environmental attributes of naturally occurring baculoviruses is that their reported host ranges are limited to a small number of invertebrates that do not include beneficial insects. The limited host range of baculoviruses has also been a deterrent to commercial development because this generally limits the potential market size.

It should be noted, however, that baculovirus host range studies with invertebrate species have been based on the signs and symptoms following exposure to the virus [for review, see ref. 3]. Viruses are capable of replicating in organisms in the absence of any overt pathological disturbance. Accordingly, virus replication in some "nonhosts" may occur. Although the inapparent wild-type virus infections could be harmless, the expression of pesticidal gene products may be deleterious to or kill these species. Therefore, a careful assessment of the effective host ranges of recombinant baculoviruses is advisable.

The ecological relationship between baculoviruses and their hosts is not well understood. Because of this, an additional concern with the release of a genetically enhanced virus is its potential to displace wild-type virus populations in nature. The displacement of the natural virus by an engineered virus in an agricultural environment would probably go unnoticed. However, in nonagricultural settings such as forests, this event could result in unanticipated ecological changes over large areas. Accordingly, in the absence of good ecological data, it would seem prudent to engineer viral pesticides in ways that ensure they will be at a selective disadvantage in nature.

An understanding of the ecology of baculoviruses is also required to assess the possibility that an engineered baculovirus may possess unanticipated properties that would warrant mitigation. Elimination of a baculovirus from nature would be highly problematic. Baculoviruses can survive in the soil for years [42]. Therefore, engineering strategies that limit survival of baculoviruses in nature are attractive.

FIELD RELEASE TESTING

Field testing of genetically engineered AcMNPV was initiated in 1986. The recombinant viruses used in these studies have not contained genes that would enhance their pesticidal activities. Rather, the testing was performed to evaluate the risks associated with recombinant baculoviruses and to test a strategy to limit the persistence of recombinant viruses in the environment.

The first field trial with a genetically engineered baculovirus was performed in 1986 by researchers at the Natural Environment Research Council's (NERC) Institute of Virology in Oxford, England [35]. Under conditions of stringent physical containment, which were equivalent to a greenhouse environment, larvae were infected with a marked strain of AcMNPV and placed in the field enclosure. The studies showed that marked virus and parental virus had identical host ranges, genetic stability in cell culture, and stability in soil.

In 1987 the NERC Institute of Virology conducted a second field trial. For this experiment a recombinant AcMPNV was used that did not express the polyhedrin gene. Insects were infected with the polyhedrin-minus virus in the laboratory and placed on plants in the field enclosure. In the absence of the polyhedrin gene, the virus particles are not occluded within polyhedra and are rapidly inactivated in the decaying insect tissues. Therefore, as anticipated, 1 week after the infected larvae died, virus infectivity could not be detected in foliage or in soil samples.

In 1989, researchers at the Boyce Thompson Institute for Plant Research conducted the first field-application release of a genetically altered baculovirus [40]. An AcMNPV isolate whose genome lacked a functional polyhedrin gene was occluded within polyhedra during replication in cells also infected with wild-type AcMNPV (co-occlusion process, discussed above). According to laboratory data [36], the polyhedrin-minus virus could not persist in the virus population because of the low probability of coinfections under natural conditions.

The co-occluded polyhedrin-minus AcMNPV was sprayed on a 0.25 acre of plants at the Cornell University's New York State Agricultural Experiment Station in Geneva, New York. Because of the level of biological containment provided in the test, physical containment was not needed. The inoculum contained 49% recombinant virus, and the progeny polyhedra isolated from field-infected larvae in 1989 contained, on average, 42% poly-minus virus particles. In 1990 and 1991, the amount of polyhedrin-minus virus isolated from infected larvae continued to decline, thereby corroborating the laboratory-based predictions.

These field release studies have been conservative in their approach toward the eventual release of genetically enhanced viral pesticides. Through these programs it is anticipated that a better understanding of the ecology of baculoviruses will evolve. This should provide data indicating how best to use recombinant viral pesticides in agricultural and forest settings and at the same time how best to minimize any risks associated with their use.

CONCLUSIONS

With the growing awareness of the real and potential problems that synthetic chemical pesticides pose to human health and to the environment as a whole, there has been increased scrutiny of chemical pesticides by regulatory agencies, both state and federal. The resulting stringent requirements for safe and effective pesticides has led to a decreased availability of new compounds for pest control. Commercial decisions to discontinue the registration of old compounds and to develope resistant pest populations are having a major impact on agriculture today. Because of these and other factors, biological control agents have become commercially attractive and will certainly play a larger role in agriculture in the future.

Because of the potential environmental and health safety benefits associated with biocontrol agents, the President's Council on Competitiveness [43] has specifically identified the development of genetically improved biocontrol agents as an area of national importance. Progress to date clearly indicates that numerous genetically enhanced viral pesticides will soon be available. A major challenge to the development of these products lies in the area of environmental risk assessments. These assessments will include evaluations of the properties of the foreign product(s) as well as ecosystem analyses.

Along with these assessments are a multitude of other commercial issues, i.e., production costs, stability, formulation, application technology, and field efficacy. In the majority of instances the commercial use of viral pesticides will be based on their cost benefit ratios as compared with synthetic pesticides. A major factor will be the industrial production of viral pesticides. Currently, only in vivo production methods are available. However, progress in the development of insect cell serum-free media and novel bioreactors should soon lead to commercially feasible in vitro production technologies [44].

REFERENCES

1. Benz, G.A. In: The Biology of Baculoviruses. Vol. 1. CRC Press, Boca Raton, FL, 1986.
2. Gehren, U. Z. Forst. Jagdwesen 24:499, 1892.
3. Groner, A. In: The Biology of Baculoviruses. Vol 1. CRC Press, Boca Raton, FL, 1986.
4. Cunningham, J.C. In: Microbial and Viral Pesticides. Marcel Dekker, New York, 1982.
5. Yearian, W.C., and Young, S.Y. In: Microbial and Viral Pesticides. Marcel Dekker, New York, 1982.
6. Kirschbaum, J.B. Annu. Rev. Entomol. 30:51–70, 1985.
7. Calderoni, P. Farm Chemicals 154:26–28, 1991.
8. Smith, G.E., Summers, M.D., and Fraser, M.J. Mol. Cell. Biol. 3:2156–2165, 1983.
9. Pennock, G.D., Shoemaker, C., Miller, L.K. Mol. Cell. Biol. 4:399–406, 1984.
10. Wood, H.A., and Hughes, P.R. In: Advanced Engineered Pesticides. Marcel Dekker, New York, (in press).
11. Carbonell, L.F., Hodge, M.R., Tomalski, M.D., and Miller, M.K. Gene 73:409–18, 1988.
12. Grishin, E.V. In: Frontiers of Bioorganic Chemistry and Molecular Biology. Pergamon Press, New York, 1980.
13. Grishin, E.V. Int. J. Quantum Chem. 19:291–298, 1981.
14. Stewart, L.M.D., Hirst, M., Ferber, M.L., Merryweather, A.T., Cayley, P.J., and Possee, R.D. Nature 352:85–88, 1991.
15. McCutchen, Choudary, P.V., Crenshaw, R., Maddox, D., Kamita, S.G., Palekar, N. Volrath, S., Fowler, E., Hammock, B.D., and Maeda, S. Biotechnology 9:848–852, 1991.

16. Maeda, S., Volrath, S.L., Hanzlik, T.N., Harper, S.A., Majima, K., Maddox, D.W., Hammock, B.D., and Fowler, E. Virology 184:777–780, 1991.
17. Tomalski, M.D., and Miller, L.K. Nature 352:82–85, 1991.
18. Tomalski, M.D., and Miller, L.K. Bio/Technology 10:545–549, 1992.
19. Aronson, A.I., Beckman, W., and Dunn, P. In: Microbiological Reviews. American Society Microbiology, Washington, DC, 1986.
20. Merryweather, A.T., Weyer, U., Harris, M.P.G., Hirst, M., Booth, T., and Possee, R.D. J. Gen. Virol. 71:1535–1544, 1990.
21. Maeda, S. Biochem. Biophys. Res. Commun. 165:1177–1183, 1989.
22. Kataoka, H., Troetschler, R.G., Li, J.P., Kramer, S.J., Carney, R.L., and Schooley, D.A. Proc. Natl. Acad. Sci. U.S.A. 86:2976–2980, 1989.
23. Hammock, B.D., Bonning, B.C., Possee, R.D., Hanzlik, T.N., and Maeda, S. Nature 344:458–461, 1990.
24. deKort, C.A.D., and Granger, N.A.A. Annu. Rev. Entomol. 26:1–28, 1981.
25. Sparks, T.C., and Hammock, B.D. Pestic. Biochem. Physiol. 14:290–302, 1980.
26. O'Reilly, D.R., and Miller, L.K. Science 245:1110–1112, 1989.
27. O'Reilly, D.R., and Miller, L.K. J. Virol. 64:1321–1328, 1990.
28. Wood, H.A., Trotter, K.M., Davis, T.R., and Hughes, P.R. J. Invertebr. Pathol. 62:64–67, 1993.
29. Eldridge, R., O'Reilly, D.R., Hammock, B.D., and Miller, L.K. Appl. Environ. Microbiol. 58:1583–1591, 1992.
30. Derksen, A.C.G., Granados, R.R. Virology 167:242–250, 1989.
31. Tanada, Y. J. Insect Pathol. 1:215–231, 1959.
32. Tanada, Y. J. Invertebr. Pathol. 45:125–138, 1985.
33. Gallo, L.G., Corsaro, B.G., Hughes, P.R., and Granados, R.R. J. Invertebr. Pathol. 58:203–210, 1991.
34. Keeley, L.L., and Hayes, T.K. Insect Biochem. 17:639–651, 1987.
35. Bishop, D.H.L., Entwistle, P.F., Cameron, I.R., Allen, C.J., and Possee, R.D. In: The Release of Genetically Engineered Microorganisms. Academic Press, New York, 1988.
36. Hamblin, M., van Beek, N.A.M., Hughes, P.R., and Wood, H.A. Appl. Environ. Microbiol. 56:3057–3062, 1990.
37. Vlak, J.M., Klingkenberg, F.A., Zaal, K.J.M., Usmany, M., Klinge-Roode, E.C., Geervliet, J.B.F., Roosien, J., and Van Lent, J.W.M. J. Gen. Virol. 69:765–776, 1988.
38. Miller, D.W. In: Biotechnology for Crop Protection. American Chemists Society, Washington, DC, 1988.
39. Shelton, A.M., and Wood, H.A. World & I 4:358–365, 1989.
40. Wood, H.A., Hughes, P.R., van Beek, N., Hamblin, M. In: Insect Neurochemistry and Neurophysiology 1989. Humana Press, 1990.
41. Tiedje, J.M., Colwell, R.K., Grossman, Y.L., Hodson, R.E., Lenski, R.E., Mack, R.N., and Regal, P.J. Ecology 70:298, 1989.
42. Thompson, C.G., Scott, D.W., and Wickham, B.E. Environ. Entomol. 10:254–255, 1981.
43. Report on National Biotechnology Policy. The President's Council on Competitiveness. February 1991.
44. Shuler, M.L., Cho, T., Wickham, T., Ogonah, O., Kool, M., Hammer, D.A., Granados, R.R., and Wood. H.A. Ann. N.Y. Acad. Sci. 589:399–422, 1990.

6

Fundamentals of Baculovirus–Insect Cell Attachment and Infection

Daniel A. Hammer, Thomas J. Wickham, Michael L. Shuler, H. Alan Wood, and Robert R. Granados

School of Chemical Engineering (D.A.H., M.L.S.) and Boyce Thompson Institute for Plant Research (H.A.W., R.R.G.), Cornell University, Ithaca, New York 14853; Genvec, Rockville, Maryland 20852 (T.J.W.)

INTRODUCTION

Baculoviruses are a family of double-stranded DNA enveloped viruses that infect insect cells. In the past two decades they have become exceedingly important in biotechnology, for two reasons. First, since they infect insects, they are candidates for insect pest control [1]. Second, certain baculoviruses, notably *Autographa californica* multiple nuclear polyhedrosis viruses (AcMNPV), have been developed as vectors for the expression of recombinant proteins in insect cell lines. Insect cell culture for the production of recombinant proteins has found widespread use in the biotechnology industry and in research and offers several advantages over recombinant protein production in bacterial or mammalian cell systems, as is thoroughly discussed in other chapters of this book.

The baculovirus life cycle is illustrated in Chapter 1. There are two forms of AcMNPV virus: the nonoccluded form (NOV), also known as *budded virus* (BV); and the occluded form (OV), viral particles from which are known as *polyhedral-derived viruses* (PDV). NOVs are single virus particles surrounded by an envelope membrane that bud from the surface of the plasma membrane of an infected insect cell. These NOVs are generated early (12–24 hours) in the replication cycle after initial infection and are the form responsible for the primary and secondary infections of insect cell tissues. OVs are generated later in the infection (after 24 hours). Many individual virus particles are found within a single occluded particle, encapsulated in a crystalline matrix of polyhedrin protein. The occluded form of the virus

Baculovirus Expression Systems and Biopesticides, pages 103–119

is responsible for primary infection of insects, as insects ingest this form, and alkaline conditions present in the insect midgut lead to polyhedrin dissolution. The resulting naked PDV particles infect midgut cells of the insect. Therefore, both the NOV and OV are infectious, but to different types of host cells. As has been illustrated elsewhere in this book, for production of recombinant proteins NOVs are particularly important, as the gene coding for a protein can be placed behind the polyhedrin promoter, which channels the replication machinery of the virus toward synthesizing the desired protein.

To infect, either the NOV or OV must get inside the appropriate host cell. Figure 1 gives a general view of virus–host cell binding and internalization. The virus must first attach to the cell surface. There are two distinct routes by which a virus may enter the cell: endocytosis or fusion. Either of these modes of entry may take place in any virus–cell system, and there is some evidence that both take place in the baculovirus–insect cell system. Although viruses must enter cells to be infectious, not every route of entry leads to infection.

Most viruses attach to cells by highly specific binding between surface proteins on the virus (viral attachment proteins) and cell surface receptor molecules on the host cell membrane (Fig. 1). The specificity of binding between viral attachment protein and receptor, as well as the limited tissue distribution of the receptor, usually

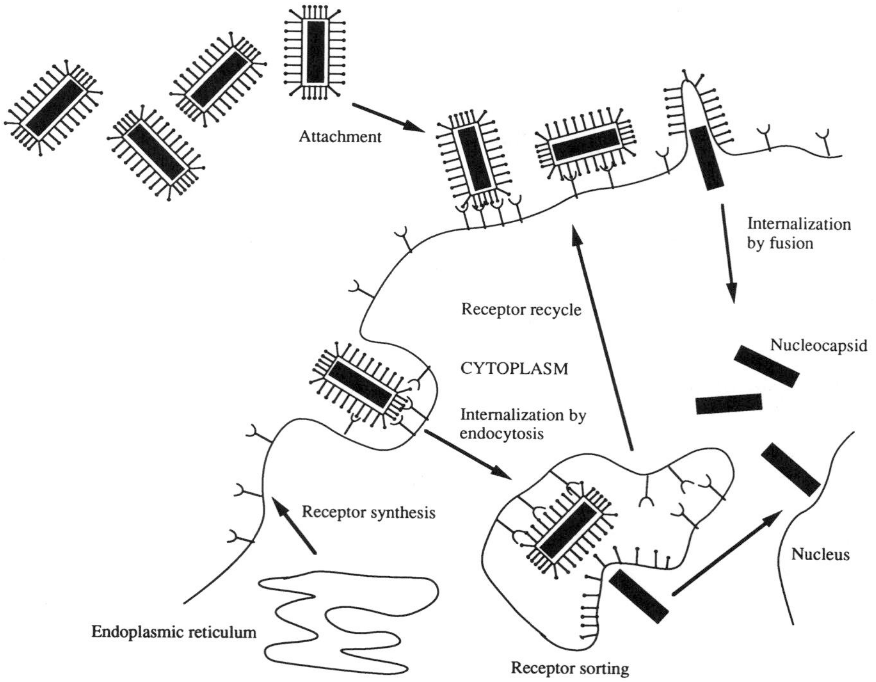

Fig. 1. A general model for the attachment and penetration of an enveloped virus involves attachment via cell surface receptors and different routes of internalization, such as fusion and endocytosis. Only some of these routes are infectious.

explains the host tropism of a virus; only hosts with the proper receptor to bind the appropriate viral surface attachment protein support the attachment of and subsequent infection by a particular virus [2].

While insect host cells are clearly infected by baculovirus, the insect cell receptor that mediates baculovirus attachment and infection has not yet been found. Although many different insect cell lines are susceptible to baculoviruses, individual baculovirus isolates tend to have a rather narrow host tropism (relatively few insect cell lines can be infected by the same virus) [3]. Do all the cell lines that can be infected by the virus have the requisite receptor? Since NOV infect cells in culture but do not infect midgut cells in larvae, while OV infect midgut cells but do not infect tissue culture cells, do NOV and PDV particles bind to different receptors? Do tissue culture cells and midgut cells express different receptors? How do structural molecules of the baculovirus cell surface interact with these receptors? Does the receptor mediate endocytosis or fusion to promote infection by the virus, or are there other cell surface molecules that perform these functions?

Elucidation of these questions will lead to natural ways to regulate the attachment and penetration of baculovirus into insect cells and ultimately optimize the productivity of insect cell cultures. While it is clear that infection of host is desirable for productive insect cell culture, the exact link between attachment and infection, or between viral infection and protein production, has not yet been made. Two considerations in making such connections are whether there is an optimum number of virus particles that can attach to the insect cell surface and whether all of the attached viral particles can enter the cell by the proper route (endocytosis or fusion) to cause infection. Once we better understand the molecular basis of attachment and entry, we will be able to quantify infection and relate it to the productivity of insect cell culture.

Baculovirologists have attempted to address some of these questions, and a great deal is now known about the structural proteins of the different forms of baculoviruses. We review what is known currently about the molecular basis of baculovirus attachment and penetration, as well as efforts to quantify viral–host cell attachment and viral internalization by insect cell cultures.

MOLECULES INVOLVED IN BACULOVIRUS ATTACHMENT AND PENETRATION

Since there are two infectious forms of virus, nonoccluded (budded) (NOV) and polyhedral derived (PDV), it is conceivable that there are different viral surface structural proteins involved in infection of the two particle types. We review the evidence for the existence of a structural protein involved in infection for each of these types of virus.

Nonoccluded Virus

As noted previously, this enveloped form of the virus involves single baculovirus particles that are responsible for primary and secondary infection of insect tissues and that are used for infection in tissue culture and bioreactors. Most of the work on structural proteins of NOV involved in infection has centered on the 64 kd protein of AcMNPV and OpMNPV.

The 64 kd Envelope Protein of Nonoccluded Virus

There is substantial evidence for a 64 kd protein on the surface of budded AcMNPV and OpMNPV [3–6]. This protein is synthesized by the infected host, can be detected as early as 6 hours postinfection with OpMNPV [7,8], and is concentrated on the plasma membrane of the host cell. Nucleocapsids become coated with envelopes containing the protein when they bud from the host cell. The protein is located at the peploma end of the rod-shaped virus, possibly in an aggregated form (see electron micrographs of Volkman [6]). After budding, these NOV particles infect susceptible hosts. The 64 kd protein has been conclusively shown to be involved in infection by NOV, as antibodies to the protein (AcV_1) neutralize the efficient entry of the virus [6]. The virus probably enters the cell largely by endocytosis, although fusion at the plasma membrane cannot be ruled out (see Biological Evidence for the Infectious Route of Entry, below). Electron micrographs of virus entry at the surface of *Trichoplusia ni* 5-B14 cells clearly show the accumulation of virus in coated pit structures [9,10]. The antibody AcV_1 appears to have little effect on attachment of the virus to the cell surface, but acts to inhibit infection. Although it has not yet been conclusively shown how gp64 mediates entry, there are several lines of evidence that show the protein is involved in the low pH induced fusion between the virus and the membrane surrounding the endosome after endocytosis [3,11]. Inhibition of AcMNPV infection in the absence of antibody can also be achieved with inhibitors that prevent acidification of the endosome, such as chloroquine and NH_4Cl [6]. Baculovirus-infected cells, which are known to express gp64 on their plasma membrane, form syncytia when placed in contact at low pH [11,12]. gp64 is necessary and sufficient for syncytia formation at low pH, as shown by virus-free expression of gp64 in insect cells and subsequent syncytia formation [11]. Also, the AcV_1 antibody does not inhibit binding of baculovirus particles to cells, and previously bound particles can be neutralized by this antibody [6]. One must be careful to preclude a role for this protein in attachment, however, since the gp64 protein might have multiple functionalities (for attachment and fusion), and AcV_1 may be directed at only one of these. In fact, Volkman and Goldsmith [13] have shown that treatment of NOV with proteases (trypsin or proteinase K) does not reduce viral infectivity, but does eliminate sensitivity of the virus to neutralization by AcV_1. This protease-treated, AcV_1-insensitive infectious virus appears to enter by low-pH fusion, since entry is sensitive to chloroquine [13]. In contrast, polyclonal antisera can block the infection of protease-treated forms of the virus. These studies suggest that an epitope not recognized by AcV_1 plays an important role in gp64-mediated infectivity. Certainly, other as yet unidentified functional epitopes can exist that mediate attachment. The hemagglutinin of influenza virus would be a model for such a multifunctional protein [14], since it is known to be both an attachment and a fusion protein.

The sequences of the 64 kd proteins expressed on the two baculoviruses, OpMNPV and AcMNPV, are known [4,5]. In the case of OpMNPV gp64, the sequence was obtained by using monoclonal antibody AcV_5 against gp64 of AcMNPV, which cross-reacts with OpMNPV [15] to screen a λgt11 expression library of OpMNPV. Inserts from immunopositive λgt11 recombinant were used to isolate the gp64 gene [5]. This gene was sequenced. A similar procedure was followed for the gp64 molecule of AcMNPV [4]. The amino acid sequence of the two proteins has been published by Blissard and Rohrmann [5]. There is 78% sequence similarity between the two proteins. The two proteins appear most different at the C and N termini, which remain hydrophobic. The C-terminus is a purported transmembrane region, which is consistent with gp64's role as a protein that can be imbedded in both the plasma membrane of the host and in the viral envelope. The N-terminus is assumed to be the signal sequence. Without considering these two termini, the sequences are 83% similar. There are two regions in the core of the protein that are dissimilar, running from amino acid positions 182–199 and 269–292. The two proteins have very similar hydropathy profiles. gp64 is known to be glycosylated; there are five conserved potential sites (arginine residues) for N-linked glycosylation in the two proteins; the gp64 from OmMNPV has two additional N-linked glycosylation sites [5]. gp64 can also be phosphorylated and contains disulfide bridges [8]. When inserted in the membrane, this protein appears to aggregate into tetramers [8,16]. This aggregate formation may be important for the function of this molecule; in influenza virus, the hemagglutinin molecule, which is also a low pH induced fusion protein, appears to form trimers [14].

Several papers describe the regulation of the expression of gp64 [3,5,17,18]. In OpMNPV, the gene for gp64 has both early and late promoters. Of four motifs available as late promoter, only two appear to be utilized. This unique combination of early and late promoters appears optimal for the continued expression of gp64 during the entire infection phase of the host, from well before the initiation of budding (6–12 hours) to well after budding of NOV is largely complete (24–36 hours), and budding has depleted gp64 from the cell membrane [5].

Recently it has been shown that the gp64 proteins from baculoviruses are related to two structural proteins found in two viruses that biochemically and morphologically resemble the orthomyxoviridae, enveloped viruses with single-stranded, negative sense RNA genomes, which include influenza virus and a number of other human pathogens [19]. These two viruses, Thogoto and Dhori, are tick-borne arboviruses. Nucleotide sequence analysis of the RNA segment of these viruses showed significant amino acid sequence identity with the gp64 proteins from baculoviruses. In particular, the distribution of cysteine residues among the four proteins (those of Thogoto and Dhori, and gp64 of AcMNPV and OpMNPV) is conserved, and the hydropathy profiles of the proteins appear similar, suggesting that the proteins have very similar functional roles in the two systems. Indeed, acidic pH cell fusion of a baby hamster kidney (BHK) cells infected with Thogoto virus has been observed [19]. Although it has not yet been directly shown that the protein with significant sequence identity to gp64 is responsible for this fusion reaction, the role of baculovirus gp64 in similar fusogenic reactions is suggestive that the functional role of the proteins has been conserved. It is not known whether baculoviruses can infect ticks or whether Thogoto and Dhori can infect insects (specifically Lepidoptera), but such cross-reactivity should be tested in the near future.

Occluded Virus (Polyhedron Derived)

The protein matrix surrounding nucleocapsids in the occluded form includes proteins such as polyhedrin and a 37 kd protein associated with the outermost edge of the occluded form [3]. These proteins represent a protective shield for the virus in the occluded form and are shed or dissolved from the virus when it enters the alkaline environment of the insect midgut. The virion envelope of the subsequently discharged PDV is the first surface of contact between the virus and midgut cell. Very little is known about the structural proteins that make up this envelope, and even less is known about the role of these proteins. It is likely that the envelope of PDV is different in composition than the envelope of the NOV particles. The envelope of NOV is derived from the plasma membrane of the host cell. Proteins in this membrane are transported to the cell surface by "normal" cellular processing for plasma membrane proteins and are deposited on the surface of the NOV when it buds through the plasma membrane. In contrast, proteins in the PDV envelope must be synthesized and assembled in the nucleus. Similar forms of proteins would exist on the two envelopes if homologous proteins were both synthesized and transported to the plasma membrane during the early phase of infection and synthesized and assembled into the envelope of the occluded form during the late phase of infection.

Structural Proteins of PDV Involved in Infection. A recent review of what is known about structural proteins of PDV envelope is given by Rohrmann [3]. Several small peptides (14.2, 18, and 26.4 kd) have been found associated with the envelope of this form of the virus, but the functions of these proteins are unknown. An additional protein, p74, has been identified in association with AcMNPV and OpMNPV occlusion bodies, either in the viral envelope of the occluded form or in the vicinity of the capsid of occluded virions [3]. The p74 protein is required for infection by the occluded form of the virus, but not the nonoccluded form, as shown by experiments in which the p74 gene is inactivated [20]. In such experiments, the virus can replicate in tissue culture, but the resulting occluded form of the virus is not infectious to insect hosts [3,20].

64 kd Protein Associated With Envelope of PDV. A 64 kd protein does appear to associate with the envelope of PDV [6], but it is not clear whether this protein is the same as the 64 kd protein known to be associated with the budded form of the virus. An early summary of evidence as to the identity of the 64 kd protein of PDV and its relationship to the gp64 molecule of NOV is given by Volkman [6].

In one instance, neutralizing antisera to NOV was also able to neutralize the associated PDV [6,21,22]; in contrast, neutralizing antisera to PDV was not able to neutralize the associated BV [6,21,22], suggesting that PDV possessed an active protein that NOV did not. In contrast, Hohmann and Faulkner [15] raised monoclonal antibodies against NOV, and four separate titers showed neutralizing activity against NOV, but not against PDV. One of these antibodies was labeled AcV_1 [6,16] and was shown to recognize the 64 kd protein of NOV. There is considerably more 64 kd protein associated with the NOV than the PDV [6], so the inability of AcV_1 to recognize a PDV 64 kd protein is either due to very small levels of expression of this molecule in PDV or to lack of identity between these proteins on the two types of virus.

The low level of expression of the 64 kd protein in PDV may explain why several groups have been able to identify 64–67 kd counterparts between BV and PDV [23,24], whereas other have not [6,25,26]. Immunological staining is similarly conflicting. Polyclonal antibodies against PDV do recognize a 64 kd protein of BV, but do not neutralize BV [6]. In contrast, AcV_1 does not recognize any PDV protein. It is not known how this PDV protein is involved in PDV particle infection, if at all.

Enhancin. A 104 kd protein, called *enhancin* (formerly referred to as *viral enhancement factor* [VEF]) contained in the granules of occluded TnGV appears to play multiple roles in enhancing the infection of polyhedral-derived nucleocapsids [27]. In three separate, distinct occluded viral systems—AcMNPV, TnSNPV, and TnGV—similar biochemical activities appear to be held within the polyhedrin protein matrix and released during the alkaline dissolution of polyhedra in the insect midgut. These released factors cause specific biochemical and structural changes in the peritrophic membrane (PM) of the insect midgut, a protective barrier that separates the insect midgut from the infection-susceptible midgut columnar cells [27]. This protein appears to comprise about 5% of TnGV polyhedra [3]; the gene for this molecule has been sequenced in TnGV, but related sequences have not been found in NPVs despite the presence of similar enzymatic activity from those systems [3,28]. However, the substrates of activity for the factors from different polyhedral-derived viral lines appear to be different; for example, the factor from TnMNPV appears to breakdown and remove a 68 kd component of the PM, whereas the protein associated with TnGV appears to break down three larger proteins (253, 194, and 123 kd) [27]. These possible differences in substrate specificity, combined with the lack of similarity between sequences, suggests there are different molecules with the same function.

Because of their ability to break down elements of the PM, and perhaps because of other effects, the released components of polyhedra enhance the infection of the PDV to the midgut columnar cells. Proteins derived from the TnGV granules or AcMNPV polyhedra enhance the infectivity of the AcMNPV in *T. ni* larvae [27]. There is additional evidence that enhancin derived from the polyhedrin matrix of TnGV specifically binds to the plasma membrane of brush border cells of *Pseudaletia unipuncta* [29], which suggests enhancin might have a dual role of breaking down the PM and serving as a specific binding site (or homing receptor) for PDV. A lipoglycoprotein with a molecular weight between 93 and 126 kd derived from the polyhedrin matrix of the Hawaiian strain of granulosis virus (GVH), called *synergistic factor* (SyF), has also been shown to bind to midgut cell membrane with a high specificity (binding with K_d = 1.5 nM), although this protein has not been shown to cleave the PM enzymatically [30]. Numerous studies implicate this protein in the enhanced infectivity of nuclear polyhedrosis virus (NPV). For example, injection of purified SyF with NPV greatly enhances the infectivity of the virus [30]. The localization and specific binding of these two factors to the plasma membrane of midgut cells suggest that these protein belong to a family, or are homologous, and act by increasing the binding or entry of virus into these cells.

Cell Surface Receptors for Baculovirus Binding

Because the molecules on the surface of NOV or PDV involved in baculovirus binding have not yet been identified conclusively, it has been difficult to identify the complementary receptors on the insect cell membrane that bind these molecules. Two lines of evidence suggest such receptors do exist. First, the virus has a narrow host range, although it enters a wide variety of cell types. Attachment is not a sufficient condition for infection, which suggests that the virus must enter properly—through a receptor-mediated route—to ensure proper replication.

There is other evidence for cell surface receptors that mediate the binding of both NOV and PDV particles. Evidence that NOV attachment is mediated by a cell surface receptor is given by Wickham and coworkers [2], in which cells treated with proteinase K do not bind nonoccluded AcMNPV, whereas untreated cells bind the virus avidly. Figure 2 shows a Scatchard analysis of AcMNPV binding to *T. ni* midgut cells with and without treatment by proteinase K. This study found that, although the virus–cell binding is rather strong, the binding between viral attachment protein and receptor is rather weak, with $K_d = 10^{-5}$ M. Apparently, there are many putative receptors available to support binding, since the virus geometrically saturates the cell surface; the number of available receptor sites was predicted to be in excess of 10^5/cell [2]. Note that this is greater than the effective number of binding sites for the virus, since the virus is multivalent. The effective number of binding sites for AcMNPV binding to TN5B1-4 and TN-F cells is 6×10^3 and 1.37×10^4, respectively [10]. Avidities of baculovirus binding can be quite strong. The affinity of AcMNPV binding to TN5B1-4 and TN-F cells is 2.35×10^{10} M^{-1} and 1.6×10^{10} M^{-1}, respectively.

There is now evidence that PDV particle binding in at least one baculovirus–insect cell system is receptor mediated [31]. PDV from *Lymantria dispar* NPV were labeled with fluorescein isothiocyanate (FITC). Binding of labeled PDV particles to both *L. dispar* LdElta cells and brush border membranes derived from *L. dispar* was measured at 4°C using fluorescent activated cell sorting. Binding of virus was saturable. Protease treatment of the host cell surface reduced virus binding as much as 70%. Unlabeled PDV could compete with binding of labeled virus. The number of PDV-specific receptor sites was measured to be 10^6/cell. This is the first study to

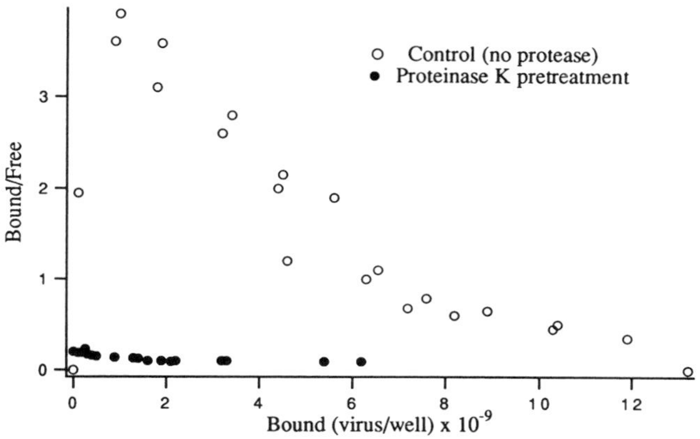

Fig. 2. Scatchard analysis of binding of AcMNPV to *T. ni* midgut cells that have been treated with and without proteinase K. Binding shows a negative cooperativity, and the reduction in binding due to proteinase K shows that a proteinaceous receptor is involved in the binding of the virus. (Modified from Wickham et al. [2], with permission of the publisher.)

present evidence for receptor-mediated binding of PDV particles to host cells or brush border membranes [31].

MEASUREMENTS OF BACULOVIRUS ATTACHMENT AND INFECTION

Clearly, attachment is a necessary condition for infection, but the exact relationship between attachment and infection has been obscure. Although there is still much to learn about the biochemical and biophysical basis of baculovirus entry and infection, the gross macroscopic relationship between attachment and infection in this system is becoming clear.

Attachment and Infection for Attached Cells

Wickham et al. [10] measured the attachment and infection of ^{32}P-labeled AcMNPV to several different cell lines attached in 96-well microtiter plates. Equilibrium binding experiments were performed for 18 hours at 4°C. Although there was some evidence for nascent internalization, as shown by electron microscopic localization of virus with coated-pit-like structures on the cell surface, internalization under these conditions was shown to be negligible [10]. Once again, treatment with three separate proteases inhibited virus–cell binding between 93% and 96%, suggesting that a cellular protein was mediating the attachment.

Simultaneously with the measurement of equilibrium attachment at 4°C, Wickham and coworkers [10] measured the rate of infection rate at 22°C. These tandem attachment and infection experiments were performed on several different cell lines from *Spodoptera frugiperda* and *T. ni,* as well as *Estigmene acrea*–88 cells derived from *Estigmene acre* [32]. The results showed that the rates of virus attachment were very different for different cell lines, and the kinetic rates of attachment correlated well with the equilibrium levels of attachment seen at 18 hours. *T. ni* cell lines in general showed greater binding than *S. frugiperda* cell lines, with TN5B1-4 cell lines showing the greatest rate of attachment. Analysis of the attachment rates of AcMNPV to TN5B1-4 cell lines shows that the binding is diffusion limited, as attachment corresponds closely with the rate of diffusion-limited encounter between the virus and cell attached on the surface of the 96-well plate [10]. Also, as evidence for the avidity of attachment, electron microscopy of highly attachment-receptive cells lines, such as TN5B1-4 shows that all available space is occupied on the cell surface, suggesting that the number of virus particles that can attach is limited by geometry rather than by numbers of cell surface receptors [10]. Whether receptors are limiting for less receptive cell lines is unclear, since space is not saturated in all systems.

Figure 3 shows the relationship between binding observed at 4°C and the rate constant for infection seen with eight cell lines. There appears to be a direct correlation between the level of attachment and the rate constant for internalization for all cell lines. TN5B1-4 cells, which showed the highest rate for binding the virus, also showed the highest infection rate, 4.4×10^9 ml/cell min. It is important to note that this rate is only slightly lower than the attachment rate observed for this virus–cell system. This suggests that infection is limited by the rate of attachment. Since attachment appears to be diffusion limited, this suggests that viral internalization and infection is determined largely by the diffusional encounter between virus and cell.

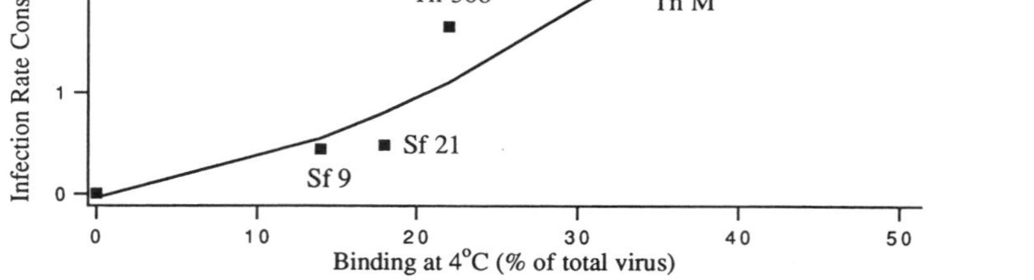

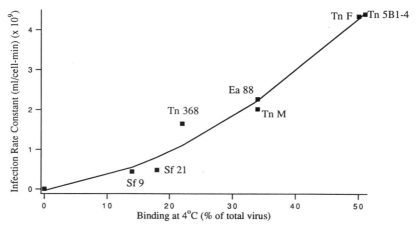

Fig. 3. Infection rates and degree of binding of AcMNPV to various cell lines show a direct correlation between the two (Data from Wickham et al. [10].)

Attachment and Infection for Cells in Suspension

The preferred bioreactor environment for insect cell culture continues to be suspension culture. An early study that attempted to measure attachment, penetration, and infection of TnMNPV by SF9 cells in suspension culture was performed by Wang and Kelly [33]. In that study, ^{32}P-labeled NOV was used to track attachment and penetration, and plaque-forming assays were used to assess cell-associated infectious virus for both NOV and PDV. The level of cell-associated NOV was maximal at 28°C and pH7.3 and required extracellular Ca^{2+} and Mg^{2+}. Binding during the first 10 minutes showed that a small but significant fraction of virus was internalized, as assessed by proteinase K–resistant levels of cell-associated virus. Internalized virus appears almost instantly after viral attachment, which supports the observation of Wickham and coworkers [10] that attachment is the rate-limiting step for internalization. Total cell-associated, internalized, and nuclear-associated viral material continued to rise over the initial 2 hours of binding. Although Wang and Kelly [33] claim that "saturation of receptor sites for the virus was not achieved," it appears that level of attachment of virus to the cell surface rapidly approaches a constant within the first 15 minutes of infection, as calculated by the difference between total cell-associated and internalized virus. Given the enormous rate of viral internalization seen in these experiments, we agree with Wang and Kelly [33] that receptor recycle (or turnover) must be rapid to maintain a constant level of attachment.

Multiplicity of infection (MOI) is defined as the number of infectious viruses added per host cell. An observation from Wang and Kelly's experiments that is not well understood is the apparent decrease in internalization efficiency as the MOI is increased. Since proteinase K was not used in studies at all MOI, it is not clear whether the attachment rate or the internalization rate could not keep pace with increases in the MOI. The possibility of defective virus particles also looms in these experiments, which might also influence the nonlinear relationship between the number of virus particles available per cell and the number of particles that actually enter the cell.

Defective Viral Particles and the Passage Effect

In serial passage of virus through a cell culture, it has been shown that the virus titer decreases in infectious activity [34]. This decrease in viral infectivity is referred to as the *passage effect;* a single passage is defined as the cycle in which NOVs bind and enter a cell and infect the nucleus and new NOVs are produced that bud from the cell membrane and enter the extracellular medium again to infect cell cultures. It is thought by many groups that the passage effect is caused by the production of a defective form of the virus that has lost a significant percentage of its viral genome, approximately 44% [35,36]. The portion of the genome lost in these cells includes that coding for the polyhedrin protein and for DNA-polymerase, so these defective viruses are themselves not infectious. However, they can attach and penetrate insect cell hosts, and, in the presence of normal, infectious NOV, can be produced in large quantities [35–37]. It is not known what surface proteins these defective nonoccluded forms possess that allows them to bind and infect insect cells, but it is likely that they use the same proteins as the normal, infectious NOV particles, as they are produced only when normal NOV particles have infected the host. Therefore, their route of entry is likely identical to that of normal NOV particles. It is suspected that, because these viruses are smaller (as has been directly shown by electron microscopy by Wickham et al. [35], more defective than normal infectious particles can be produced per unit time after infection. An understanding of the mechanisms of attachment and infection of normal and defective particles can be used to devise strategies for channeling cell culture away from the production of defective particles.

Biological Evidence for the Infectious Route of Entry

In general, viruses exhibit two possible routes of entry, either of which hold potential for being infectious. In infection by adsorptive endocytosis, a route pursued by influenza virus, the virus attaches to the cell surface, usually through binding cell surface receptors. The virus is then internalized through endocytosis, which may occur randomly on the cell surface (pinocytosis) or at specialized cell surface structures designed for internalization called *coated pits* (Fig. 1). After internalization, endosomes fuse with each other and other larger intracellular lysosomes. At some point in the internalization cascade, the virus nucleocapsid exits the endosome, either by fusion of the outer leaflet of the virus with the membrane of the endosome (in the case of enveloped viruses) or by biochemical manipulation and lysis of the endosome membrane (in the case of picornaviruses). In the case of enveloped virus fusion with the endosome membrane, the fusion is often catalyzed by the low pH environment of the endosome. Once in the cytoplasm, the nucleocapsid, or the genetic material it contained, makes its way into the nucleus of the cell to begin replication.

In infection by fusion, a virus membrane will fuse directly with the plasma membrane of the host cell after binding. The human immunodeficiency virus is an example of this class. These viruses usually possess a surface fusion molecule that can promote membrane fusion at neutral pH. Once again, in the cytoplasm, the nucleocapsid, or the genetic material it contained makes its way into the nucleus of the cell to begin viral replication. However, the genetic material has a much greater distance to travel to the nucleus.

Unfortunately, controversy exists over the actual route that causes viral infectivity in the baculovirus–insect cell system; for example, both surface fusion and endocytosis, either passive (pinocytosis) or adsorptive, have been implicated in infectivity of NOV particles [6]. Part of this controversy stems from trying to understand the infectious route by using techniques that cannot distinguish between infectious and noninfectious particles, such as electron microscopy [6,38] or ^{32}P labeling [33]. These studies, which implicate fusion as the infectious route, may not properly deal with the extremely low frequency of infectious virus in an virus titer

[6]. Until a fusion molecule is found and blocked, it will not be easy to prove fusion as an infectious route for this virus.

Evidence is mounting for infection by NOVs by the route of adsorptive endocytosis. First, electron microscopy of the virus on the cell membrane has shown that the virus localizes in coated pits [see ref. 10]. Second, as noted previously, gp64 appears to be a low pH fusion protein, as demonstrated by the ability of baculovirus infected cells to fuse at low pH [12] and the ability of the AcV_1 antibody, which is specific to gp64, to block the hemolysis of red blood cells at pH lower than 5.8 [6]. Furthermore, lipophilic amines such as chloroquine and NH_4Cl can block the infection of baculoviruses, reducing it to levels seen by treating the virus with AcV_1, which neutralizes gp64 [39]. These results suggest that NOVs enter by the endocytic pathway and fuse with endosomes at low pH using gp64 [6].

In contrast, there is evidence that PDV particles enter cells by fusion at the plasma membrane [31]. Using a standard fluorescent dequenching assay in which virus envelopes are labeled with a self-quenching concentration of octadecyl rhodamine-B (R18), virus was observed to fuse with LdElta cells and brush border membranes of *L. dispar* cells by measuring the dequenching of R18 fluorescence. Fusion was greatest at 27°C, but also occurred significantly at 4°C. Viral fusion occurred at all values of pH between 4 and 11, but was greatest at the alkaline pH of the insect midgut. Techniques commonly used to block infection by endocytosis, such as treatment with lysomotrophic agents such as chloroquine, has no effect on PDV infectivity [31]. These results suggest that PDVs enter and infect largely by fusion, whereas NOVs infect largely by endocytosis. It is interesting that baculoviruses, which, unlike other viruses, have two separate portions of their viral life cycle, might employ two separate mechanisms for entry in each portion of the virus life cycle.

Mathematical Models for Cell Attachment and Infection

Two groups have recently introduced detailed mathematical models of the processes of viral infection and viral production [37,40]. Licari and Bailey [40] extended their model to account also for the production of recombinant protein in cell lines. There are overt and subtle differences between the modeling of these two groups, yet each make useful contributions to understanding baculovirus–insect cell reactor dynamics or to optimizing insect cell culture for protein production. In this chapter, we point out how attachment and infection is modeled in these cases. The role of these models in predicting protein production and for designing insect cell bioreactors is addressed in Chapter 8.

In the model of de Grooijer and coworkers [37], the concentration of insect cells and the concentrations of several different forms of the virus are calculated for a number of different bioreactor configurations. The forms of the virus considered are normal virus (i-NOV, i = infectious), defective NOV (d-NOV, d = defective), and abortive virus (a-NOV, a = abortive). The last of these are virus particles that can bind to the cell surface and be internalized but do not infect due to some failure to maneuver properly through the intracellular machinery. Additionally, infectious virus particles are tracked by passage number. In their companion experiments, infection was assessed visually by observing the density of cells infected with polyhedral forms of the virus. Such observation was possible since the polyhedrin gene was not replaced with the code for a recombinant protein.

In this model, three modes of infection are considered. In the first, i-NOV and a-NOV simultaneously enter a target cell, resulting in the production of many i-NOV of greater passage number, as well as some a-NOV and d-NOV. (Passage number is augmented by one through each passage.) In the second, cells are infected with all three types of virus, which results in the production of a large number of d-NOV, as well as some i-NOV and a-NOV. In the third, cells are infected with a-NOV and d-NOV alone, which results in no virus being produced. At each time, the fraction of cells infected in each of the different modes can be calculated, based on the fact that the cell has a certain number of attachment sites available and the number of those sites occupied by each virus type depends on the concentration of each type.

In their model, infection was treated simply. Attachment of virus of any type to the cell surface was assumed to occur instantly on the time scale of the reactor experiment. Furthermore, it was assumed the number of available attachment sites for virus was less than the total number of suspected receptors for baculovirus particles [2] and that those numbers of sites did not change during infection. However, we suspect that the number of those sites changes; at early times, the number of sites is likely given by the geometric availability of space on the cell surface, due to the likely excess in receptors. At later times, the number of available sites will depend on the relative rates of receptor internalization and recycle [41]. In this model, both the level of infection and the time to reach maximum infection were very sensitive to the number of available attachment sites.

In the modeling of Licari and Bailey [40], the focus was on the generation of recombinant protein in batch cultures of attached cells. The infection strategy of their experiments was to expose cells to a brief, 1 hour dose of virus at various MOIs at different times after plating and observe the production of recombinant β-galactosidase. In experiments performed on this system, Licari and Bailey [42] found that when cells were infected early in the growth cycle, the production of protein with MOIs went through a maximum at intermediate MOI, but was overall a weak function of MOI. When infected late in the exponential growth phase, protein production increased with increasing MOI. The experimental results suggested that for this simple experiment there was an optimal relationship between MOI and protein production, and the objective of the modeling was to determine if that optimal relationship could be determined mathematically.

In the model of Licari and Bailey, a single form of the virus—the infectious form—was considered. Since the experiments were performed in batch culture, it seems reasonable to assume that, provided the initial inocula was healthy (free of substantial quantities of defective or abortive NOVs), there would be few passages before the culture became inactive due to overwhelming infection, suggesting that there will be little passage effect. The key quantity in this model is $n(t,\tau, NV)$, where $n(t,\tau, NV)dt$ is the concentration of cells between t and $t + dt$ infected with

NV virions at $t - \tau$. The number of infected cells at any time in the culture can be calculated by performing a double integration of this function over both all possible previous times of infection and all possible numbers of virions, NV. A Poisson distribution is used to calculate the number of viruses of a given type that can infect a single cell, and maximum number of viruses that can infect a cell is assumed. The maximum number that is assumed, and that appears to fit the experimental data, is well below the number of such viruses that can attach to the cell [2,10]; nevertheless, it is possible that infection may be saturable at a much lower level than attachment might be. To predict levels of protein production, the level of infection of each cell was related to the time for extracellular virion synthesis, recombinant protein synthesis, and cell lysis.

As a final note on the modeling, there appears to be some controversy about modeling infection itself. In attached cell cultures, Wickham and coworkers [10] measured infection as a function of cell density and virus concentration and showed that the rate of infection was first order in both virus and cell concentration (second order overall). This expression is similar to that proposed by De Grooijer and coworkers [43] to describe infection, who assumed infection was first order in cell concentration when virus was in excess. Licari and Bailey [40] find that such an expression is inadequate to describe the kinetics of infection, and later, of viral and protein production, presumably because too many viruses are allowed to infect in the second order formalism, leading to saturation of the system at higher MOIs that is not observed experimentally. We believe that the discrepancy can be resolved by distinguishing infection, as observed by plaque-forming assay with polyhedra-producing virus, and production of NOV and protein, which may be saturable. Therefore, many viruses that infect can go on to form plaques, but only some of the early infectious virus can go on to produce NOV and proteins. Furthermore, it seems clear that although the virus can enter the cell by different routes (fusion and endocytosis), only one of these routes may be infectious. The relative fluxes of virus through each of these routes need to be considered in formulating a model that relates extracellular virus concentration to the intracellular concentration of truly infectious virus. The relative contributions of different viral forms to infection, viral generation, and protein production still need to be sorted out by detailed analysis of the relative rates of these processes at the cell level for the most accurate mathematical description of the viral infection process.

SUMMARY

In this chapter, we summarized the recent progress in understanding the attachment and penetration of baculoviruses into insect cell. This included a summary of what is known about structural proteins that seem to be involved in baculovirus attachment and infection, evidence for routes of entry and infection of the virus, and mathematical modeling to predict the processes of viral cell attachment and infection. Significant work remains to be done in elucidating the detailed molecular and cellular mechanisms of baculovirus attachment and entry. Further elucidation of this phenomenon will facilitate the development of robust models of cell culture and bioreactor behavior and allow the intelligent manipulation of baculovirus attachment and infection for maximum productivity at the cell level.

ACKNOWLEDGMENTS

The authors gratefully acknowledge the valuable assistance of Dr. Gary Blissard in providing several key references, reading the manuscript, and providing useful information and ideas for this work. This effort was supported by the National Science Foundation through BCS-9111091.

REFERENCES

1. Wood, H.A., and Granados, R.R. Annu. Rev. Microbiol. 45:69–87, 1991.
2. Wickham, T.J., Granados, R.R., Wood, H.A., Hammer, D.A., and Shuler, M.L. Biophys. J. 58:1501–1516, 1990.
3. Rohrmann, G.F. J. Gen. Virol. 73:749–761, 1992.
4. Whitford, M., Stewart, S., Kuzio, J., and Faulkner, P. J. Virol. 63:1393–1399, 1989.
5. Blissard, G.W., and Rohrmann, G.F. Virology 170:537–555, 1989.
6. Volkman, L.E. Curr. Top. Microbiol. Immunol. 131:103–118, 1986.
7. Bradford, M.B., Blissard, G.W., and Rohrmann, G.F. J. Gen. Virol. 71:2841–2846, 1991.
8. Blissard, G.W., and Rohrmann, G.F. Ann. Rev. Entomol. 35:127–155, 1990.
9. Wickham, T.J. Ph.D. Thesis, Cornell University, Ithaca, NY, 1991.
10. Wickham, T.J., Shuler, M.L., Hammer, D.A., Granados, R.R., and Wood, H.A. J. Gen. Virol. 73:3185–3194, 1992.
11. Blissard, G.W., and Wenz, J.R. J. Virol. 66:6829–6835, 1992.
12. Leikina, E., Onaran, H.O., and Zimmerberg, J. Acidic pH induces fusion of cells infected with baculovirus to form syncytia. FEBS Lett. 304:221–224, 1992.
13. Volkman, L.E., and Goldsmith, P.A. Virology 166:285–289, 1988.
14. Doms, R.W., White, J., Boulay, F., and Helenius, A. In: Membrane Fusion. (Wilschut, J. and Hoekstra, D., eds.). Marcel Dekker, New York, pp. 313–336, 1991.
15. Hohmann, A.W., and Faulkner, P. Monoclonal antibodies to baculovirus structural proteins: determination of specificities by Western blot analysis. Virology 125:432–444, 1983.
16. Volkman, L.E., and Goldsmith, P.A. Virology 143:185–195, 1984.
17. Blissard, G.W., and Rohrmann, G.F. J. Virol. 65:5820–5827, 1991.
18. Blissard, G.W., Hogan, P.H., Wei, R. and Rohrmann, G.F. Site CA Virology 190:783–793, 1992.
19. Morse, M.A., Marriott, A.C., and Nuttall, P.A. Virology 186:640–646, 1992.
20. Kuzio, J., Jaques, R., and Faulkner, P. Virology 173:759–763, 1989.
21. Volkman, L.E., Summers, M.D., and Hsieh, C.-H. J. Virol. 19:820–832, 1976.
22. Roberts, P.L. Arch. Virol. 75:147–150, 1983.
23. Smith, G.E., and Summers, M.D. Virology 89:517–527, 1978.
24. Maruniak, J.E., and Summers, M.D. Virology 109:25–34, 1981.
25. Dobos, P., and Cochran, M.A. Virology 103:446–464, 1980.
26. Stiles, B., and Wood, H.A. Virology 131:230–241, 1983.

27. Derksen, A.C.G., and Granados, R.R. Virology 167:242–250, 1988.
28. Hashimoto, Y., Cosaro, B.G., and Granados, R.R. J. Gen. Virol. 72:2645–2651, 1991.
29. Wang, P., Hammer, D.A., and Granados, R.R. J. Gen. Virol, in press, 1994.
30. Uchima, K., Harvey, J.P., Omi, E.M., and Tanada, Y. Insect Biochem. 18:645–650, 1988.
31. Horton, H.M., and Burand, J.P. J. Virol. 67:1860–1868, 1993.
32. Granados, R.R., and Naughton, M. In: Invertebrate Tissue Culture, Applications in Medicine, Biology and Agriculture (Kurstak, E., and Maramorosch, K., eds.) Academic Press, New York, pp. 379–389, 1976.
33. Wang, X., and Kelly, D.C. J. Gen. Virol. 66:541–550, 1985.
34. Faulkner, P. In: Pathogenesis of Invertebrate Microbial Diseases (Davidson, E.A., ed.) Osmun & Co., Allanhead, NJ, pp. 3–37, 1981.
35. Wickham, T.J., Davis, T., Granados, R.R., Hammer, D.A., Shuler, M.L., and Wood, H.A. Biotech. Lett. 13:483–488, 1991.
36. Kool, M., Voncken, J.W., van Lier, F.L.J., Ramper, J., and Vlak, J.M. Virology 183:739–746, 1991.
37. De Grooijer, C.D., Koken, R.H.M., Van Lier, F.L.J., Kool, M., Vlak, J.M., and Tramper, J. Biotechnol. Bioeng. 40:537–548, 1992.
38. Bassemir, J., Miltenburger, H.G., and David, P. Cell Tissue Res. 228:587–595, 1983.
39. Volkman, L.E., and Goldsmith, P.A. Virology 139:185–195, 1985.
40. Licari, P., and Bailey, J.E. Biotechnol. Bioeng. 39:432–441, 1992.
41. Shuler, M.L., Cho, T., Wickham, T., Ogonah, O., Kool, M., Hammer, D.A., Granados, R.R., and Wood, H.A. Annals N.Y. Acad. Sci. 589:399–422, 1990.
42. Licari, P., and Bailey, J.E. Biotechnol. Bioeng. 37:238–246, 1991.
43. DeGooijer, C.D., van Lier, F.L.J., Van den End, E.J., Vlak, J.M., and Tramper, J.A. Appl. Microb. Biotechnol. 30:497–501, 1989.

7

Development and Evaluation of Host Insect Cells

Thomas R. Davis and Robert R. Granados

Boyce Thompson Institute for Plant Research, Cornell University, Ithaca, New York 14853

INTRODUCTION

Since the development of the first established cell lines susceptible to baculoviruses in the mid 1970s, progress in cell line establishment has grown at an accelerated rate. The establishment of new insect cell lines is no longer an "art," and numerous investigators have developed procedures for the isolation of novel cell lines from species in Lepidoptera, Hymenoptera, Orthoptera, Hemiptera, Diptera, Homoptera, and Coleoptera [1].

Several important advances in insect cell culture methods that were responsible for the rapid growth in this area included 1) the development of new media formulations based on Grace's medium [2]; 2) a greater appreciation of the physical/chemical factors (pH, temperature, serum requirement, and osmotic pressure of media) that may influence cell growth [3]; and 3) the preparation of primary tissue explants directly from larval and/or embryonic tissue [3,4].

Until recently, few established lepidopteran cell lines had been evaluated for baculovirus production or expression of recombinant proteins. The cell line established from *Spodoptera frugiperda* (IPLB-SF21) and a cloned isolate (SF9) have become standards and are widely used in baculovirus research. Recent studies have demonstrated that other cell lines may be equal or superior to SF21 and SF9 for baculovirus production, and in this chapter we describe methods for establishing and evaluating novel insect cell lines for baculovirus replication and recombinant protein production.

DEVELOPMENT OF CELL LINES

Goodwin [4] and others described methods for the establishment of lepidopteran cell lines for virus research that were based on the dissection of larval or adult

Baculovirus Expression Systems and Biopesticides, pages 121–130

tissues as a source of insect cells. Although these methods were suitable for initiating cell lines, the procedures were somewhat tedious and slow and the success rate low. For details on tissue culture methods and their use in virus research, the reader is referred to Chapter 2.

Miltenburger et al. [5] and Naser et al. [6] introduced a new method for the rapid and easy establishment of embryonic cell lines from Lepidopterous species. This procedure is outstanding because of its simplicity, and researchers with minimal training in insect cell culture can readily carry out this procedure successfully. This procedure has been used in our laboratory to develop numerous embryonic cell lines from *Trichoplusia ni,* and *Pieris rapae* [7,8]. This method, outlined in Table 1, is particularly useful for establishing cell lines in a short period of time and screening them for virus susceptibility and recombinant protein expression [9,10].

In general, after initiation of primary, embryonic cell cultures using the procedure outlined in Table 1, cultured cells will begin dividing within 4 to 10 days, and the first subculture will be possible within 2–3 weeks. The following procedure for the screening of novel cell lines for susceptibility to *Autographa californica* multiple nuclear polyhedrosis virus (AcMNPV) and expression of model recombinant viruses has been used successfully in our laboratories.

1. The donor insect species for new cell lines is usually one whose larval stage is highly susceptible to AcMNPV. Also, the insect species should have a relatively short life cycle and be easily reared on artificial diet. Examples of such insect species include *T. ni, S. frugiperda, Plutella xylostella,* and *Anticarsia gemmatalis.*
2. New cell lines are subcultured in serum-containing medium for 15–20 passages in 25 cm^2 T-flasks prior to initial testing. During this time, cell cultures are selected for properties such as their ability to grow as anchorage-dependent or suspension cultures and short doubling time.
3. The cell lines, prior to forming a monolayer, are screened for susceptibility to AcMNPV by inoculating cells in multiwell plates (duplicate wells per cell line) at a multiplicity of infection (MOI) of 10 plaque-forming units (pfu) per cell. Other viruses can also be tested at this time. Infection levels are determined by phase contrast examination of the inoculated cells at 2–3 days postinoculation and counting the percentage of cells containing polyhedra or occlusion bodies (OBs).
4. Since it is well known that some cell lines may support OB formation with

TABLE 1. Procedure for Establishment of New Embryonic Cell Lines From Lepidopterous Eggs

1. 300–400, 24-hour-old eggs are sterilized in 2% chlorox, 70% ethanol, and rinsed in GTC-100 tissue culture medium
2. With a rubber policeman the eggs are crushed through a 100 μm sieve into fresh medium
3. The homogenate is centrifuged at 200*g* for 5 minutes and the pellet resuspended in 5 ml of TNM-FH tissue culture medium [11]
4. The cells are seeded into a 25 cm^2 tissue culture flask and incubated at 28°C

Reproduced from Granados et al. [7], with permission of the publisher.

little or no occlusion of infectious virions, it is important to conduct larval bioassays with OBs purified from infected cells [12]. This step is important if the main goal is to select cell lines for production of viral pesticides.

5. Cell lines that exhibit over 90% susceptibility to AcMNPV and are infectious to larvae are selected for further screening, and three to six ampules of each selected cell line are frozen in liquid nitrogen. The remaining lines are discarded. At this time (after about 30 passages) selected cell lines are adapted to serum-free media. Depending on the cell line, adaptation to a serum-free medium could take an additional 5–20 passages.

6. Screening of cell lines for virus and/or recombinant protein production usually will occur at about passages 40–50. Wickham et al. [9], Davis et al. [10], and Wang et al. [12] described a rapid screening method for recombinant proteins using multiwell tissue culture plates. A similar method can be used for evaluating OB production in selected insect cell lines [12].

7. Following step 6, only the high producing cell lines are kept and 10–15 ampules are stored in liquid nitrogen. The remaining cell lines are discarded.

8. Cell lines can be further improved and stabilized if they are cloned. This step will require re-evaluation of the cloned cell lines to produce high levels of virus or recombinant proteins. Further characterization of the lines would include isozyme analysis, karyotyping, determining cell doubling times, and growth curve analysis.

Serum-containing medium is generally used for the establishment of cell lines. However, it was recently reported that serum-free medium could be used to initiate continuous cell lines from embryonic tissue of the cotton boll weevil *Anthonomus grandis* [13]. McKenna and Granados (unpublished data) have also been successful in initiating cell lines from *T. ni* eggs in a serum-free medium supplemented with egg yolk. These observations suggest that an enriched serum-containing medium (i.e., TNM-FH) may not be necessary for establishing cell lines. Initiating lines directly in serum-free media would eliminate the sometimes difficult adaptation of specific cell lines to these media. In addition, the use of this medium for initiation of cell cultures might provide a selection mechanism for cell types that have the capacity to grow readily in other types of serum-free media.

EVALUATION OF CELL LINES

Much of the research to increase the amount of recombinant protein expressed in the baculovirus expression vector system (BEVS) has centered around manipulating the transcriptional signals of the viral promoter [14–16]. An avenue of research integral to recombinant protein yield that has gone virtually unexplored is the examination of those elements relating to the host of the baculovirus infection, the insect cell line. In all but a few examples, recombinant protein expression has been in the *S. frugiperda* SF9 or SF21 cell line. Viral and/or recombinant protein expression and post-translational modification has been examined in relatively few of the established insect cell lines.

In 1980, Lynn and Hink [17] compared the replication of AcMNPV in *Estigmene acrea* (BTI-EAA); *Lymantria dispar* (IPLB-LD64BA); *Mamestra brassicae* (IZD-MB0503); *S. frugiperda* (IPLB-SF1254); and *T. ni* (TN368) cell lines. There were significant differences among the cell lines with respect to both nonoccluded virus (NOV) and polyhedra yield and susceptibility of the cell lines to virus infection. The *M. brassicae* and *T. ni* cell lines produced significantly more polyhedra than any other cell line tested. The same two cell lines also had a higher sensitivity to AcMNPV infection as determined by the plaque assay procedure. However, the *E. acrea* cells released significantly more NOV than the *S. frugiperda, M. brassicae, T. ni,* and *L. dispar* cell lines.

In a more recent study, McIntosh and Ignoffo [18] investigated the virus yield, susceptibility, and growth rate of AcMNPV in lepidopteran cell lines from *T. ni* (TN-CL1); *S. frugiperda* (IPLB-SF21); *Heliothis virescens* (BCIRL-HV-AM1); *P. xylostella* (BCIRL-PXZ-HNV3); and *A. gemmatalis* (BCIRL-AG-AM). The cell doubling times during exponential growth for all five cell lines did not differ significantly; however, the *A. gemmatalis* and *H. virescens* cell lines had approximately a 48 hour lag prior to exponential growth. Maximum viral titers were obtained at 72 hours postinfection (pi) in the *S. frugiperda* and *T. ni* cell lines, whereas maximum titers in *H. virescens, P. xylostella,* and *A. gemmatalis* were obtained at 96 hours pi. The *T. ni* and *H. virescens* cell lines had significantly higher titers than the other cell lines, but both had a 10-fold decrease in titer from 96 to 120 hours pi.

The *P. xylostella* cell line produced a minimum of two-fold more OBs than any of the other cell lines. However, the OB size was measured and found not to differ significantly from the most productive to least productive cell lines. The OBs from *S. frugiperda, T. ni, A. gemmatalis,* and *P. xylostella* gave greater than 86% mortality when fed to *T. ni* larvae. However, an equivalent number of OBs from the *H. virescens* cell line only caused 16% mortality. While no explanation for this result was given, it was noted that the restriction endonuclease digests of the AcMNPV DNA from the *T. ni* and *H. virescens* cell lines were identical, thus suggesting that no host modification of the viral DNA had occurred.

In 1991, King et al. [19] compared replication and production of one wild-type and two recombinant AcMNPV encoding for β-galactosidase and chloramphenicol acetyl transferase in *S. frugiperda* (SF21) and *M. brassicae* cell lines. The *M. brassicae* cell line expressed approximately two- to threefold more polyhedrin or recombinant protein and produced more polyhedra per cell than the *S. frugiperda* cells. However, the *M. brassicae* cell line had a significantly lower extracellular virus titer for both the wild-type and β-galactosidase recombinant viruses. The authors suggested that because both viruses were produced and amplified in the *S. frugiperda* cells, the life cycle had become biased toward the production of extracellular virus in the *S. frugiperda* cells.

Ogonah et al. [20] compared recombinant protein production capabilities of *T. ni* (TN368) and *S. frugiperda* (SF9) cells in suspension culture. The *T. ni* cell line produced 50% more recombinant protein (β-galactosidase) on a per cell basis than did the SF9 cell line. However, the TN368 cells were twice as large as the SF9 cells and appeared to be more sensitive to the shear forces of the spinner flask.

Hink et al. [21] compared recombinant AcMNPV-expressing pseudorabies virus glycoprotein (gp50T), human plasminogen (HPg), and β-galactosidase in cell lines from *A. gemmatalis* (UFL-AG-286); *Choristoneura fumiferana* (IPRL-CF-1); *E. acrea* (BTI-EAA); *H. virescens* (IPLB-HvTl); *Heliothis zea* (BCIRL-HZ-AM1); *L. dispar* (IPLB-LdElta), (IPLB-LdEltf); *M. brassicae* (IZD-MB0503), (NIAS-MaBr-92), (NIAS-MB-25), (SES-MaBr-1), (SES-MaBr-3), (SES-MABr-4); *Manduca sexta* (CM-1), (UCR-SE-1a); *P. xylostella* (BCIRL-PX2-HNV3); *Spodoptera exigua* (UCR-SE-1), (UCR-SE-1a); *S. frugiperda* (IPLB-SF21AE), (IPLB-SF21AE-15), (SF9), (IPLB-SF1254); and *T. ni* (IPLB-TN-R^2), (TN368). Amounts of recombinant protein were evaluated during late viral infection in each of the cell lines.

The IZD-MB0503 cell line had the highest yield of HPg, while four other cell lines, IPLB-TN-R^2, IPLB-SF-1254, IPLB-LdElta, and CM-1, had expression levels above SF9 cells. The pseudorabies glycoprotein gp50T had the greatest expression in the IPLB-HvT1, IPLB-SF21AE, IPLB-SF21AE-15, and IPLB-SF1254 cell lines. Equally high levels of β-galactosidase were expressed in the SF9, IZD-MB0503, and BCIRL-PX2-HNV3 cell lines.

Wild-type virus infection was also evaluated in the same cell lines at 96 hours pi by counting the number of cells containing polyhedra. The percentage of cells infected with wild-type virus ranged from 0 to 95%.

In 1992, Wang et al. [12] compared virus replication and recombinant protein (β-galactosidase) production in the cell lines IPLB-SF21, SF9, and TN368 that were adapted to serum-free medium. Production of β-galactosidase was greatest in SF21 cells, followed by TN368 and SF9 cells. All three cell lines produced a similar maximum extracellular virus titer; however, the rate of virus production in SF9 cells was significantly slower. The number of polyhedra per cell for the SF9 cell line was also 45% less than either the SF21 or TN368 cells. However, bioassay data indicated that there were not significant differences in the infectivity of the polyhedra produced in the three cell lines.

Wickham et al. [22] compared the production of AcMNPV NOV and β-galactosidase in cell lines from *S. frugiperda* (SF9), (IPLB-SF21AE); *T. ni* (BTI-TN-M), (BTI-TN5B1-4), (BTI-TN-AP_2), (TN368); *M. brassicae* (IZD-MB0507); and *E. acrea* (BTI-EA-88). After accounting for the differences in cell size, the BTI-TN-M and BTI-TN5B1-4 cell lines produced five- and twofold more β-galactosidase, respectively, than the SF9 or SF21 cells. However, in experiments examining NOV production, the TN-M cell line was one of the lowest producers. Experiments with the *M. brassicae* and SF21 cell lines yielded results similar to those of King et al. [19] in that the production of the β-galactosidase recombinant protein from the *M. brassicae* cell line was threefold greater than the SF21 cells after differences in cell size were taken into account. However, the SF21 cell line produced greater numbers of NOV than the *M. brassicae* cell line.

Davis et al. [10] expressed a truncated human placental alkaline phosphatase (SEAP) protein in cell lines derived from *T. ni* (BTI-TN5B1-4), (BTI-TN-AP_2), (BTI-TN-MG-1), (TN368); *S. frugiperda* (SF9), (IPLB-SF21AE); *M. brassicae* (IZD-MB0507); and *E. acrea* (BTI-EAA). The SEAP protein is a secreted glycoprotein modified with N-linked oligosaccharide. When SEAP was expressed in all cell lines at the optimal cell density, the BTI-TN5B1-4 cells produced a minimum of ninefold more SEAP on a per cell basis and sixfold more on a per milliliter basis than any other cell line.

FACTORS AFFECTING RECOMBINANT PROTEIN EXPRESSION

There are many factors that have been identified that affect foreign gene expression in insect cell cultures. Factors such as choice of media and its additives, cell culture conditions, the recombinant protein being expressed, and the particular vector selected are all components that affect recombinant protein expression and are described in greater detail in other chapters of this book. Other factors such as multiplicity of infection, cell density, and the presence of defective interferring particles in the viral inoculum also may have an affect on recombinant protein production. In order that optimal protein expression can be achieved for each cell line in a comparison study, it is important that all of these factors be thoroughly examined.

Multiplicity of Infection

The concensus of most studies investigating the affect of the multiplicity of infection (MOI) is that recombinant protein production is insensitive to MOI [23,24]. Murhammer and Goochee [25] demonstrated that β-galactosidase synthesis in middle to late exponential growth phase SF9 cells in gas-sparged bioreactors was relatively consistent over a range in MOI from 1 to 5. Maiorella et al. [26] used a recombinant AcMNPV to infect SF9 cells in exponential growth phase for the production of human macrophage colony-stimulating factor (M-CSF). Recombinant protein expression was similar in either serum-supplemented or protein-free medium in a variety of sparged culture vessels with a range of MOI from 0.5 to 10.

King et al. [27] assessed virus production and chloramphenicol acetyl transferase (CAT) expression in *S. frugiperda* cells using a temperature-sensitive baculovirus. Flasks of cells infected with an MOI of 0.02, 0.2, or 2.0 were incubated at 33°C (the nonpermissive temperature) for 3–4 days, after which the temperature was decreased to the permissive temperature of 27°C and aliquots taken daily for both CAT and virus assays. Maximum virus titer and CAT expression were obtained at an MOI of 0.02. At the higher MOIs of 0.2 and 2.0, there was less virus and CAT expression. King et al. also performed a parallel experiment at a constant temperature of 27°C that resulted in the highest virus and CAT titers at an MOI of 2.0. They hypothesized that the contrasting results were due, in part, to the exposure of cultures to the higher nonpermissive temperature (33°C). They further suggested that at high MOI all the cells became infected and these infected cells were more susceptible to thermal damage than uninfected cells. At lower MOI, all the cells were not synchronously infected, thus giving the uninfected cells time to adapt to the elevated temperature.

Licari and Bailey [28] evaluated the influence of MOI on recombinant protein expression by infecting SF9 cells with a β-galactosidase recombinant AcMNPV at MOI values ranging from 0 to 100 and monitoring the expression of β-galactosidase with time. They found that the final β-galactosidase titer, as a function of MOI, was dependent on the growth phase of the cells at the time of infection. If cells were infected in the early-exponential growth phase, β-galactosidase production was relatively insensitive to MOI. However, the higher the MOI used to infect late-exponential growth phase cells, the greater the amount of β-galactosidase produced. The authors point out that these results suggest the importance of cell density and growth phase of the cells, as well as the importance of the MOI.

Research related to these MOI experiments has shown that undiluted serial passage of baculoviruses in cell culture provides an opportunity for the formation of spontaneous mutants. The few polyhedra (FP) mutant identified by several groups [29–32] contains small insertions or deletions at specific sites within the baculovirus genome that result in the altered FP phenotype [33,34]. These mutant viruses can have altered protein expression levels, as was demonstrated with the isolation of a "spontaneous mutant" that had an FP phenotype that was deficient 1) in its ability to synthesize normal levels of polyhedrin protein and 2) in viral occlusion body formation [35].

Defective Interfering Particles

Production of wild-type or recombinant (β-galactosidase) AcMNPV in a bioreactor system resulted in a decrease in productivity of NOV, β-galactosidase production, number of polyhedra per cell, and fraction of cells that contained polyhedra after only a few weeks of operation [36]. Kool et al. [37] found that the mechanism responsible for this decrease in productivity was a deletion mutant of AcMNPV. These defective interferring (DI) particles were generated in continuous cell culture under conditions of high MOI. Restriction enzyme analysis of DI particle DNA revealed they had 43% of their genome deleted compared with standard AcMNPV virus particles. The deleted region was colinear and extended from map position 1.7 to 45; thus it included the polyhedrin and DNA polymerase genes. The DI particles were also found to interfere with standard virus replication, thus explaining the reduction in AcMNPV productivity and the large decreases in expression of foreign genes placed under control of the polyhedrin promoter.

Wickham et al. [38] described the generation of DI particles by serial passage of AcMNPV in SF21 cells grown in tissue culture flasks. The DI particle–containing viral innoculum interferred with β-galactosidase and infectious virus production in three insect cell lines. The authors speculated that this mutant may explain the results of some indicating that there is an MOI effect on protein expression. At high MOI, DI particle–containing preparations are able to coinfect cells with both standard virus and DI particles. The resulting interference because of the presence of DI particles results in reduced protein expression. At low MOI the cells would receive either standard virus or DI particles separately. Since the DI particles interfere with recombinant protein expression, the cells receiving normal virus alone would produce higher levels of protein than cells coinfected with DI particles and standard virus.

Cell Density

Factors associated with the cell line are also important to virus replication/protein production. Several studies have reported that infection of cells with recombinant virus should occur during the exponential growth phase in order to achieve maximum foreign protein expression [39,40]. However, no exact cell density was ever given in order to obtain this optimal expression. Wood et al. [41] found that confluent monolayers of SF21 and TN368 cell lines produced fewer OBs and less NOV than cell cultures at less confluent cell densities. This inhibition was found to be due to cell–cell contact and not to the cells becoming refractive to viral infection, depletion of nutrients, or diffusable cell-associated factor(s). Also, DNA hybridization and thymidine incorporation studies indicated that cell–cell contact reversibly inhibited both cellular and viral DNA synthesis.

In a recent study, Wickham et al. [22] found that some cell lines appear to exhibit density-dependent inhibition of virus and recombinant protein production due to cell–cell contact. Upon reaching confluency on tissue culture plates, β-galactosidase production would decrease three- to sixfold in five of the eight cell lines tested. A nutrient depletion or an accumulation of a toxic by-product were not suspected as the cause for the inhibition; however, oxygen limitation could not be completely ruled out.

Davis et al. [10] also found that cell density significantly influenced production of a secreted glycoprotein. A recombinant *A. californica* baculovirus expressing SEAP was used to evaluate secretion and post-translational processing. Four of the eight cell lines tested exhibited a decrease in SEAP production with increases in cell density above the optimum. Furthermore, the commonly used SF9 and SF21 cell lines were among the four that did not exhibit dramatic cell density inhibition.

CONCLUSIONS

The procedures for establishing insect cell cultures have been advanced rapidly in recent years, making for a wealth of novel cell lines. Studies comparing protein expression in these cell lines have shown that they differ in their abilities to synthesize and modify viral and/or recombinant proteins. Many of the same studies have also shown that the BEVS is capable of expressing foreign proteins that are, in many ways, similar to the natively expressed proteins. However, very few of these novel cell lines have been screened for superior expression or post-translational processing qualities. As Luckow and Summer [15] have noted, "there is much that can be learned studying the expression, processing, targeting, and transport of recombinant proteins in cell lines other than *S. frugiperda*." The aforementioned comparative studies support this statement and indicate that the host cell line chosen for BEVS expression can influence recombinant protein expression. Therefore, future efforts to develop novel insect cell lines and improve the expression levels obtained by the BEVS should include studies that will screen these novel cell lines for their inherent ability to synthesize and post-translationally modify various foreign proteins. Along the same lines, the factors that influence foreign gene expression in the BEVS must continue to be identified. Undoubtedly, further understanding of the basic concepts of the BEVS will lead to the discovery of more of these production-altering conditions.

REFERENCES

1. Lynn, D.E. Proc. 8th Int. Conf. Invertebr. Fish Tissue Culture. Anaheim, CA, June 16–20, pp. 1–6. Tissue Culture Assoc. Pult. 1991.
2. Grace, T.D.C. Nature 195:788–789, 1962.
3. Lynn, D.E. J. Tissue Culture Methods 12:23–29, 1989.
4. Goodwin, R.H. In Vitro 11:369–378, 1975.
5. Miltenburger, H.G., Naser, W.L., and Harvey, J.P. Z. Naturforsch. 39:993–1002, 1984.
6. Naser, W.L., Miltenburger, H.G., Harvey, J.P., Huber, J., and Huger, A.M. FEMS Microbiol. Lett. 24:117–121, 1984.
7. Granados, R.R., Derksen, A.C.G., and Dwyer, K.G. Virology 152:472–476, 1986.
8. Dwyer, K.G., Webb, S.E., Shelton, M., and Granados, R.R. J. Invertebr. Pathol. 52:268–274, 1988.
9. Wickham, T.J., Shuler, M.L., Hammer, D.A., Granados, R.R., and Wood, H.A. J. Gen. Virol. 73:3185–3194, 1992.
10. Davis, T.R., Wickham, T.J., McKenna, K.A., Granados, R.R., Shuler, M.L., and Wood, H.A. In Vitro Cell. Dev. Biol. 29A:388–390, 1993.
11. Hink, W.F. Nature 226:466–467, 1970.
12. Wang, P., Granados, R.R., and Shuler, M.L. J. Invertebr. Pathol. 59:46–53, 1992.
13. Stiles, B., McDonald, I.C., Gerst, J.W., Adams, T.S., and Newman, S.M. In Vitro Cell. Dev. Biol. Z8A:355–363, 1992.
14. Matsuura, Y., Possee, R.D., Overton, H.A., and Bishop, D.H.L. J. Gen. Virol. 68:1233–1250, 1987.
15. Luckow, V.A., and Summers, M.D. Bio Technology 6:47–55, 1988.
16. Luckow, V.A., and Summers, M.D. Virology 170:31–39, 1989.
17. Lynn, D.E., and Hink, W.F. J. Invertebr. Pathol. 35:234–240, 1980.
18. McIntosh, A.H., and Ignoffo, C.M. J. Invertebr. Pathol. 54:97–102, 1989.
19. King, L.A., Mann, S.G., Lawrie, A.M., and Mulshaw, S.H. Virus Res. 19:93–104, 1991.
20. Ogonah, O., Shuler, M.L., and Granados, R.R. Biotechnol. Lett. 13:265–270, 1991.
21. Hink, W.F., Thomsen, D.R., Davidson, D.J., Meyer, A.L., and Castellino, F.J. Biotechnol. Prog. 7:9–14, 1991.
22. Wickham, T.J., Davis, T., Granados, R.R., Schuler, M.L., and Wood, H.A. Biotechnol. Prog. 8:391–396, 1992.
23. Furlong, A.M., Thomsen, D.R., Marotti, K.R., Post, L.E., and Sharma, S.K. Biotechnol. Appl. Biochem. 10:454–464, 1988.
24. Schopf, B., Howaldt, M.W., and Bailey, J.E. J. Biotechnol. 15:169–173, 1990.
25. Murhammer, D.W., and Goochee, C.F. Bio Technology 6:1411–1418, 1988.
26. Maiorella, B., Inlow, D., Shauger, A., and Harano, D. Bio Technology 6:1406–1410, 1988.

27. King, G., Kuzio, J., Daugulis, A., Faulkner, P., Allen, B., Wu, J., and Goosen, M. Biotechnol. Bioeng. 38:1091–1099, 1991.
28. Licari, P., and Bailey, J.E. Biotechnol. Bioeng. 37:238–246, 1991.
29. Potter, K.N., Faulkner, P., and McKinnon, E.A. J. Virol. 18:1040–1050, 1976.
30. Fraser, M.J., and Hink, W.F. Virology 117:366–378, 1982.
31. Burand, J.P., and Summers, M.D. Virology 119:223–229, 1982.
32. Hink, W.F., and Vail, P.V. J. Invertebr. Pathol. 22:168–174, 1973.
33. Fraser, M.J., Smith, G.E., and Summers, M.D. J. Virol. 47:287–300, 1983.
34. Kumar, S., and Miller, L.K. Virus Res. 7:335–349, 1987.
35. Wood, H.A. Virology 105:338–344, 1980.
36. Kompier, R., Tramper, J., and Vlak, J.M. Biotechnol. Lett. 10:849–854, 1988.
37. Kool, M., Voncken, J.W., Van Lier, F.L.J., Tramper, J., and Vlak, J.M. Virology 183:739–746, 1991.
38. Wickham, T.J., Davis, T., Granados, R.R., Hammer, D.A., Schuler, M.L., and Wood, H.A. Biotechnol. Lett. 13:483–488, 1991.
39. Stockdale, H., and Gardiner, G.R. J. Invertebr. Pathol. 30:330–336, 1977.
40. Vaughn, J.L. J. Invertebr. Pathol. 28:233–237, 1976.
41. Wood, H.A., Johnston, L.B., and Burand, J.P. Virology 119:245–254, 1982.

8

Overview of Issues in Bioreactor Design and Scale-Up

Ronald A. Taticek, Daniel A. Hammer, and Michael L. Shuler

School of Chemical Engineering, Cornell University, Ithaca, New York 14853

INTRODUCTION

The commercial exploitation of the insect cell–baculovirus expression system for the production of heterologous proteins or biopesticides requires an efficient large-scale cultivation method. Recombinant proteins or bioinsecticide baculoviruses can be produced in vivo by infecting whole insect larvae or in vitro by infecting insect cells grown in culture. Although the in vivo approach can result in higher yields per gram cell weight than achieved in vitro, the in vitro method has a number of advantages. Unlike insects, cell cultures do not need to be cultured continuously as cells can be grown and then frozen until needed. Being able to start from a common stock of cells minimizes the variations in cells that can result from long-term passaging in culture. This should result in a more consistent yield over time. Also, cell lines can be carefully selected for their production capability. During production, optimum conditions can be maintained by controlling important environmental parameters. The in vivo approach is usually more expensive as it requires special pathogen-free, insect-growing facilities. The process is very labor intensive, time consuming, and potentially difficult in terms of product recovery. The initial product obtained is impure and can be highly allergenic as it contains microorganisms, insect cuticle, and protein. The most important disadvantage is that it is difficult to scale-up economically. As a broadly applicable production strategy, the cell culture approach is more suitable for large-scale production of recombinant proteins or bioinsecticides. Early attempts at large-scale insect cell culture were motivated by the possibility of large-scale production of viral insecticides. The technology already exists for large-scale animal cell culture and is used by many companies for the production of viruses for veterinary vaccines or recombinant proteins.

Baculovirus Expression Systems and Biopesticides, pages 131–174

The insect cell–baculovirus expression system has significant industrial potential. The strategy for using the baculovirus as an expression vector or for the production of viral insecticides involves three steps (see Chapter 4). First, the insect cells are grown to mid- or late-exponential growth phase [1–3]. Second, the cells are then infected with the baculovirus. Third, a sufficient period of time is allowed for protein or virus production, and then the product is harvested and purified. To date, very little attention has been focused on the engineering considerations involved with this system. This chapter will describe important characteristics of the system and how they can be incorporated into efficient bioreactor production systems.

Often a process that works well at the bench scale (T-flasks and spinner flasks) works very poorly at larger scales (bioreactors). Since it is the same medium, the same cells, and the same environmental conditions (temperature and pH), it may not be intuitively obvious to the reader why the system's performance should change so much upon scale-up. The answer usually lies in a switch in the rate-determining step from a kinetic or intrinsic limitation in the biological system to a physical limitation due to changes in shear, mixing times, surface-to-volume ratios, or ability to transport two reactive components together. In principle, an understanding of the physical environment in the bioreactor and the intrinsic biological kinetics from bench-scale experiments should allow the prediction of large-scale behavior. We do not yet understand these factors well enough to make such predictions with certainty. The purpose of this chapter is to bridge the material in the earlier chapters with actual experience at larger scales.

WHY IS SCALE-UP PROBLEMATIC?

Problems associated with scale-up have always plagued chemical engineers, and biochemical engineering is no different. If a cylindrical reactor vessel is made larger, we note that the wall surface area and the gas–liquid surface area change at a different rate than the volume (see Fig. 1). As volume increases, both of the surface area to volume ratios decrease. The decrease in relative wall-surface area is important if either virus or cells adsorb to it. The gas–liquid surface to volume ratio is critical if oxygen transfer is to occur through head space aeration. At low volumes this ratio is high, and aeration through the head space is satisfactory. Oxygen supply is proportional to the gas–liquid interfacial area, while demand is proportional to the volume. At high interface to volume ratios (corresponding to low volumes), aeration without gas sparging is adequate. But if we scaled-up using head space aeration only, we would reach a critical volume (with insect cell cultures, a relatively low volume of 100–500 ml) in which cell growth and product formation are limited by oxygen availability. That is, we would have gone from a kinetic limitation (at low volumes) based on the intrinsic rates in the biological system to a transport or physical limitation based on our ability to supply oxygen at higher volumes.

The physical conditions in a large bioreactor can never exactly duplicate those in

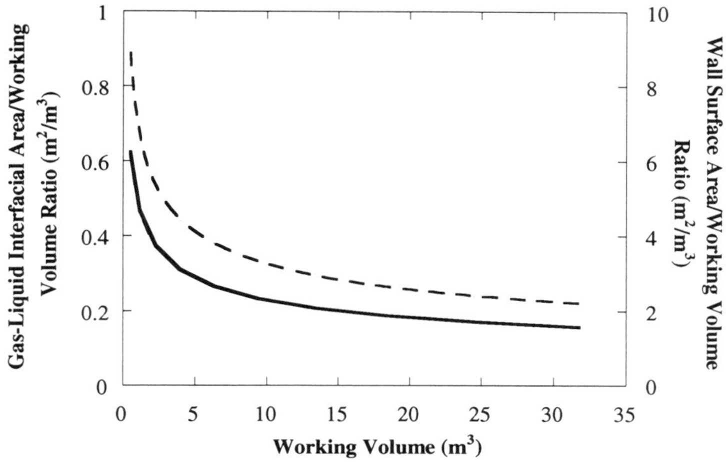

Fig. 1. Changes in the wall surface area to volume ratio (———) and gas–liquid interfacial area to volume ratio (– – –) as a function of working volume for a cylindrical vessel. A working volume of 75% and a height to diameter ratio of 2 are assumed in these calculations. The working volume is that occupied by liquid, and the remainder is occupied by gas.

a smaller bioreactor even if geometric similarity is maintained. Because shear, fluid-mixing, gas-bubble size, and mixing times all have different dependencies on impeller rotation rate, N, and impeller diameter, D_i, in geometrically similar stirred tank systems, the physical environment must change upon scale-up. Also, different rules for scale-up can be employed and can give very different results. Keeping the power input per unit volume (P/V) constant upon scale-up implies a constant oxygen transfer rate, while a constant agitation rate implies a constant mixing time. The maximum impeller tip speed determines the maximum shear rate; hence, keeping the tip speed constant would imply a constant shear. In the case described in Table 1, an 80 liter stirred tank bioreactor has been scaled-up to 10,000 liters by increasing the diameter by 5 and keeping the height to diameter ratio the same. The differences obtained by applying each of the different scale-up rules are illustrated. Because the physical environment will change upon scale-up, there will be some size in which either a change from a kinetic to a transport limitation will occur (which may be due to the altering of the distribution of chemical species in the reactor) or the biological system itself will change (e.g., cell lysis due to shear; see Chapter 9). As a result, the metabolic response of the culture will differ from one scale to the other.

Many novel bioreactor systems are available. Each of these will have an ultimate maximum useful size due to transport limitations. These designs will also vary in terms of scalability. Some also promote a heterogeneous environment that, while suitable for some cells, does not appear to be attractive for the insect cell–baculovirus system. The heterogeneity could be problematic in achieving efficient

TABLE 1. Interdependence of Scale-Up Parameters[a]

			Large bioreactor (10,000 liters)			
Scale-up criterion	Dependence	Small bioreactor (80 liters)	Constant P/V[b] (OTR)	Constant Re[b] (Flow patterns)	Constant N (Mixing times)	Constant ND_i (Shear)
Diameter, D		1.0	5.0	5.0	5.0	5.0
Volume, V		1.0	125	125	125	125
Impeller diameter, D_i		1.0	5.0	5.0	5.0	5.0
Impeller rotation rate, N		1.0	0.34	0.04	1.0	0.2
Maximum shear rate, γ	$\gamma \propto ND_i$	1.0	1.7	0 2	5.0	1.0
Liquid recirculation rate, Q	$Q \propto ND_i^3$	1.0	42.5	5.0	125	25
Mixing time constant,[c] t	$t \propto \frac{1}{N}$	1.0	2.94	25	1.0	5.0
Maximum stable bubble size[d], D_b	$D_b \propto \frac{1}{N^{1.2}D_j^{0.8}}$	1.0	1.0	13.1	0.28	1.9

[a] Adapted from Oldshue [139].

[b] $(P/V) \propto N^3D_i^2$ and $Re \propto \rho_{liq}ND_i^2/\mu_{liq}$.

[c] Shuler and Kargi [5].

[d] Bailey, J.E. and D.F. Ollis, *Biochemical Engineering Fundamentals*, 1986, eq. 8.54, p. 480.

infections, and heterogeneous culture conditions may result in heterogeneous protein post-translational processing resulting in a product of lower quality [4].

Many novel reactor configurations yield high cell densities. High cell density, through the buildup of toxic byproducts, depletion of nutrients, or "contact inhibition" on cell division can be detrimental to productivity. Since high volumetric productivities (g product/liter·day) are important to process economics and are a goal of many novel bioreactor systems, understanding such limitations is critical when making a reactor choice. These issues are generic to all of bioprocessing, although the fragility of animal cells imposes more severe constraints than with most bacterial systems. See Shuler and Kargi [5] for a more detailed discussion of aspects of scale-up.

BIOENGINEERING CONSIDERATIONS APPLIED TO THE INSECT CELL–BACULOVIRUS SYSTEM

Introduction

There are a number of factors that are unique to the insect cell–baculovirus expression system that must be considered in addition to the more general considerations discussed in the previous section. The cells must be infected with virus in an efficient manner. The point in the growth cycle at infection, the cell density at infection, and the amount of virus used to infect the cells are very important parameters. As the system is lytic, the harvest time is much more crucial than in other systems. To simplify downstream processing and ensure the stability of the product, it should be harvested before significant cell lysis has occurred. Figure 2 illustrates how all of these factors are interrelated.

As is the case with mammalian cells, the design and scale-up of bioreactors for the mass cultivation of insect cells is constrained by supplying adequate oxygen for the cells to grow and synthesize products without stressing the cells. The traditional approach of using submerged aeration to supply oxygen may be of limited applicability due to the shear sensitivity of insect cells. The rates of oxygen mass transfer required for large-scale production require higher agitation and aeration rates, and these can result in cell damage or death. The low shear tolerance of insect cells, their susceptibility to damage and death when exposed to air bubbles, and their high oxygen demand relative to mammalian cells are the main reasons why special bioreactor designs may be necessary for insect cells. Fluid shear stresses may also influence the ability of the insect cell to produce protein from recombinant DNA. For example, Folkman and Mascona [6] found that rates of DNA synthesis, important for viral replication, depend on cell shape and adhesive interactions.

Suspension versus Attached Culture

Insect cells can grow attached to the surface of a culture vessel or some other solid support or can grow suspended in the liquid medium. The growth of insect cells in

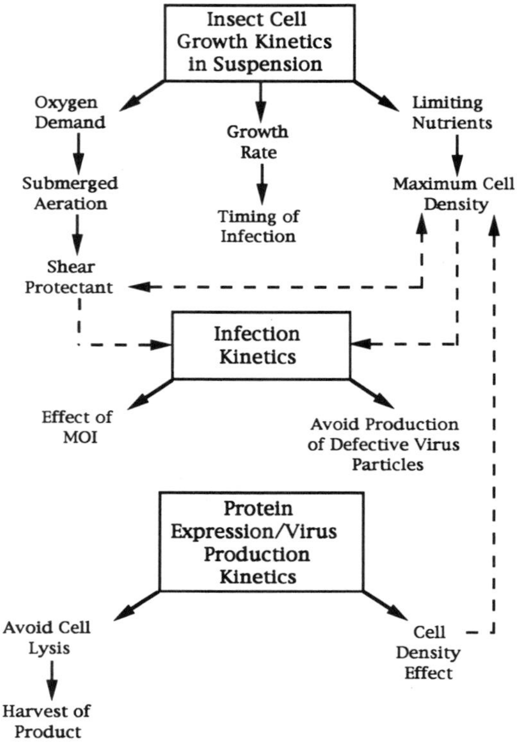

Fig. 2. Diagram illustrating relationships between factors affecting application of the insect cell–baculovirus expression system.

attached culture is difficult at large scales due to the decrease in surface to volume ratio with increasing scale of operation if T-flask or roller bottle culture is used. There would be a corresponding decrease in the profitability of the process.

The use of microcarriers in stirred tank bioreactors can maintain a more constant surface to volume ratio. A microcarrier is a small bead to which cells attach and grow. Microcarrier culture has been used successfully with mammalian cell systems. A major problem with microcarriers is heterogeneity in cell loadings, as some beads may have a few or no cells while cells have grown to confluency on others. Since productive infections in insect cell cultures are obtained at confluencies of 60%–75%, many of the beads would give suboptimal performance. Furthermore, bead-to-bead contact during stirring can damage cells [7]. The damage due to bead contact can be avoided in packed bed systems, although the problem of heterogeneity may remain. Packed bed systems have been used with insect cell systems [8,9], but neither of these designs have been well explored. With no well-accepted large-scale system for attached cells, most investigators favor suspension cultures if at all possible. Since most insect cell lines can be manipulated to grow either in suspension or attached, larger scale efforts have focused on suspension culture.

Only if a compelling reason exists, such as vastly superior production or enhanced product quality due to more complete post-translational modification in attached culture, would attached cell reactors be considered. Lanford [10] reports three to five times higher specific production in adhered cells than in suspension culture [10], while Maiorella et al. [11] report a threefold improvement in M-CSF production going from attached to suspension culture. Murhammer and Goochee [12] obtained similar β-galactosidase expression per cell in attached (T-flask) and suspension systems (unsparged bioreactors). In our experience, whether suspension or attached culture produces more or less target protein may depend more on the protein. With β-galactosidase, we obtain a two- to threefold increase in specific production going to suspension culture, while with a secreted alkaline phosphatase we observe a threefold decrease going to suspension culture [9,13,14].

Suspension systems provide a more homogeneous and controllable environment. Also, the condition of the culture (cell number and viability) are more easily assessed. The cultivation of insect cells in suspension uses engineering technology and bioreactors that are in use with microbial and mammalian cells. Large-scale insect cell culture is in many ways analogous to large-scale mammalian cell culture, but there is significantly less published data on the behavior of insect cells growing in bioreactors.

Adapting Insect Cells to Suspension Culture

Some anchorage-dependent cell lines can be made to grow in suspension after a period of adaptation. The period of adaptation varies with the cell line. The variables that are important in the adaptation of cells to growth in suspension are the tendency for the cells to grow attached (which can be affected by the agitation rate), their shear sensitivity (which is also affected by the agitation rate), and biochemical factors (such as medium components). Some insect cell lines, such as *Spodoptera frugiperda,* can easily be grown in suspension culture. In our laboratory, stable and reproducible growth of *S. frugiperda* (IPL-SF21AE) cells in suspension can be obtained after only a few passages in Ex-Cell 400 serum-free medium (usually <10). A period of 3–4 weeks was required to adapt *Drosophila melanogaster* cells to suspension culture [15]. Other groups report it taking about 6 weeks (corresponding to six passages) to adapt SF cells to suspension culture from static culture [16]. The same group reports little success in adapting *Trichoplusia ni* cells to suspension culture in shake flasks. The cells grew from 1.5×10^5 to only 5×10^5 cells/ml, indicating that the *T. ni* cells may not be able to tolerate shear to the same extent as other cell lines or may require different biochemical factors to aid in their initiation into suspension. In our laboratory, it has taken 30 passages to adapt *T. ni* cells to suspension culture in Ex-Cell 405 serum-free medium.

Medium composition can be an important factor in establishing a suspension culture. *Aedes albopictus* was adapted to suspension culture with Eagle's minimal essential medium (MEM) supplemented with 1% nonessential amino acids, 1% lactalbumin hydrolysate, and 10% FBS [17]. The cells were first allowed to adapt to the new medium in stationary culture for several months before being introduced into suspension. It took several additional months before a well-adapted suspension culture was obtained. Lengyel et al. [15] added 0.5% bactopeptone and 0.1% lactalbumin hydrolysate to Dulbecco's medium in order to adapt *D. melanogaster* to suspension culture [15]. Initially a cell density of 3×10^6 cells/ml was achieved in roller bottle. Eventually, densities as high as 1×10^7 cells/ml were achieved in spinner flasks.

The inoculation cell density is important when growing insect cells in suspension. Miller et al. [18] and Summers and Smith [19] suggest that insect cells will grow very slowly or not at all if they are inoculated at densities below 1×10^5 cells/ml. A higher inoculum, such as 5×10^5 cells/ml, may be necessary for more delicate cells. Wu et al. [16] found that SF9 cells grew at densities as low as 0.7×10^5 cells/ml providing that the inoculum came from a culture in exponential growth. Neutra et al. [20] report that optimal growth and cell yield were obtained if SF9 cells were seeded at densities of 8×10^5 cells/ml or higher. Somewhat lower cell yields were obtained with an inoculum of 5×10^5 cells/ml. Lazarte et al. [21] report being able to maintain SF9 cells in exponential growth for longer using inocula of 8×10^5 cells/ml or higher. An inoculum size $1–5 \times 10^5$ cells/ml was recommended for *D. melanogaster* cells, although growth was normal at cell densities greater than 6×10^4 cells/ml [22]. Little or no growth was obtained at $2–3 \times 10^4$ cells/ml. SF cells (IPL-SF21AE) grown in our laboratory exhibit no lag in growth if inoculated at densities greater than 3×10^5 cells/ml in 50 ml spinner flasks stirred at 85 rpm and 100 ml spinner flasks stirred at 120 rpm. The existence of a critical inoculum may be explained by a cooperative effect among the cells. Cultured cells may leak intermediate metabolites and growth factors that are essential for cell growth. To have an adequate amount of these metabolites in the medium, a minimal cell density is required [23].

Agitation rates must be chosen to maintain the cells in suspension and to provide them with adequate oxygen without shearing them. Using *Bombyx mori* cells, Stavroulakis et al. [24] found that at low agitation rates (30 rpm) cells either attached to the flask wall or aggregated to form large clumps. Increasing the agitation rate to 60 rpm and adding 0.1% methylcellulose increased the growth rate, but it was still significantly less than observed in static culture. Methylcellulose may alter the effect of shear on these cells (see Chapter 9 for a more detailed discussion). After five passages, the growth rate improved dramatically and differed little from static culture. When the agitation rate was further increased to 80 rpm, the concentration of methylcellulose had to be increased to 0.3% to achieve good growth. Miyake et al. [22] grew *D. melanogaster* cells in suspension culture in shake flask. To keep the cells in suspension, agitation rates of greater than 180 rpm were necessary. Cameron et al. [25] recommend agitation rates of around 100 rpm for suspension culture and observed oxygen limitation problems in spinner flasks at volumes larger than 1,000 ml. We have noted similar problems with SF21 cells in 250 ml spinner flasks stirred at 75 rpm (see Fig. 3). Increasing the agitation rate to 95 rpm had no significant effect on the growth of the cells in the 250 ml spinner flask. Caron et al. [26] observed oxygen limitation in 250, 500, and 1,000 ml spinner flasks agitated at 135 rpm. Agathos and coworkers varied the agitation rate of 100 ml spinner flasks from 60 to 120 rpm and found that 100 rpm was optimal for the growth (both rate and cell yield) of *A. albopictus* cells [27].

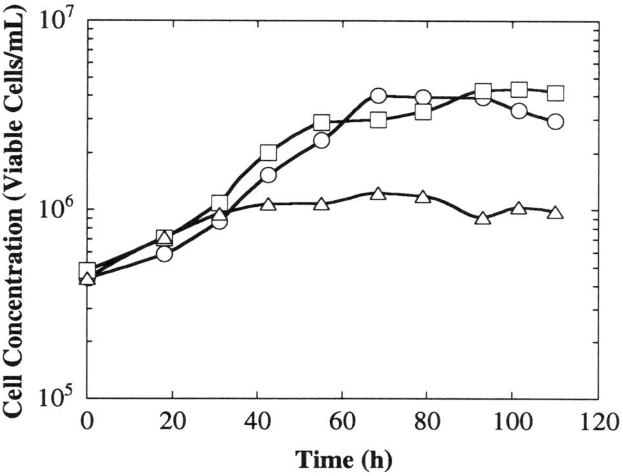

Fig. 3. Viable cell concentration as a function of time for three different capacities of spinner flasks: ○, 50 ml (40 ml working volume); □, 100 ml (80 ml working volume) △ 250 ml (200 ml working volume). *Spodoptera frugiperda* (IPL-SF21AE) cells grown at 85 rpm.

The working volume used also affects how well cells grow in suspension. This effect is most likely just an oxygen transfer limitation problem. Using a 125 ml shake flask agitated at 100 rpm, Neutra and coworkers [20] varied the working volume and found that the best growth and highest cell number were achieved using a working volume of 40 ml. Larger volumes (60 and 80 ml) gave very poor growth, as mixing was not adequate enough to keep the cells suspended. Much smaller volumes ($<$10 ml) gave no growth.

Cell aggregation in suspension culture can make initiating a culture difficult. *T. ni* were grown in suspension in modified TNM-FH medium by Hink and Strauss [28]. To alleviate the problem of cell clumping, 0.1% methylcellulose was added to the medium. Other additives, such as heparin, have been effective in particular cases [29]. Allowing the large aggregates to settle out and subculturing only the single cells and small aggregates have also given good results with *T. ni* cells [30].

Table 2 summarizes the growth rates and maximum cell densities achieved for various cell lines introduced into suspension either in shake flask or spinner flask. Doubling times range from about 16 to 50 hours, which are comparable to values for most mammalian cells. The maximum cell densities achieved range from 4 to 6 $\times$ 10^6 cells/ml, which is about twice as high as densities achieved in mammalian cell culture.

Insect Cell Sensitivity to Shear and Bubbling

Insect cells are sensitive to shear because of their lack of a cell wall and their large size. High shear due to agitation or aeration will destroy the cells. Moderate rates of

TABLE 2. Growth Data for Various Insect Cell Lines Established in Suspension Culture in Small-Scale Bioreactors

Bioreactor	Working volume (ml)	Agitation rate (rpm)	Cell line	Medium	Growth rate (h^{-1})	Maximum cell density (cells/ml)	Reference
Shake flask	5	180	*D. melanogaster*	D20 + FBS[a]	0.043	$5–10 \times 10^6$	[22]
Shake flask	25	100	*S. frugiperda* (SF9)	Grace's + FBS	0.020	5×10^6	[16]
Shake flask[b]	25	100	*S. frugiperda* (SF9)	SF900	0.024	8.7×10^6	[31]
				Grace's + FBS	0.034	3.5×10^6	
				IPL-41 + FBS	0.034	10.6×10^6	
Shake flask	25–40	100	*S. frugiperda* (SF9)	TNM-FH + FBS	0.035	5×10^6	[20]
Shake flask	20	120	*S. frugiperda* (SF21)	TNM-FH + FBS			[32]
Spinner flask	NA	NA	Schneider *Drosophila*	DMEM + FBS	0.023	10×10^6	[15]
Spinner flask	100	100	*A. albopictus*	MEM + FBS	0.040	5.2×10^6	[27]
Spinner flask	<1,000	NA	*S. frugiperda* (SF9)	Serum-free	0.028–0.035	5×10^6	[25]
Spinner flask	100	75–100	*S. frugiperda* (SF9)	IPL-41 + FBS + PF68	0.029–0.039	5.5×10^6	[11]
Spinner flask	200	80	*Bombyx mori*	IPL-41 + FBS + MCL	0.013	4×10^6	[24]
Spinner flask	40	75	*S. frugiperda* (SF21)	Ex-Cell 400	0.031–0.034	$4–5 \times 10^6$	[14]

[a]Abbreviations: FBS, fetal bovine serum; MCL, methylcellulose; PF68, Pluronic-F68; NA, not available.
[b]Growth experiment carried out at 33°C.

agitation have a minimal effect on insect and animal cells grown in suspension in the absence of bubbles. Sparging, cavitation, and bubble incorporation via vortex formation have been shown to cause damage to insect and animal cells grown in suspension (see Chapter 9) [33–39]. Hink and Strauss [28] report that *T. ni* cells stopped growing when the stirrer speed in 2 and 3 liter stirred tank bioreactors equipped with marine impellers was increased to 200 rpm in medium containing 0.1% methylcellulose. This would correspond to a critical shear stress of 1.5 N/m^2. Tramper et al. [40] attempted to assess the shear sensitivity of insect cells by monitoring the effect of stirrer speed on the viability of *S. frugiperda* cells grown in a 1 liter round-bottom bioreactor equipped with a marine impeller. It was found that cells died at agitation rates over 220–510 rpm, corresponding to shear rates of 1.5–3 N/m^2 (after 3 hours of exposure). Agathos et al. [41] estimated the critical shear stress for *Aedes albopictus* cells to be about 1 N/m^2 [41]. The value obtained by Tramper et al. [40] may be higher as the medium they used contained 0.1% methylcellulose. They did not test lower shear stresses. Goldblum et al. [42] found that laminar shear stresses of 0.1 and 0.59 N/m^2 caused significant cell lysis of *T. ni* 368 and *S. frugiperda* (SF9) cells, respectively (within 5 minutes, without use of protectants) [42]. It is not clear how a cell's ability to withstand laminar shear translates into its ability to survive the wide range of hydrodynamic stresses encountered in bioreactors.

In bioreactors, cell viability has been shown to be affected dramatically by the use of submerged aeration (bubbling). Microscopic visualization has shown that cells adsorb to rising bubbles, and when these bubbles reach the liquid surface of the reactor they burst [43]. A number of research groups have hypothesized that it is the interaction of animal cells with these bursting bubbles that shears the cells [35,36,40,44]. It is clear that damage is not occurring in the region of rising bubbles [35,44], but it is still debated as to whether the interaction of cells with bubbles disengaging from the sparger also shears the cells. Murhammer and Goochee [37] conclude that damage does occur and that it is dependent on the design of the sparger. To minimize cell damage in this region it is important to minimize the pressure drop across the sparger [37]. Tramper et al. [40] saw a decrease in viability with increasing air flow rate. Similar results are reported by Wu et al. [45] in an airlift bioreactor. In both of these cases, increased air flow rates would result in increased interactions between cells and bubbles at both the sparger and the liquid surface.

A number of polymers have been investigated as "shear protectants" with great success. Methylcellulose, dextran, and the nonionic surfactant Pluronic-F68 and various other reverse Pluronic polyols (BASF) have been used to increase cell resistance to shear [12,33,34,35,37,42,46]. It has been hypothesized that the protective effect is due to the interaction of pluronic with the cell membrane or as a result of reduced adhesion of cells to rising bubbles which results in a reduced interaction between cells and bursting bubbles at the liquid surface (for more details, see Chapter 9).

Hink and Strauss [47] found that by increasing the concentration of methylcellulose from 0.1% to 0.3% that *T. ni* cells grown in a stirred tank bioreactor equipped with marine impellers and with submerged aeration exhibited a much more normal morphology. Agathos et al. [41] also found that adding methylcellulose improved the growth of *Aedes albopictus* cells. Murhammer and Goochee [37] found that concentrations of Pluronic-F68 in the range of 0.3%–0.5% provided good protection with submerged aeration in medium containing serum. Neutra et al. [20] found that adding 0.05% Pluronic-F68 improved the growth of SF9 cells in shake flasks agitated at 150 rpm. The addition of 0.1% Pluronic-F68 increased the cell viability of SF9 cells grown in serum-containing medium in spinner flasks agitated at 200 rpm [26]. Research done in our laboratory growing SF21 cells in serum-free medium in a 1.2 liter stirred-tank bioreactor with submerged aeration showed that a much higher concentration of Pluronic-F68 was required to "protect" the cells (see Fig. 4) [14]. This trend is in agreement with Caron et al. [26], who report having to increase the amount of pluronic from 0.1% to 0.3% in surface aerated cultures when switching from serum to serum-free medium.

Another strategy to avoid contact between cells and bubbles is to use novel aeration systems such as those employing semipermeable tubing to supply oxygen [48–50]. Unfortunately, the use of such aeration systems is limited to smaller scale reactors. The length of tubing necessary to supply oxygen to very large volumes of liquid becomes prohibitive. The concentration gradient will decrease to zero. Other possible solutions involve the separation of the oxygenation and growth areas of the reactor. One disadvantage of such a system is that it is more complicated to scale-up.

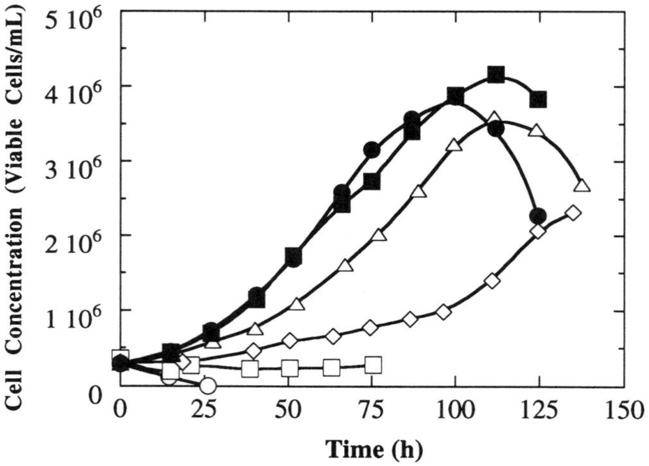

Fig. 4. Effect of Pluronic-F68 concentration on growth of *Spodoptera frugiperda* (SF21) cells in Celligen bioreactor (1.2 liter working volume, submerged aeration at 17 ml/min, agitation at 85 rpm; maximum amount of antifoam added was 0.08%): ○, 0.1%; □ 0.3%; ◇, 0.5%; △, 0.7%; ●, 0.9% pluronic; ■, control (50 ml spinner with working volume of 40 ml, 0.1% pluronic).

Oxygen Requirements

Oxygen demand is a key design parameter for any aerobic biological process. Oxygen can be difficult to supply especially in larger volume bioreactors. Some data have been published on the oxygen uptake rates for insect cells and the effect of dissolved oxygen concentration on cell growth. As can be seen from Table 3, typical maximum oxygen uptake rates range from 1.3 to 1.5 mmol l^{-1} h^{-1}, while specific oxygen uptake rates range from 0.15 to 0.55 μmol (10^6 cells) $^{-1}$ h^{-1}. The specific uptake rates are comparable with those observed for mammalian cells. The higher total oxygen uptake rates are a reflection of the higher cell densities achieved by insect cells. Typical total oxygen uptake rates range from 1 to 3.5 mmol l^{-1} h^{-1} for plant cells, 9 to 15 mmol l^{-1} h^{-1} for yeast cells, and 5 to 90 mmol l^{-1} h^{-1} for bacterial cells. Hink and Strauss [47] found that *T. ni* cells grew equally well at 50% dissolved oxygen (sparging with air) as at 100% dissolved oxygen (sparging with oxygen) [47]. Cells grown at 15% dissolved oxygen were vacuolated by 120 hours, and a rapid decrease in cell number followed. Oxygen utilization depends on the physiological state of the culture. Figure 5A shows the total and specific oxygen uptake rates as a function of time for SF21 cells grown in a 150 ml stirred-tank bioreactor with submerged aeration. The maximum uptake rates occur late in the exponential phase of growth.

Most research groups have observed changes in oxygen requirements when cells are infected with virus. The exact magnitude varies. The increases observed can be

TABLE 3. Oxygen Requirements of Insect Cells Compared With Mammalian, Plant, and Microbes

Cell type	Cell line	Maximum specific oxygen demand μmol(10^6 cells)$^{-1}$ h^{-1}	Maximum total oxygen demand (mmol l^{-1} h^{-1})	Reference
Insect	*S. frugiperda* (SF9)			
	Exponential phase	0.15		[11,51]
	Stationary phase	0.05		
	S. frugiperda (SF9)	0.23		[52]
		0.29	1.5	[53]
		0.20		[54]
	S. frugiperda (SF21)	0.55	1.3	[14]
	T. ni (TN368)	0.41		[55]
Mammalian	Hybridoma	0.25		[56]
	Hybridoma	0.34		[57]
	Hybridoma	0.54		[58]
	Mammalian cells		0.053–0.59	[59]
Plant			1–3.5	[60]
Yeast			9–15	[61]
Bacteria			5–90	[61]

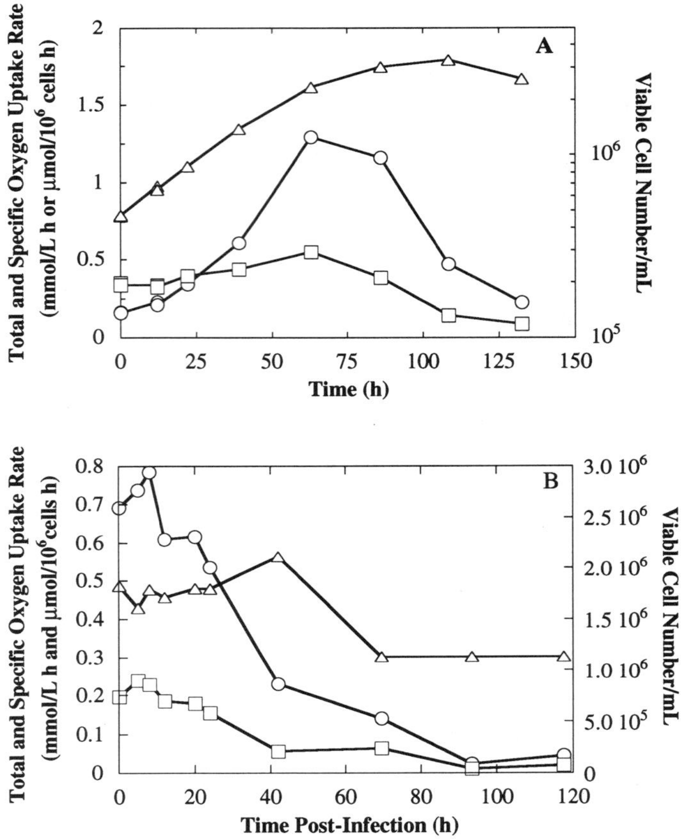

Fig. 5. Oxygen uptake rates for uninfected and infected *Spodoptera frugiperda* (SF21) cells. **A:** Uninfected cells grown in a 150 ml stirred tank bioreactor with submerged aeration: ○, Total oxygen uptake rate; □, specific oxygen uptake rate; △, viable cell number/ml (85 rpm, 0.9% Pluronic-F68, 0.08% antifoam C). **B:** Infected cells grown in 100 ml spinner flasks: ○, Total oxygen uptake rate; □, specific oxygen uptake rate; △, viable cell number/ml (120 rpm).

explained by the increased metabolic rate as a result of virally induced macromolecule biosynthesis. Street and Hink [55] infected *T. ni* cells (TN368) with wild-type AcMNPV and monitored oxygen uptake and compared it with that of uninfected cells. Early during the infection, the specific uptake rates for the uninfected and infected cells were similar (corresponding to the period before the initiation of viral replication). By 14 hours postinfection, the oxygen requirements had increased from 0.36 to 0.72 µmol (10^6 cells) $^{-1}$ h^{-1} (corresponding to the initiation of NPV nucleocapsid production), representing a 76% increase over the maximum specific uptake for uninfected cells. The oxygen consumption then decreased rapidly until the infected cells showed little or no respiratory activity 96 hours postinfection. Schopf et al. [52] also observed that the maximum respiratory activity occurred 16 hours postinfection and was followed by a sharp decrease in oxygen requirements.

Other research groups have not observed as dramatic an increase in oxygen requirements. Inlow et al. [51] and Maiorella et al. [11] report the uptake rate increasing from 0.15 to 0.17 μmol (10^6 cells)$^{-1}$ h^{-1} after viral infection. Kamen et al. [53] observed an increase in maximum uptake from 0.27 to 0.32 μmol (10^6 cells)$^{-1}$ h^{-1} and a 30%–40% increase in respiration coefficient during infection. Schopf et al. [52] infected SF9 cells in T-flasks with recombinant β-galactosidase AcMNPV at multiplicities of infection (MOI) of 1 and 10 and observed an increase from 0.14 to 0.31 μmol (10^6 cells)$^{-1}$ h^{-1}. This is a 35% increase over the maximum specific uptake rate that was measured for uninfected cells (see Table 3). The MOI had no effect on the increase in oxygen uptake. Weiss et al. [62] observed a 25% increase in oxygen uptake when SF21 cells were infected with AcMNPV. King and coworkers [54] observed a 20% decrease in specific oxygen uptake 24 hours after SF9 cells were infected in a 14 liter bioreactor. Higher glucose and glutamine consumption rates have also been observed with infected cells [53]. Wang et al. [63] report that infected cells utilize sucrose once glucose is depleted and hypothesize that it is the switch from direct uptake of glucose to the metabolizing of sucrose that results in a higher oxygen requirement. Although glucose was not depleted in the medium, Reuveny and coworkers [64] observed a doubling in oxygen consumption after infection. Scott et al. [65] only saw large increases in oxygen uptake postinfection when the dissolved oxygen concentration in the medium was allowed to drop to zero; otherwise, only minimal increases were observed. In our laboratory, we have seen about a 15% increase in specific uptake of oxygen after infection (see Fig. 5B).

As a result of the increased oxygen requirements of infected cells, a higher oxygen transfer capacity is required. Since infected cells are more shear sensitive than uninfected cells [12], the oxygen transfer capacity should be increased but not in conjunction with increased shear (i.e., as a result of increased aeration or agitation rates). A novel approach to eliminating oxygen limitation after infection would be to decrease the operating temperature after the cells are infected, as decreasing temperature increases oxygen solubility and oxygen transfer. Reuveny et al. [64] found that decreasing temperature to as low as 22°C had no effect on protein yield at low oxygen demand levels and were able to alleviate oxygen limitations at high cell densities.

Infection Parameters

There are a number of factors to consider when attempting to obtain an efficient infection process. They include the amount of virus added per cell, the timing of the infection (growth phase of the cells and cell density), the medium condition postinfection (spent vs. fresh), and the "age" of the virus (the number of passages in vitro).

Multiplicity of Infection. The number of plaque-forming units used to infect each cell is called the *multiplicity of infection* (MOI). There is considerable disagreement in the literature as to whether the amount of virus used to infect cells has an effect on virus or recombinant protein yield and in what way. It is important to note that comparing MOIs between different research groups is difficult as the assays used to determine virus titer are far from accurate and production levels vary with virus stock. Some work with wild-type and recombinant virus does show an "MOI effect." Brown and Faulkener [1] infected *T. ni* cells with low-passage TnMNPV at MOIs ranging from 0.01 to 500. Below 4.0, the yield was about 20 polyhedra per cell. Above 4.0, a marked increase in yield was observed, with a maximum of about 60 polyhedra per cell occurring at MOIs between 20 and 30. At MOIs higher than 30, the yield fell off. Neutra et al. [20] varied MOI from 0.1 to 20 and found that the yield of β-galactosidase per cell increased from 13 to 152 U/(10^6 cells), with the increase of MOI from 0.1 to 5 and decreased to 123 U/(10^6 cells) at an MOI of 20. Lazarte et al. [21] found a similar decrease in the production of rCD4 at high MOI. Increasing the MOI from 10 to 580 resulted in a decrease in expression from 11.1×10^4 to 5.2×10^4 molecules per cell. The trend could be reversed by resuspending the cells in fresh medium at the time of infection. The decrease in production with increase in MOI in this case is due to the increased carry over of spent medium with the virus stock so that cells become nutrient limited more quickly. Using the higher MOI, the yield per cell was increased to 33.8×10^4, still indicating an MOI effect (though in the reverse direction). King et al. [31] report decreases in CAT (chloramphenicol acetyltransferase) and virus titers/ml as MOI was increased from 0.02 to 2. The magnitude of the decrease was found to vary with the medium used.

Other research groups report relatively constant production for a wide range of MOI values. Wickham [13] observed no decrease in protein expression with increasing MOI when low-passage virus was employed. The production of M-CSF (human macrophage colony-stimulating factor) was not significantly affected when the MOI was varied from 0.5 to 10 [11]. Equivalent production was achieved in sparged culture vessels ranging in size from 6 to 36 liters. Murhammer and Goochee [12] obtained a relatively constant yield of β-galactosidase over the range of MOIs of 1–10. Gardiner and coworkers [66] found no MOI effect with AcMNPV production (both OV and NOV) in TN368 cells for MOIs between 0.001 and 500. Using wild-type and recombinant AcMNPV, Schopf et al. [52] found that MOI did not have a dramatic effect on virus or protein expression. Licari and Bailey [67] report that the relationship between product yield and MOI was dependent on the growth phase of the cells prior to infection. The product yield from cells infected in the early exponential phase was relatively independent of MOI, while the product yield from cells infected in the late exponential phase increased with MOI in contradiction to the results of King et al. [31].

The effect of MOI may be a kinetic one: the lower the MOI, the slower the infection process and the later the expression. Lazarte et al. [21] report that only 70% of cells were infected after 50 hours with an MOI of 10, while 91% were infected with an MOI of 580. The percentage of infected cells could be increased to 97.6% with the addition of some fresh medium after infection, indicating that nutrient limitation is playing a role. Cell growth postinfection is also a good indicator of the efficiency of infection. More growth is expected with a lower percentage of infected cells. At low MOI's of 0.02 and 0.2, King and coworkers [31] observed cell growth for 1–2 days postinfection, while no growth was seen at an MOI of 2.0. Licari and Bailey [68] found that cell death occurred more rapidly at higher MOI. They also observed less growth postinfection with increasing MOI. Cells infected with an MOI of 0.01 grew from 0.55×10^6 to almost 2×10^6 cells/ml, while cells infected at an MOI of 100 grew only to 0.9×10^6 cells/ml. Neutra et al. [20] found a similar inverse relationship between MOI and cell growth postinfection for MOIs below 1. For MOIs greater than 1, a decrease in viable cell number was observed. Schopf and coworkers [52] found that SF9 cells infected with wild-type AcMNPV at an MOI of 1 grew at the same rate as uninfected cells for 24 hours postinfection. Significantly less growth was observed for the MOI 10 case. Cell viability rapidly dropped 24 hours later for the MOI of 1 compared with the MOI of 10. The peak in β-galactosidase activity was shifted 24 hours later by decreasing the MOI from 10 to 1. Caron et al. [26] found that decreasing the MOI to less than 1 from 1 or greater shifted peak production of VP6 (bovine rotavirus nucleocapsid protein) from 48 to 72 hours. Wu and colleagues [69] studied baculovirus-induced cell death in insect cells as a function of MOI and a number of other factors. Decreasing MOI prolonged cell viability. The delay time (time until viability begins to drop) increased, while the rate of cell death decreased.

Generation of Defective Virus Particles In Vitro. Reductions in yields at high MOI may be the result of the presence of defective virus particles in the virus stock. A number of research groups have found aberrant virus particles present in AcMNPV that has undergone numerous serial passages in culture [70,71]. Purifying the virus stock eliminated the decline in production observed with increasing MOI [71]. MacKinnon et al. [72] found that continuous passage of virus in cell culture leads to a loss of polyhedra formation in infected cultures as a result of slow changes in the replication of the virus. The polyhedra produced were normal for the first 10 passages. Between 10 and 20 passages in vitro, various incomplete forms of viral replication were observed. Some cells did not develop virions. Others did, but few or no polyhedra formed. As a result, there was a reduction in the average number of polyhedra formed per cell from about 28 to less than 5. By passage 43, the polyhedra formed were no longer infectious. They attributed the drop off to the production of aberrant particles. These particles, which contain only part of the viral genome, replicate preferentially (as they are smaller), need an intact virus as a helper, and inhibit the replication of the standard virus. During continuous culture conditions, Kool et al. [73] observed that a mutant lacking 40% of the viral genome became predominant. More of these defective particles are present in higher passage virus and at higher MOIs and thus have more of an effect. Gardiner et al. [66]

observed interference phenomena at MOIs greater than 100. de Gooijer and coworkers [74] found an order of magnitude decline in specific NOV production between passage 4 virus and passages greater than 9. Passaging the virus at low MOI would reduce the accumulation of defective virus particles.

Other mutant viruses have been identified. Prolonged passage of NOV also results in the appearance of a variant known as the "FP" mutant [75,76]. This variant is characterized by the formation of less than 15 polyhedra per cell as opposed to the standard virus known as the "MP" variant, which produces more than 30 per cell. "FP" variant polyhedra contain few normal virions and have a lower virulence for susceptible larvae. The variants could be the result of the selection of a strain of the wild virus that is more suited for in vitro replication. Potter and coworkers [77] concluded that "FP" NOV production is favored over "MP" virus synthesis in vivo. The passage of virus results in the accumulation of defective particles similar to the defective interfering particles found in other viruses when they are repeatedly cultured in vitro.

Effect of Cell Density, Stage of Cell Growth, and Medium Condition at Infection. Another characteristic of the insect cell–baculovirus expression system that limits its industrial applicability is that specific productivity of both virus and recombinant protein (i.e., production per cell) decreases with increasing cell density at infection, with increasing culture "age," and with increasing medium "age." The so-called cell density effect (really a combination of all three of these factors) has been observed with virus production in attached culture since the mid-1970s [3,13,66,78–81]. Stockdale and Gardiner [80] found that cell density, phase of growth, and medium condition all affect the yield of virus in attached culture. Rapidly growing TN368 cells (at low cell density) produced more polyhedra than those approaching stationary phase (near confluency). When freshly started monolayer cultures were overlain with spent media removed from different aged cultures, they found that the inhibition of virus production increased with the "age" of the medium. Medium from the stationary phase almost completely suppressed virus production. With any given age of medium, a decrease in polyhedra per cell was still observed with increasing cell density. Stationary phase cells were resuspended at densities from 1×10^5 to 3×10^6/ml. Upon infection, the number of polyhedra produced per cell dropped rapidly at densities above 4×10^5/ml. They speculated these results were due to a nutrient becoming limiting and/or the exhaustion of a precursor. Varying cell density from 4×10^5 to 1.75×10^7/ml, Hink et al. [79] observed a 14-fold decrease in virus production per cell with increasing cell density, but did not find that culture age had a significant effect on the yield of polyhedra.

In an attempt to understand the effect of cell density on virus replication in attached culture, Wood et al. [81] infected cells at low cell density (LCD, $1.4 \times 10^5/cm^2$) and high cell density (HCD, $5.7 \times 10^5/cm^2$) and monitored viral and cellular DNA replication [81]. Less than 0.1% of cells from HCD cultures contained polyhedra compared with greater than 90% for cells from LCD cultures. At HCD, cells that did contain polyhedra were in areas where the monolayer was interrupted. LCD cells when exposed to conditioned medium (from HCD cultures) exhibited no inhibition. When infected HCD cells were diluted to LCD conditions, the production of virus was restored to LCD levels. Viral DNA replication was significantly reduced at HCD and could be reinitiated when the cells were diluted. They concluded that the inhibition of viral and cellular DNA replication was reversibly mediated by cell-to-cell contact rather than by a depleted medium component or diffusable extracellular inhibitor. Lynn and Hink [82] synchronized cultures of TN368 and infected them with AcMNPV during different phases of the cell cycle. Their results showed that cultures exposed to virus during the mid- and late-S phase of the cell cycle had higher percentages of infected cells than did cultures inoculated with virus in the G_2 phase. Although more cells were infected, there was not a significant difference in the production of polyhedra. They concluded that S-phase cells are more susceptible to infection and speculated that it may be due to the condition of the cell membrane that allows for increased adsorption and/or penetration of the virus. Volkman and Summers [78] found that cells infected in log phase produced the highest yield of NOV. Stationary phase cells (confluent and primarily in G_0) may be less susceptible to infection.

More recently, the effect of cell density at infection on recombinant protein expression has been investigated in suspension culture. Maiorella et al. [11] saw a dramatic decrease in yields of M-CSF if cells were infected during stationary phase compared with cells in exponential growth. Caron and colleagues [26] studied the effect in SF9 cells in TNM-FH supplemented with 10% FBS, by removing 70 ml aliquots from a 4 liter Celligen bioreactor at various points in the growth cycle and infecting them in 100 ml spinner flasks at an MOI of 1. Densities between 1.3×10^6 and 4.6×10^6 cells/ml were used. They observed a 10-fold decrease in VP6 production between 2.4 and 3.2×10^6 cells/ml. Production was barely detectable at 4×10^6 cells/ml. Taking cells from the early stationary phase (4.6×10^6 cells/ml), infecting them for 1 hour, and resuspending the cells in fresh media restored protein production per cell to maximum levels. They found that addition of just glucose and glutamine was not sufficient to restore production. They did not try medium replacement over the whole range of cell densities, nor did they account for cell growth postinfection. Furthermore, the cells were in very different physiological states at the time of infection. Using SF9 cells growing in suspension in TN-MFH with serum and infecting them at an MOI of 1 with β-galactosidase-AcMNPV, Neutra et al. [20] were able to study the effect of the timing of infection on the yield per cell. Growth was inversely correlated with the time of infection. The earlier the infection, the lower the extent of growth after infection. The highest yield on a volumetric and per cell basis was obtained for the 24 hour infection (corresponding to cells in the early exponential phase). The 0 and 48 hour infections respectively achieved 32% and 62% less product per cell than the 24 hour infection. The 72 hour infection (corresponding to late exponential phase) produced only 2% as much product as the 24 hour infection.

The hypothesis that the cell density effect in suspension is the result of nutrient limitations has been further confirmed by Lindsay and Betenbaugh [83], who studied the effects of infection cell density using SF9 cells infected with two different recombinant AcMNPVs (VP4 outer capsid protein of porcine-OSU strain of group A rotavirus and β-galactosidase). Their results indicate that if spent media is replaced with fresh medium and if the spinner flask cultures are aerated through the side arms, the cell density at infection had no influence on specific protein production. It must be noted that the difference between what they considered LCD and HCD was small (3–5 × 10^5 versus 0.9–1.4 × 10^6 cells/ml). In our laboratory, β-galactosidase has been expressed in 40 ml spinner cultures of SF21 cells grown in Ex-Cell 400 serum-free medium [14]. Cells from midexponential growth phase were infected at an MOI of 10 for 2.5 hours and then resuspended in fresh medium at various cell densities. The caps on the side arms were replaced with porous foam closures to allow for increased oxygen transfer. Figure 6A shows the results on a

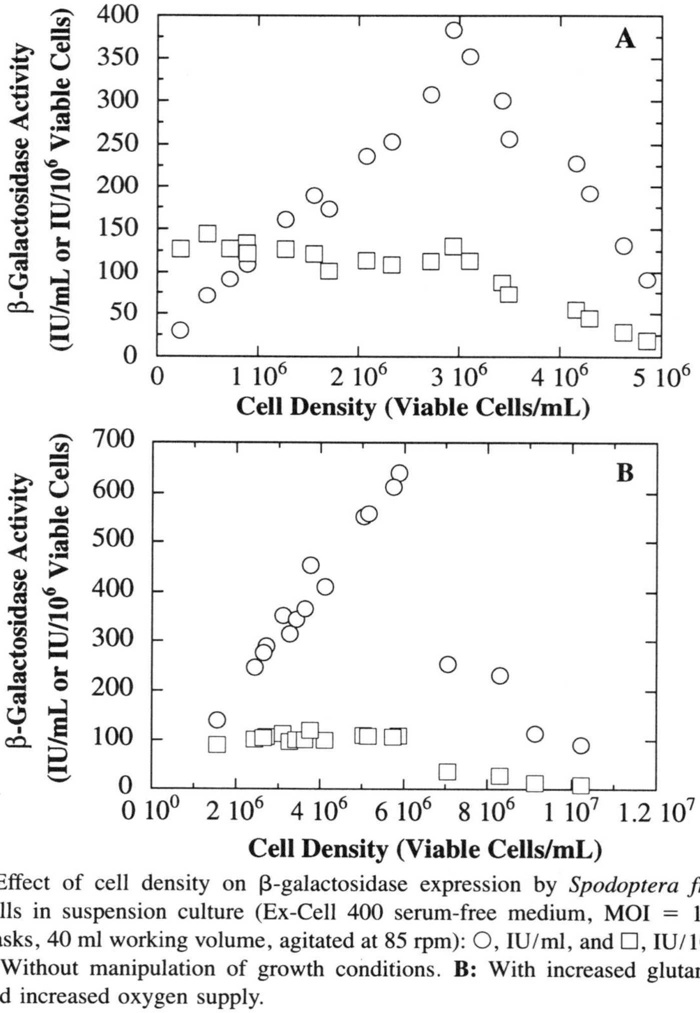

Fig. 6. Effect of cell density on β-galactosidase expression by *Spodoptera frugiperda* (SF21) cells in suspension culture (Ex-Cell 400 serum-free medium, MOI = 10, 50 ml spinner flasks, 40 ml working volume, agitated at 85 rpm): ○, IU/ml, and □, IU/10^6, viable cells. **A:** Without manipulation of growth conditions. **B:** With increased glutamine (1.5 g/liter) and increased oxygen supply.

volumetric and per cell basis. Protein production is independent of cell density for densities below 3×10^6 cells/ml. The kinetics of expression on a per cell basis for densities below 3×10^6 cells/ml were very similar. Significant lags were obtained at densities above 3×10^6 cells/ml, indicating that a change occurs in the rate-controlling step. For the densities above 3×10^6 cells/ml, the volumetric yield continued to increase until day 9 postinfection, while for densities below 3×10^6 cells/ml the maximum was reached much earlier (between days 4 and 6). Increasing the oxygen supply to the cells fourfold (by increasing the oxygen in the incubator from 20% to 80%) and increasing the glutamine in the medium by 50% allowed the per cell productivity to be maintained until 6×10^6 cells/ml. As can be seen from Figure 6B, this resulted in a 75% increase in volumetric yield. Increasing oxygen or glutamine independently resulted in some increase, but both are required to sustain production to 6×10^6 cells/ml. Further increases in either oxygen or glutamine had no effect.

A number of research groups have now attempted to control dissolved oxygen concentrations postinfection and supplement the medium with key nutrients. Using a 6 liter air lift bioreactor operated with dissolved oxygen controlled at 50% of saturation, Lazarte et al. [21] were able to triple the yield of CD4 at $4–5 \times 10^6$ cells/ml with the addition of 1 liter of fresh medium after infection. If oxygen is controlled so that it does not become limiting during the infection stage, it appears that glucose becomes the controlling nutrient. Reuveny and coworkers [84] studied the production of two proteins at high cell density using SF9 cells in a bioreactor controlled at a dissolved oxygen (DO) concentration of 65% of saturation. By replacing the medium prior to infection, glucocerebrosidase expression was tripled at 5×10^6 cells/ml. In both cases, the maximum level of expression corresponded to when glucose was depleted. By supplementing the fresh medium with glucose, production was increased to 4 from 3 mg/ml. Further supplementing the medium with glucose, glutamine, and yeastolate (2 g glutamine and 17 g yeastolate per 20 g glucose), the expression level was increased to 6 mg/ml. The cell density remained at or just above 5×10^6 cells/ml in each case, so that there was an increase in specific productivity but not quite to the level seen at 0.5×10^6 cells/ml. Similar trends were observed with β-galactosidase.

A more detailed study of the effect of oxygen, glucose, and glutamine supply on recombinant protein expression has been done by Bentley's group [63,85]. In a bioreactor infection with SF9 cells at 6×10^5 cells/ml, the dissolved oxygen fell to zero and stayed there for 90 hours without DO control. Lactate accumulated and reached a maximum during the time oxygen was limiting. With DO controlled at 35%, the cells grew postinfection, lactate remained below 1 mM, and the level of glucose dropped from 14 to 8 mM. Specific productivity was increased 100%. In an attempt to increase production further, glucose and glutamine were fed to cells after infection in spent medium in spinner flask. An additional 100% increase in specific production was obtained. Specific productivity at the high cell density (4×10^6 cells/ml) with feeding was almost the same as that obtained for cultures resuspended at 1.1×10^6 cells/ml in fresh medium after infection. The pH remained stable with feeding or dilution, but increased in spent medium. Little has been published on insect cell metabolism after infection. Wang et al. [63] report that none of the amino acids was determined to be rate-limiting after infection at the low densities used (1.2–2.4 × 10^6 cells/ml).

It is clear from the above discussion that the best infection strategy is one that infects actively growing cells (early to midexponential phase) with a low-passage virus at moderate MOI. Production can be optimized by controlling dissolved oxygen and supplementing key nutrients postinfection. For the insect cell–baculovirus expression system to become an economically viable method of producing recombinant protein, yields must be further improved. Insect cells can be grown to cell densities as high as 11 × 10^6 cells/ml [31]. If the environment in the bioreactor could be manipulated so that specific productivity could be maintained at such high cell concentrations, product yields could be increased 5–10-fold. Up to now, the variability observed with the limiting nutrients is likely due to the use of different media, different bioreactors, and the variability between clones of the same cell line. Certainly, more work needs to be done at higher cell densities (greater than 5 × 10^6 cells/ml) to better determine what nutrients are rate-controlling.

KINETIC DATA AND THEIR ROLE IN SCALE-UP

The rational scale-up of processes from bench-scale observations to large-scale bioreactors would require a quantitative ability to predict cellular response to operating conditions. Unlike most other bioprocesses, the insect cell–baculovirus system requires the design of an infection strategy. Efficient use of virus will be critical, and large-scale suspension cultures, in particular, may require very precise delivery of virus. Laboratory systems such as multiwell plates, T-flasks, spinner flask, and microcarrier-containing spinner flasks are efficient systems for developing kinetic data. Mathematical expressions for key steps in the process for production of the target protein will be needed. Such descriptions include the kinetics of virus attachment, of infection, of protein production, and of cell lysis. For attached cell systems, the kinetics of cell attachment may also be important. Ideally these expressions can be used to select optimal strategies for reactor operation. Relatively little work has been done in this area with insect cell cultures, but the opportunities are significant.

The first step in the process must be the attachment of virus to the cell surface. As described in Chapter 6, a receptor appears to be involved in the attachment of baculovirus to the insect cell surface. Since receptor numbers and configuration might vary somewhat from cell line to cell line, a cell line–dependent susceptibility to attachment and infection might be expected. Indeed this is observed [86] and is described in more detail in Chapter 6.

Using multiwell plates with 3.5 cm diameter wells, Wickham and his coworkers [86] measured infection kinetics for several cell lines. All experiments were done at moderate cell densities (<1.5 × 10^6 cells per well) with dilute virus stock ($\ll$ 1 pfu/cell). Under these conditions, the infection rate could be expressed as

$$r_i = k_i \,(C)(V) \tag{1}$$

where r_i is the infection rate (infections/ml/min), C is the cell concentration (cells/ml), V is the virus concentration (pfu/ml), and k_i is the infection rate constant (ml/cell min). Table 4 summarizes the measured infection rate constants for a variety of cell lines. These constants vary 10-fold. The value of k_i for *T. ni* 5B1-4 cells was 4.4×10^{-9} cell/ml/min, which is very close to the measured attachment rate of 5.2×10^{-9} ml/cell. The similarity of these values suggests that virus attachment is the rate-limiting step and infection by bound virus is very efficient. The rates of attachment were close to the theoretical predictions for the thermally driven collision rate of virus to the cell surface, indicating that for *T. ni* 5B1-4 cells, attachment was diffusion limited.

de Gooijer et al. [87], working with suspension cultures, modeled infection as first order in cell concentration and zero order with respect to virus when virus was in excess. Since Eq. 1 was determined at low virus concentrations, these two results are not necessarily in disagreement. In fact, one could speculate that a more general form of an infection equation (where infection is defined as virus uptake into the cell) could be

$$r_i = k_i'\{C\}\left\{\frac{V}{K_V + V}\right\} \tag{2}$$

where K_v is a saturation-like parameter with units of pfu/ml, and k_i' would have units of infections/ml/min. Equation 2 would suggest that at low virus concentrations the infection rate is first order in both virus and cell concentration, but becomes zero order in virus concentration when $V \gg K_v$. However, at high cell densities not all infections are productive; hence virus and protein production will be saturable in cell density, i.e.,

$$r_p' \propto \left\{\frac{C}{K_C + C}\right\} \tag{3}$$

TABLE 4. Infection Rate Constants Measured for Various Cell Lines[a]

Cell line	Infection rate constant[b] (10^9 ml/cell/min)
S. frugiperda (SF9)	0.44
S. frugiperda (SF21)	0.48
T. ni (TN368)	1.65
T. ni (TN-M)	2.02
E. acrea (EAA)	2.27
T. ni (TN-F)	4.35
T. ni (TN5B1-4)	4.40

[a]For a full description of the cell lines, see Wickham [13].
[b]Measured at a cell density of 10^6 cells/ml and an MOI of 0.005.

where r'_p is the volumetric rate of production of virus or protein, and K_c is a saturation-like parameter with units of cells/ml.

A more complete study of the kinetics of infection and effects on protein production has been reported by Licari and Bailey [68]. Their model is constructed for attached cell growth and validated with T-flask data. Their model assumes that virus, once attached, will infect the cell with negligible delay. Protein production and lysis are assumed to occur at a fixed time after infection. They tested Eq. 1 and found it unsatisfactory for MOIs greater than 1. They developed a model accounting for the influence of multiple virions infecting a single cell. A Poisson distribution was used to calculate the probability of a cell being infected with one or more virions at a given time. The model had the capacity to account also for secondary infections. The model accurately predicted that for infections of early exponential phase cultures (also low density) the optimal MOI is about 1, although total production (units/ml) is relatively constant for $0.1 < \text{MOI} < 10$. For late exponential cells (also at high density as a consequence), production is low at low MOIs (below 1) but equals that of the early exponential phase culture at a high MOI (between 10 and 100). These results are the consequence of the importance of secondary infections at low cell densities.

de Gooijer and colleagues [74] have presented another model to help explain the decline in virus production in a cascade of continuous bioreactors running in series where the first bioreactor is the growth vessel and the subsequent vessels are infection vessels. They incorporate several different forms of virus (normal, defective, and abortive) in their model. The abortive form of virus can be internalized by the cell but does replicate within the nucleus. In this model, three modes of infection can be distinguished. The first is by infectious and abortive virus, which results in mostly infectious virus being produced (a little abortive and defective virus is also produced). The second involves infection by all three types of virus and leads to the production of a large amount of defective virus and some of the other two types of virus. The third route of infection is by abortive and defective virus only and results in no virus being produced. In a simulation of the two-stage configuration, when the infection level is 50% of the steady-state level, most of the virus is of passage 8 or higher. Virus of such high passage number produces fewer virions and would be unable to sustain infection in the continuous bioreactors. The authors suggest that the optimal mode of operation is repeated batch, as the production of virus is predicted to be maintained much longer than in continuous culture. In repeated batch mode, the virus is kept at a low concentration thereby establishing a low MOI favoring infection by the infectious form of the virus.

Power et al. [88] use a model that divides the cell population into viable, nonviable, infected, and noninfected cell populations. The number of viable infected cells is a function of the number of viable noninfected cells and the amount of virus present (similar to Eq. 1). At MOIs greater than 2, infection is assumed to be synchronous. Postinfection, cells are virus-producing cells and/or protein-producing cells. No decomposition of virus or protein is included in the model. Infection parameters were taken from runs that were free of any nutrient limitations, but unfortunately they could not completely confirm their results as they had no measurement of the number of infected cells. Good agreement was obtained between the model and experiment for virus titer. The agreement was reasonable for protein expression. Once the cells become nutrient-limited, the model predictions and experimental results diverge, suggesting that a high enough MOI must be used so as to infect the cells quickly so they do not grow and become nutrient limited. This is similar to the result of Licari and Bailey [68].

Conceptually, it should be possible to extend the Licari and Bailey model to predicting optimal infection strategies in large-scale attached cell reactors. Kumar [89] uses the concepts from the Licari and Bailey model to simulate the infection of *T. ni* cells attached to glass beads in the downcomer of a split-flow air lift bioreactor (see Fig. 7 for a schematic of the bioreactor). The extension to suspension culture reactors is less straightforward as the state of confluency of the culture, which is

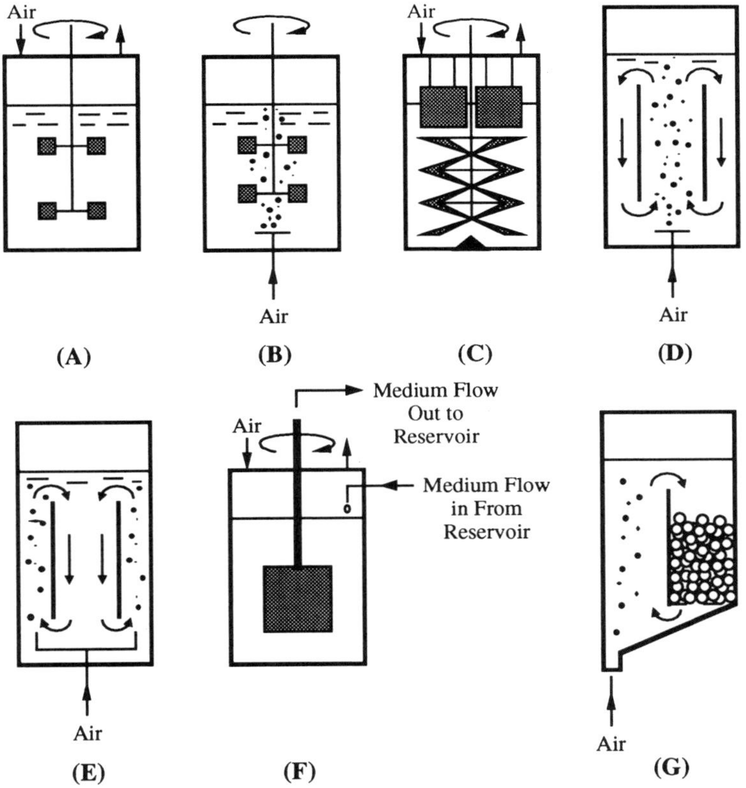

Fig. 7. Schematic diagrams illustrating the various bioreactor configurations that have been employed successfully to grow insect cells. **A:** Stirred tank with surface aeration. **B:** Stirred tank with submerged aeration. **C:** Stirred tank with helical-ribbon impeller. **D,E:** Draught-tube air lifts. **F:** Spin-filter perfusion reactor. **G:** Split-flow airlift bioreactor with cells attached to glass beads in downcomer.

defined for attached cell systems, is not well-defined in suspension cultures. In suspension cultures, the nutritional status of the culture may need to be considered more explicitly. Also attached cell systems present another problem in large scale systems: the seeding and distribution of attached cells. If microcarriers were to be used, some beads might receive no cells, while others would receive many. After a period of growth, some beads would have populations near confluency (where a high MOI would be necessary) and others with a low density (where a lower MOI would be optimal). Such heterogeneity in physiological states would complicate the infection strategy. Mathematical models for cell attachment and optimal strategies for maximizing attachment have been developed for mammalian cells [90,91]. Presumably these models could be adapted for use with insect cells.

BIOREACTORS FOR LARGE-SCALE INSECT CELL CULTURE

Laboratory-scale cultivation of insect cells is carried out in T-flasks and roller bottles (for attached culture) and in spinner and shake flasks (for suspension culture). Simply increasing the number of vessels used to achieve a larger volume (which has been done with roller bottles [92]) is not the most practical route for large-scale cultivation. Instead, bioreactors (fermenters) of varying configurations and volumes have been used to scale-up growth of insect cells and the production of virus and recombinant protein. A bioreactor is a vessel designed to provide conditions that promote cell growth and/or product formation. During the late 1960s and early 1970s, it was natural for animal cell culturers to use microbial fermenter technology. There were no specialized bioreactors then available. Since then companies have developed new systems, such as high cell density perfusion bioreactors, that attempt to address the low productivity and efficiency characteristic of animal cell culture (compared with microbial systems). There is some reluctance to adopt these systems for large-scale manufacturing because they are considered over-sophisticated, unreliable, and difficult to scale-up. This reluctance is diminishing with the reports of these systems being used at large scale to produce interferon and TPA[93,94]. A wide range of different types of bioreactors have been used with insect cells, including perfusion type bioreactors. Figure 7 illustrates some of these configurations. Kinetic data for various insect cell lines grown in larger scale bioreactors is given in Table 5.

Since mechanically agitated bioreactors serve as the work horses of the pharmaceutical industry, it is not surprising that a large number of research groups have used mechanically stirred vessels to culture insect cells. With this type of bioreactor, achieving good mixing and homogeneous environmental conditions is difficult because of variable shear (high near the impeller and lower near the walls). Hink and Strauss [28] grew *T. ni* cells in suspension in four types of stirred tank reactors: a 100 ml spinner flask, a 400 ml glass bioreactor equipped with a vibromixer, a 400 ml Bioflo stirred tank bioreactor, and a 2 liter stirred tank bioreactor (MF-205) [28]. The inoculation cell density was $1–2 \times 10^5$ cells/ml. Growth was significantly slower in the Bioflo and MF-205 bioreactors compared with the spinner flask. The

TABLE 5. Growth Kinetic Data for Various Insect Cell Lines Grown in Suspension Culture in Large-Scale Bioreactors

Bioreactor	Working volume (liters)	Cell line	Medium	Aeration	rpm	Growth rate (h^{-1})	Maximum cell density (cells/ml)	Reference
Airlift	21	*S. frugiperda* (SF9)	IPL-41 + FBS + PF68	DO controlled at 20%	—	0.027	5×10^6	[11]
Spin-filter perfusion	3	*S. frugiperda* (SF21)	IPL-41 + FBS	Submerged with DO controlled at 93%	—	—	—	[95]
Spin-filter perfusion	14	*S. frugiperda* (SF9)	SF900	Submerged	150	0.034	1.45×10^7	[96]
Helical-ribbon impeller STR	11	*S. frugiperda* (SF9)	TNM-FH + FBS	Surface with O_2	60–75	0.028–0.031	5.4×10^6	[26]
			IPL-41 SFM					[53]
STR (Celligen with marine impellers)	4	*S. frugiperda* (SF9)	TNM-FH + FBS	Surface with O_2	100	0.030–0.035	4.7×10^6	[26]
			IPL-41 SFM			0.029–0.032	5.5×10^6	[26]
Airlift	10	*S. frugiperda* (SF21)	IPL-41 + FBS	DO controlled at 20%	—	—	3×10^6	[95]
STR	5–10	*M. brassicae*	Serum-containing	Silicone tubing	NA	0.035–0.039	3×10^6	[48]
STR (Celligen with marine impellers)	2	*S. frugiperda* (SF9)	SF900 II	Submerged with DO controlled at 50%	130	0.033	$6.8–8.0 \times 10^6$	[97]

STR, stirred tank reactor; FBS, fetal bovine serum; PF68, Pluronic-F68.

cells in the bioreactor equipped with a vibromixer grew even more slowly and seemed to be damaged. Methylcellulose at 0.1% was added to minimize cell clumping. Aeration (either submerged or surface) increased cell growth. More recently, Caron et al. [26] reported being able to grow SF9 cells to high cell densities in a CelliGen bioreactor using surface aeration with oxygen and a marine impeller [26]. Barkhem and coworkers [98] successfully scaled-up the production of human thyroid receptor β_1 from 2 to 100 liters in a stirred tank reactor. Others have had success growing cells by using novel impellers or aeration methods. Kamen and coworkers [53] report growing SF9 cells in an 11 liter bioreactor equipped with a helical ribbon impeller. The impeller offers excellent oxygen transfer capacity while minimizing shear effects. Silicone tubing has been used to supply oxygen to insect cells grown in stirred tank bioreactors, avoiding the shear associated with submerged aeration [48,49]. Graf and Schugerl [50] found that using a reciprocating silicone membrane tubing aeration system compared with one that did not reciprocate resulted in a higher cell yield by extending the exponential growth phase. Using a bubble-free aeration system may also increase protein expression. Murhammer and Goochee [12] obtained 73% higher expression of β-galactosidase per cell in unsparged bioreactors than in sparged ones.

Pneumatically agitated bioreactors have proven to be successful for growing shear-sensitive cells, such as hybridoma and plant cells. Air lifts and bubble columns provide environments with relatively uniform mechanical shear. They could have limited applicability for insect cells due to the shear effects associated with the cell–bubble interactions, but the use of shear protectants has reduced the severity of the problem (see earlier discussion and Chapter 9). Maiorella et al. [11] report growing *S. frugiperda* cells to cell densities in excess of 5×10^6 cells/ml in a 20 liter airlift reactor. Weiss et al. [99] report growing *S. frugiperda* cells in 5, 10, and 40 liter airlift bioreactors [99]. Cell densities achieved ranged from 2.4×10^6 cells/ml in the 40 liter airlift to 6.4×10^6 cells/ml in the 5 liter airlift. King and coworkers [54] were able to grow SF9 cells to 1×10^7 cells/ml in serum-free medium in a 14 liter draught-tube air lift bioreactor.

A series of novel bioreactors have been investigated for use with insect cells. A rotating biological disk bioreactor was found to be unsuitable for the cultivation of anchorage-dependent *T. ni* cells [8]. Cell growth was abnormal over most of the surface area of the disks. Only cells near the center of the disk grew normally. A packed bed reactor using 3 mm glass beads was also used to grow *T. ni* cells [8]. A separate unit was employed for gas exchange, and the oxygenated medium was circulated through the packed-bed reactor. The growth rate obtained was comparable with that observed in stationary culture. Infection with an AcMNPV virus containing the β-galactosidase gene resulted in protein production ranging from 500 to 900 μg/10^6 cells, which is slightly lower than the 1,100 μg/10^6 cells obtained in T-flask. A split-flow air-lift bioreactor has been used to express a secreted human alkaline phosphatase in *T. ni* cells attached to 3 mm glass beads in the downcomer [9,100]. The idea of separating oxygenation from growth was retained, but the bioreactor system is simplified.

Recently, a number of perfusion-type systems have been evaluated for the expression of recombinant protein at high cell densities [95,101–105]. A perfusion bioreactor is operated by continually adding and removing medium from the growth section of the vessel. Also contained in the reactor or external to it is a separation apparatus that allows medium to be removed from the bioreactor while retaining the cells in the reactor. Perfusion culture reactors are used to maintain high levels of nutrients and to dilute out metabolic inhibitors. They may prove effective in eliminating the decrease in specific productivity observed with the insect cell–baculovirus expression system. Weiss et al. [95] report infecting SF21 cells with a recombinant virus for human β-interferon in a spin-filter perfusion bioreactor system. Subsequently, they reported being able to achieve a cell density of 1.45×10^7 cells/ml after 12 days of perfusion [96]. The perfusion rate had to be increased to 0.04 from 0.01 h^{-1} to achieve that density. Kim et al. [105] used a spin-filter bioreactor to compare the growth of and β-galactosidase expression by SF9 cells in Grace's, TNM-FH, and IPL-41 media supplemented with FBS and 0.3% Pluronic-F68. Cells were allowed to reach densities between 5 and 10×10^5 cells/ml before perfusion mode was begun. Perfusion rates of 0.01 to 0.06 h^{-1} were used. In all cases, β-galactosidase production per cell decreased dramatically as the perfusions progressed. Whether this is the result of the generation of defective interfering particles or is due to a "cell density" type phenomenon is unclear. Cell densities increased to about $3–3.5 \times 10^6$ cells/ml within about 5 days. Jager et al. [101] used a conventional stirred tank bioreactor equipped with a double membrane stirrer to run in perfusion mode. The double membrane provided bubble-free aeration and allowed for cell retention. A maximum cell density of 5.5×10^7 cells/ml was achieved. Cells at concentrations between 1 and 3×10^7 cells/ml were infected with recombinant virus. Product was harvested continuously with spent medium, which did not contain cells or cell debris. Significantly higher cell specific productivity was obtained compared with batch culture.

External filtration devices have also been used in perfusion systems. Cavegn et al. [102] used an external microporous hollow fiber cartridge to separate cells from spent medium in their perfusion system. Two different proteins were expressed using this reactor. The volumetric productivity of the intracellular protein, CD23 (B cell surface antigen) was increased from 4 to 14 μg/(1 hour) when going to perfusion mode. An eightfold higher volumetric productivity was obtained for the secreted protein. Another research group has also used an external hollow fiber cartridge to separate the cells [104]. Cell densities as high as 20×10^6 cells/ml and β-galactosidase yields of 1 g/liter were achieved. Massie and coworkers [103] used a tangential flow filtration system in conjunction with a 4 liter stirred tank bioreactor operated at a perfusion rate of 0.04 h^{-1} to achieve cell densities of at least 15×10^6 cells/ml.

MODES OF CULTURE

Batch, Fed-Batch, and Continuous Bioreactors

Batch culture is the traditional and often preferred production technology for animal cells. Vaccine manufacturing using lytic systems made batch culture the only option. The simplest scale-up approach was to adopt microbial-type stirred tanks operated in batch mode, as it has a proven record of reliability as a manufacturing tool. Unfortunately, batch operation has a low productivity because conditions change continually, with inhibition becoming more problematic with culture time due to the accumulation of waste products. One way of reducing this problem is to use fed-batch or semicontinuous modes of operation. Fed-batch involves adding fresh medium to the bioreactor, while semicontinuous involves removing a portion of the medium and replacing it with fresh medium. Both of these modes of operation result in increased cell densities and productivity by lengthening the culture period and allowing for repeated harvests. They are also more economical than batch as the downtime involved (e.g., for clean-up and sterilization) is reduced, but culture conditions still vary with time and toxic materials still accumulate (though to a lesser extent).

Semicontinuous culture has been used with insect cells. Hink and Strauss [47] were able to use a series of stirred tank bioreactors to produce virus. Cells were grown up in the stock vessel and then allowed to settle out. Fifty percent of the medium was removed and put into a second bioreactor. Fresh medium was added to both reactors. Cells in the second reactor were infected for 24 hours, and the cells were then allowed to settle out. Again, 50% of the medium with cells was removed and replaced with fresh medium. This was repeated five more times. High percentages of infection and yields of polyhedra were obtained with the first four batches of harvested cells. Yields in the fifth and sixth harvests were reduced. Zhang et al. [106] employed a two-stage bioreactor system in which the first vessel is used for growth of the cells and the second is used for infection. Cells are grown to a high cell density (3.5×10^6 cells/ml), and then 90% of the volume is transferred to the infection vessel. The remaining 10% is used as the inoculum for the next cycle in the first bioreactor. Virus is added to the second vessel, and a sufficient time is allowed for viral adsorption. The medium is exchanged for fresh medium via a porous membrane. At the end of the production phase, infected cells and virus are pumped out of the second vessel, and the whole process is repeated. Consistent yields of 250 mg/liter of chloramphenicol transferase were obtained for four cycles of operation lasting more than 600 hours. A similar system was used by Kloppinger et al. [107] to produce AcMNPV in SF9 cells. They exchanged medium prior to infection, and 5% of the volume in infection vessel was retained to infect the next batch of cells. They report being able to sustain production for several weeks without a decline in virus yield.

Continuous flow systems (without cell retention) operate by adding fresh medium to the bioreactor and removing spent medium with cells. Such a mode of operation allows the system to be operated at a single steady state that can be controlled to be optimal for the desired process (i.e., product formation). Since a continuous flow process can be run for weeks or months there is a saving in downtime. Continuous processes are not readily acceptable to manufacturers, as they are more complex and are more difficult to get licensed. Each batch of product must be tested and certified before it is sold. Exactly what constitutes a batch from a continuous process is difficult to define. But because the kinetics of insect cell growth, infection, and virus or protein production are different, a continuous flow system could allow for the optimization of each step, resulting in an efficient bioreactor system.

Applicability of Multiple Stage Bioreactor Systems

Attempts have been made to run continuous flow bioreactors with insect cells but with limited success. Using a two-stage continuous system (the first vessel for cell growth and the second for baculovirus replication), Kompier et al. [108] were able to sustain continuous polyhedra production for about 25 days. A decrease in productivity began thereafter, and virtually no polyhedra were being produced by day 56. The reduction observed was attributed to the so-called passage effect (degradation of virus as the number of passages increase). Using a computer model developed earlier [40], it was calculated that virus passage numbers greater than 10 (above which the passage effect is severe) become significant in a continuous flow stirred tank reactor (CFSTR) after about 1 month.

In an attempt to make the infection process more efficient, they replaced the second stage (virus replication vessel) by two reactors, each containing half the volume of the original vessel [87,109]. The most efficient infection of cells occurs in a plug-flow reactor, and this system approaches plug-flow conditions more closely. The fraction of cells containing polyhedra increased, but a decrease in productivity was still observed. This time the decrease occurred more rapidly (after 15 days). The number of cells with polyhedra, the number of polyhedra per cell, and the number of infectious NOVs were all found to decrease significantly. They attributed the faster decrease to an acceleration in the occurrence of virus of higher passage number in the cascade of reactors.

A similar decrease in production has been observed with the production of a recombinant protein. The same group has done work producing β-galactosidase using a multistage continuous reactor system [32,73,110]. Cell concentrations in the production reactors were slightly higher than in the growth reactor, indicating that not all of the cells were immediately infected. β-Galactosidase, as measured by its activity, was produced at a relatively constant rate for about 25 days (for three replicates) and then dropped to a negligible amount by day 40. The decrease in production coincided with the decline in infectious virus (from 10^2–10^4 pfu/ml to essential zero). To determine if the decrease was the result of a decrease in production or due to inactivation of the enzyme, SDS-PAGE analysis was carried out. The results indicated that the total amount of β-galactosidase decreased with time (not an inactivation), while the production of most viral proteins remained constant. They report that during the process, a defective virus (which lacks about 40% of the viral genome) becomes predominant and is the cause of the decrease in production. This is in agreement with the observations of a number of research groups [70,71].

The work of Tramper's group has not exploited the differences in kinetics for the three steps. Using knowledge of the kinetics of growth, infection, and production, it might be possible to design a better multistage continuous flow reactor system whose three component reactors are optimized with respect to cell growth, infection, and expression (see Fig. 8). Medium enters the first vessel inoculated with cells, and more cells are produced. In the second vessel, low passage, standard

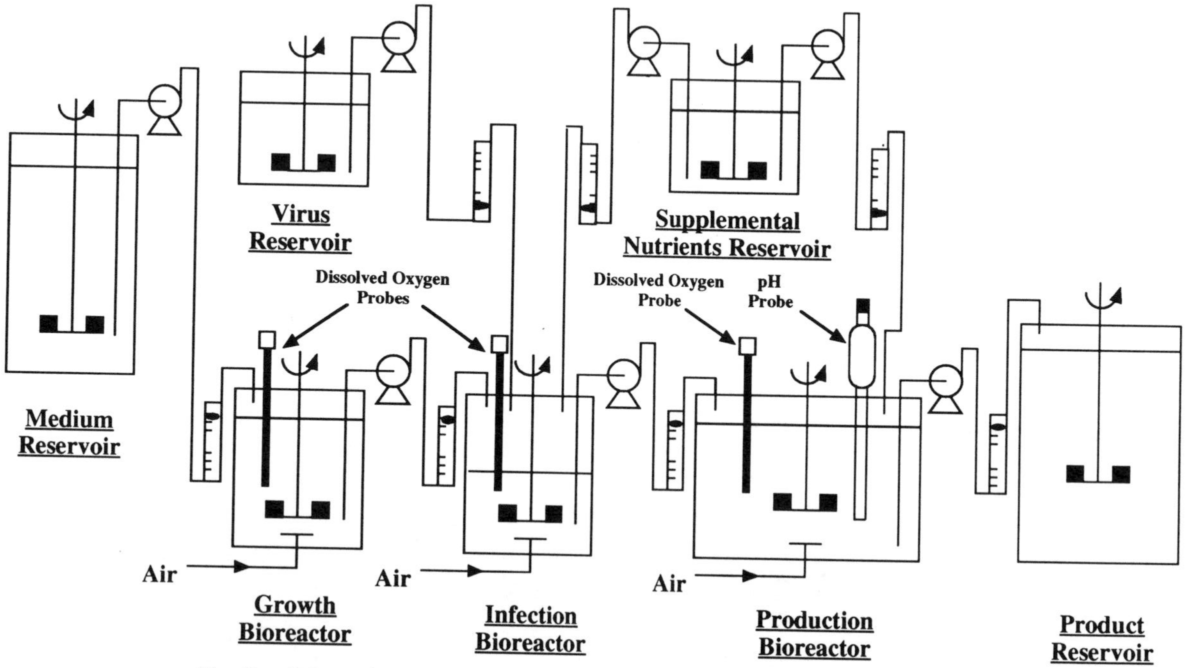

Fig. 8. Schematic diagram of the three-stage continuous flow bioreactor system.

virus is added to infect the entering cells. Infected and uninfected cells enter the third vessel, where virus particles replicate and the protein is expressed. This system will avoid the detrimental effects of defective virus particles by separating the initial infection of the cells from the subsequent virus and protein production. The dilution rate (flow rate/volume of reactor) of the second stage is kept high enough so that cells become infected with standard virus and then move to the third stage before producing progeny virus particles. Cycling of the virus is thus avoided so that defective virus particles are not able to proliferate. Potentially, defective virus particles can be generated in the third stage. Experiments have shown that defective particles entering a cell 9 hours after infection with standard virus have little or no effect on standard virus or protein production [71]. Thus, by choosing a sufficiently long residence time in the second reactor, defective particle production will be greatly inhibited in the third stage. Even if defective particles are produced in the third stage, they should have little effect on protein production.

Large-Scale Insecticide Production

In the near future, the global market for biopesticides will be quite large, especially considering the concerns regarding the environmental safety of chemical pesticides. The application of baculoviruses as pesticides is an area of research that is gaining popularity. Reviews on the mass production of viruses using both in vivo and in vitro approaches have been written by Hink and Strauss [47] and by Shapiro [111]. For baculoviruses produced in vitro to be economical biopesticides, they need to be produced in large quantities using low-cost serum-free medium (a price range of $2–$5/liter compared with the present $20–$30/liter [97]), and they need to kill the insect more rapidly. A detailed discussion of the development of improved baculovirus biopesticides is given in Chapter 5. One way of producing large quantities of viruses would be using a two-stage continuous bioreactor system (similar to the system discussed in Batch, Fed-Batch, and Continuous Bioreactors, above). In fact, Pollard and Khosrovi [112] designed a two-stage, continuous-flow, vertical, tubular bioreactor system for virus production. No experimental work was done to verify the feasibility of their design. In their calculations, they found that the cost of serum-containing medium to be prohibitive.

A number of research groups have developed inexpensive serum-free medium targeted for the production of viruses in vitro (see Chapter 2) [97, 113]. Godwin and his coworkers [97] have developed a prototype medium called Biopesticide-SFM and found that growth and virus production comparable to serum-containing medium was obtained with SF9 cells. TN368 cells achieved almost twice the cell density in Biopesticide-SFM compared with serum-containing medium.

IMMOBILIZED SYSTEMS

Whole cell immobilization involves physically confining or localizing intact cells to a certain defined region of space with the preservation of the desired catalytic activity. In most applications of whole cell immobilization, the objectives are to increase volumetric productivity, to simplify downstream processing, and to "reuse" expensive biomass. However, the lytic nature of the insect cell–baculovirus system does not allow cell reuse. In some cases, surface attachment is necessary for cell growth or optimal productivity. The higher cell densities that are achieved usually result in higher volumetric reaction rates. Confinement of the cells facilitates their separation from the product in solution, and the product concentration may increase because of the higher volumetric productivity. Cell immobilization allows for a broader range of options when considering the mode of operation of a bioreactor (medium exchange, perfusion, and so forth). Methods of immobilization include surface attachment (both natural and chemical), entrapment within porous matrices (gel entrapment and preformed supports), and containment behind a barrier (phase entrapment and synthetic membranes) (see Fig. 9). A more complete discussion of immobilization techniques and their application to whole cells is given by Karel et al. [114], Scheirer [115], and Mattiasson [116]. Radovich [117] details mass transfer effects that are important to immobilized cell systems. As surface attachment (on microcarriers, glass beads, and preformed matrices) and encapsulation are the techniques that have been used with insect cells, the advantages and disadvantages associated with these methods are discussed below.

Advantages and Limitations

The ideal cell immobilization technique is one that is simple and mild such as surface attachment to microcarriers or preformed matrices. Adsorption of the cells to the supports can take place under conditions that are not harmful to the cells. In situ encapsulation methods are much more complex than any of the other methods, as the polymeric capsules are formed around the cells, and conditions and chemicals that are used must be biocompatible. The use of microcarriers and some preformed matrices (e.g., collagen beads) are relatively easy to scale-up as they can be used in suspension with conventional bioreactors (with minor modifications). With all of the methods, products can be easily separated from the cells and the medium can be easily exchanged. Assessing cell growth on microcarriers is easily done by removing the cells from the microcarriers using collagenase (for collagen-coated beads) and RDB (for DEAE-based beads) [118]. With other methods, like encapsulation and use of glass fibers, quantification of cell number is more difficult. Another advantage of immobilization is the protection of the cells against shear, although the degree varies with the method. Although the cells are attached to the surface of microcarriers, they still experience shear and abrasion when in suspension [7,119,120]. If cells are grown on larger beads (e.g., glass) or on preformed matrices in a packed-bed configuration, the cells will not experience extreme shear, but may experience nutrient limitations due the difficulty of achieving adequate mixing. As cells grow inside the preformed matrices, they will also protect the cells from shear when used in suspension and from compression that can occur in packed beds.

Achieving locally high cell densities in immobilized systems may not improve

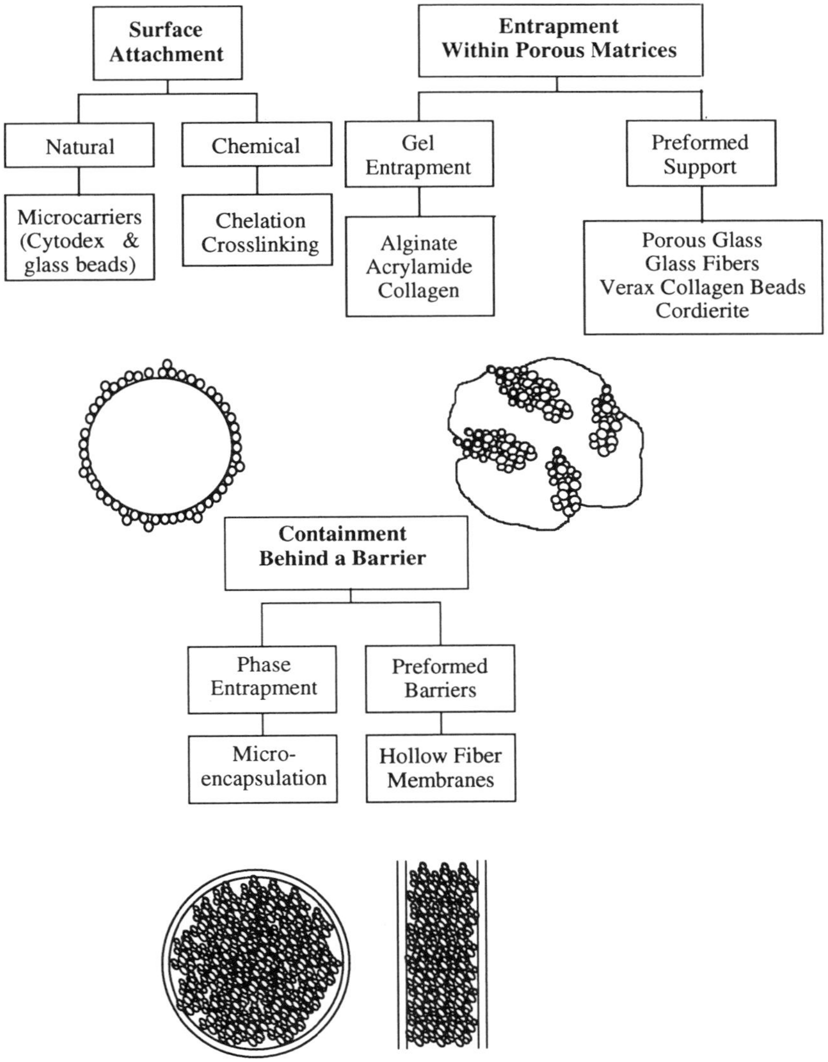

Fig. 9. Classification of immobilization methods with schematics illustrating the different methods.

production of virus or recombinant protein using the insect cell–baculovirus system because of the cell density effect. Using immobilized insect cells in a perfusion reactor may eliminate the cell density effect if it is due to a nutrient limitation, but long-term production is impossible because of the lytic nature of the expression system. The efficient delivery of virus to the immobilized cells may be more difficult if the cells are in a packed bed or encapsulated.

Immobilized Insect Cell Systems

Some of the earliest attempts at growing insect cells in larger scale "immobilized" culture used roller bottles [3,92,121]. More recently, researchers have employed microcarriers [27,122], glass beads [8,9,100], and microencapsulation [123].

Insect cells have been successfully grown on a number of different types of microcarriers. Wickham and Nemerow [122] tested the suitability of two different types of microcarriers (a DEAE-based microcarrier called Dorma cell and a collagen-coated microcarrier called Cytodex 3) for the growth of a *T. ni* cell line (TN5B1-4). The cells would attach and grow on the DEAE microcarriers, but when the beads were introduced into suspension they clumped together within 1 hour of inoculation (as a result of cell bridging). Cells would begin lysing almost immediately so that by 24 hours postinoculation cell viability was very low. When the microcarriers were attached to the walls of polystyrene roller bottles, the cells grew preferentially between the beads and did not cover the entire area of the beads. When the cells were infected with β-galactosidase-AcMNPV, increased production was not obtained as the cells did not use the increased area available for growth. Expression was still found to be highly sensitive to cell density. The collagen-coated beads were much more suitable for use in suspension. Cell bridging was minimal (only small aggregates of about five beads formed compared with as many as 30 with the DEAE-based microcarrier). Cells grew to confluency, and a high viability was achieved. Cells growing on the DEAE-based beads in a roller bottle, on collagen-based beads in suspension, and in T-flask were all infected with a virus expressing Epstein-Barr viral attachment protein (EBV gp105). All infections were done at optimal cell densities. The levels produced by the cells attached to beads came very close to that produced in T-flask, indicating that microcarriers are a possible solution for scaling up attachment-dependent cell lines like TN5B1-4.

Morita and Igarashi [124] grew *Aedes albopictus* cells on collagen Cytodex 1 beads suspended in 1 liter spinner flasks agitated at 60 rpm and infected the cells with encephalitis virus. Virus containing fluid was harvested daily between days 2 and 8 postinfection by allowing the microcarriers to settle out, withdrawing most of the fluid, and then replenishing the flask with fresh medium. Virus titers between 10^7 and 10^8 pfu/ml were obtained. Lazar and coworkers successfully grew *A. aegypti* on cellulose-based microcarriers and scaled it up to 8 liters [118]. Semicontinuous production of virus was conducted over a 16 day period. Virus titers obtained were higher than those obtained in T-flask. Using collagen-based Verax microspheres to immobilize *A. albopictus* cells, cell densities of about 2×10^7 cells/ml were achieved [27]. Cell densities in suspension are normally around 5×10^6 cells/ml [41].

Microencapsulation entraps cells within a semipermeable polysaccharide/polycation membrane. The capsule membrane is selective and allows small molecules such as nutrients and oxygen to diffuse through while preventing large molecules and cells from diffusing out. Microencapsulation has been used commercially to produce monoclonal antibodies from hybridoma cells [125,126]. King et al. [123] were able to achieve intracapsular cell densities of SF21 as high as 8×10^7 cells/ml using alginate/poly-L-lysine (PLL) encapsulation. Maximum suspension culture densities were at least 10-fold lower. The cells, when infected with a temperature-sensitive AcMNPV, produced a virus titer of 1×10^9 pfu/ml in the capsules. The virus titer in the supernatant was about 300 times lower, indicating that essentially all of the virus was retained within the capsules. The molecular weight cut off could be controlled by varying the molecular weight of the PLL, the reaction time, and the PLL concentration. King et al. [123] were able to develop a multiple membrane microcapsule system that reduced the alginate concentration in the capsule by 20% [123]. Biocompatibility problems were observed at concentrations of sodium alginate greater than 0.8%. Losses in cell viability were observed with *S. frugiperda* cells [127].

Another method of growing cells attached involves the use of man-made fibers. Kompier et al. [128] successfully grew SF21 cells in a 50 ml column packed with Fibra-cell carrier (polyester fabric laminated to a polypropylene screen, surface treated, and precoated with poly-D-lysine). TNM-FH medium supplemented with fetal calf serum was circulated through the column from a 300 ml reservoir. Oxygenation was carried out via thin-walled silicone tubing in the medium reservoir. The perfusion rate was varied between 90 and 120 ml/h (for 15–20 minutes residence time). Seeding densities were varied between 1×10^4 and 5×10^5/ml and seemed to have no effect on cell growth rate or rate of attachment. When the cell density reached 6×10^6 cells/ml packed bed (8 days postinoculation), the cells were infected with β-gal-AcMNPV at an MOI of 10. Fourteen hours postinfection, the medium in one of the two reactor systems was changed to serum-free (TNM-FH with 1% ADC-1). β-galactosidase production reached a maximum on day 13, while the maximum virus titer was measured on day 17 postinfection. Protein and virus yields were slightly less in serum-free than in complete medium. Protein yields/ml were comparable to the maximum achieved by others in attached culture [12,13].

Cell–Substratum Adhesion in Insect Cell Culture

The molecular biology of cell adhesion of insect cells of the lepidopteran lineage is not well understood. Most biotechnological processes involving anchorage-dependent lepidopteran insect cells, such as bioreactors, have employed some form of nonspecific adhesion for cell immobilization. For example, Shuler and coworkers [8,9,100] were able to grow *T. ni* 5B1-4 cells on nonporous borosilicate glass (Pyrex) supports in a packed bed column. No special treatment of the glass surfaces was required to support adhesion. Other materials that have been tested as nonspecific adhesive supports for *T. ni* cell lines include hydrophilic and hydrophobic polypropylene, polyethylene, polycarbonate, polystyrene, and boro-alumino-silicate glass. Adhesion, cell viability, and cell growth rate were superior on Pyrex and polystyrene substrates [8].

Clearly, little attention has been paid to cell adhesion, especially mediated by cell surface receptors, as a way of manipulating insect cell performance in bioreactors. There are reasons to believe that such an effort might be fruitful. The *T. ni* 5B1-4 cell line is a subclone of *T. ni* that was selected by repeated culture of the most adherent cell subpopulation. These cells are more susceptible to viral infection and produce more protein on a per cell basis, especially at low cell densities [129]. TN5B1-4 cells are more difficult to adapt to suspension culture due to cell aggregation. It may be that cellular machinery that makes the cells more adhesive might be related to cellular machinery that controls viral infection and replication and that efforts to select more adhesive cell lines might further prove fruitful in generating more productive host cell lines. There are numerous reports from mammalian systems that adhesion, especially mediated by cell surface receptors, and the structure of the cytoskeleton, are intimately related. The "stick and grip" hypothesis offered by Rees and coworkers [130] suggests that there is a dynamic, reciprocal interaction between adhesion and cytoskeleton, where receptor–ligand binding can influence cytoskeleton, and cytoskeleton can in turn influence adhesion strength. This concept has been confirmed by both experiment and substantial theoretical calculation [131,132]. This reciprocal relationship between adhesion and cytoskeleton suggests that adhesion will affect all processes that depend on the structure of the cytoskeleton.

Indeed, the cytoskeleton has been directly implicated in the ability of lepidopteran insect cells, principally *S. frugiperda* (SF21), to produce properly infectious virus. In a series of papers, Volkman and coworkers [133] have clearly established that microfilaments (f-actin) are necessary for the assembly of proper nucleocapsids that are infectious to other cells in culture; treatment of SF21 cells with cytochalasin D, a microfilament assembly inhibitor, results in the production of noninfectious nonoccluded virus with defective nucleocapsids. The role of microfilaments can be traced to nucleocapsid assembly in the nuclear region, as viral DNA replication and synthesis are unaffected [134]. Since infection is transmitted by infectious nonoccluded virus in culture, cytoskeleton may play a crucial role in sustaining long-term infectivity in culture. Microfilaments also are necessary for the formation of proper polyhedral in SF21 cells [135]. Volkman and Zaal [136] have shown that microtubules are important regulators of insect cell shape and that depolymerization of the microtubule network may be a necessary requirement for viral replication. When microtubule depolymerization was inhibited with taxol, viral replication was inhibited in both rate and extent [136]. The proper assembly of the insect cells cytoskeleton will be critical for the proper production of proteins from recombinant baculovirus; cytochalasin D treatment caused either a delay in or an inhibition of the production of several viral proteins. In particular, it caused a delay in the amplification of the production of polyhedrin by 8 hours. Since most recombinant proteins are placed upstream of the polyhedrin promoter, counting on the normal amplification of that protein to increase recombinant protein synthesis, a proper functioning cytoskeleton is quite important for a properly functioning insect cell culture.

The above argument rests on the assumption that cell–substrate adhesion in insect cell systems will show similarities to cell substrate adhesion in mammalian systems. This has already been shown in *D. melanogaster,* which has been intensely studied as a model to elucidate the relationship between cell adhesion and morphogenesis and development in simpler organisms [137]. Three of the four major families of cell adhesion molecules have been found in *Drosophila,* including the immunoglobulin family, single-chain adhesion molecules with looping repeat structures; the integrins, heterodimers that in mammalian systems link extracellular ligands to the cytoskeleton and require cations for proper function; and cadherins (transmembrane calcium-dependent homophilic adhesion molecules), which are involved in vertebrate morphogenesis [137]. A comprehensive review of cell adhesion molecules in *Drosophila* is given by Hortsch and Goodman [137]. These adhesion molecules are very similar in structure to their mammalian counterparts. Furthermore, *Drosophila* integrins such as PS2, when transfected into cells that lack this receptor, promote spreading on both vitronectin and fibronectin, as well as the RGD peptide that is found on both of those extracellular matrix molecules [138]. Therefore, in cytoskeleton, cell surface adhesion receptors, and extracellular matrix molecules, there are similarities between insect and mammalian systems, so one would expect technologies for manipulating mammalian cell culture through adhesion ultimately to be applicable to lepidopteran insect cell culture. This is an area of promise that deserves further investigation.

CONCLUSIONS

The insect cell–baculovirus expression system has great potential for producing a wide range of biological agents. Effective and economical cultivation of insect cells on a large scale is an important step toward the commercial production of biochemicals from insect cell culture. Although an increasing number of researchers are investigating engineering aspects of insect cell culture and developing novel bioreactor systems, many questions remain. These questions will have to be addressed before the full potential of the insect cell-baculovirus expression system will be realized.

ACKNOWLEDGEMENTS

The authors acknowledge the support of the National Science Foundation through grant BCS-91111091 and NSERC through a Centennial Science and Engineering postgraduate fellowship to R.A.T.

REFERENCES

1. Brown, M., and Faulkner, P. J. Invertebr. Pathol. 26:251–257, 1975.
2. Hink, W.F. In: Microbial and Viral Pesticides (Kurstak, E., ed). Marcel Dekker, New York, 1982.
3. Vaughn, J.L. J. Invertebr. Pathol. 28:233–237, 1976.
4. Goochee, C.F., and Monica, T. Bio/Technology 8:421–427, 1990.
5. Shuler, M.L., and Kargi, F. Bioprocess Engineering: Basic Concepts. Prentice-Hall, Englewood Cliffs, NJ, 1992.
6. Folkman, J., and Moscona, A. Nature 273:345, 1978.

7. Croughan, M.S., Hamel, J.-F.P., and Wang, D.I.C. Biotechnol. Bioeng. 32:975–982, 1988.
8. Shuler, M.L., Cho, T., Wickham, T., Ogonah, O., Kool, M., Hammer, D.A., Granados, R.R., and Wood, H.A. Ann. N.Y. Acad. Sci. 589:399–422, 1990.
9. Chung, I.S., Taticek, R.A., and Shuler, M.L. Biotechnol. Prog. 9:1007–1012, 1993.
10. Lanford, R.E. Virol. 167:72–81, 1988.
11. Maiorella, B., Inlow, D., Shauger, A., and Harano, D. Bio/Technol. 6:1406–1410, 1988.
12. Murhammer, D.W., and Goochee, C.F. Bio/Technology 6:1411–1418, 1988.
13. Wickham, T.J. Baculovirus–Insect Cell Interactions in Producing Heterologous Proteins: Attachment, Infection and Expression in Different Cell Lines. Ph.D. Thesis, Cornell University, Ithaca, NY, 1991.
14. Taticek, R.A., and Shuler, M.L. Unpublished results.
15. Lengyel, J., Spradling, A., and Penman, S. In: Methods in Cell Biology (Prescott, D.M., ed.). Academic Press, New York, 1975.
16. Wu, J., King, G., Daugulis, A.J., Faulkner, P., Bone, D.H., and Goosen, M.F.A. J. Ferment. Bioeng. 70:1–5, 1990.
17. Spradling, A., Spinger, R.H., Lengyel, J., and Renman, S. In: Methods in Cell Biology (Prescott, D.M., ed.). Academic Press, New York, 1975.
18. Miller, D.W., Safer, P., and Miller, L.K. In: Genetic Engineering: Principles and Methods (Setlow, J.K., and Hollaender, A., eds.). Vol. 8. Plenum Press, New York, 1986.
19. Summers, M.D., and Smith, G.E. A Manual of Methods for Baculovirus Vectors and Insect Cell Culture Procedures. Texas Agricultural Station, Bulletin No. 1555, 1987.
20. Neutra, R., Ben-Zion, L., and Shoham, Y. Appl. Microbiol. Biotechnol. 37:74–78, 1992.
21. Lazarte, J.E., Tosi, P.-F., and Nicolau, C. Biotechnol. Bioeng. 40:214–217, 1992.
22. Miyake, T., Saigo, K., Marunouchi, T., and Shiba, T. In Vitro 13:245–251, 1977.
23. Paul, J. Cell and Tissue Culture. Livingstone, London, 1970.
24. Stavroulakis, D.A., Kalogerakis, N., Behie, L.A., and Iatrou, K. Can. J. Chem. E. 69:457–464, 1991.
25. Cameron, I.R., Possee, R.D., and Bishop, D.H.L. TIBTECH 7:66–70, 1989.
26. Caron, A.W., Archambault, J., and Massie, B. Biotechnol. Bioeng. 36:1133–1140, 1990.
27. Agathos, S.N., Jeong, Y.-H., and Venkat, K. Ann. N.Y. Acad. Sci. 589:372–398, 1990.
28. Hink, W.F., and Strauss, E.M. In: Invertebrate Tissue Culture (Kurstak, E., and Maramorosch, K., eds.). Academic Press, New York, 1976.
29. McKenna, K., and Granados, R.R. Personal communication.
30. Montgomery, T. JRH Biosciences, personal communication.
31. King, G., Kuzio, J., Daugulis, A., Faulkner, P., Allen, B., Wu, J., and Goosen, M. Biotechnol. Bioeng. 38:1091–1099, 1991.
32. van Lier, F.L.J., van der Meijs, W.C.J., Grobben, N.G., Olie, R.A., Vlak, J.M., and Tramper, J. J. Biotechnol. 22:291–298, 1992.
33. Gardner, A.R., Gainer, J.L., and Kirwan, D.J. Biotechnol. Bioeng. 35:940–947, 1990.
34. Handa, A., Emery, A.N., and Spier, R.E. Develop. Biol. Standard. 66:241–253, 1987.
35. Handa-Corrigan, A., Emery, A.N., and Spier, R.E. Enzyme Microb. Technol. 11:230–235, 1989.
36. Jobses, I., Martens, D., and Tramper, J. Biotechnol. Bioeng. 37:484–490, 1991.
37. Murhammer, D.W., and Goochee, C.F. Biotechnol. Prog. 6:391–397, 1990.
38. Oh, S.K.W., Nienow, A.W., Al-Rubeai, M., and Emery, A.N. J. Biotechnol. 12:45–62, 1989.
39. Tramper, J., Smit, D., Straatman, J., and Vlak, J.M. Bioproc. Eng. 3:37–41, 1988.
40. Tramper, J., Williams, J.B., Joustra, D., and Vlak, J.M. Enzyme Microb. Technol. 8:33–36, 1986.

41. Agathos, S.N., Jeong, Y.-H., and Venkatasubramanian, K. Effects of Serum and Supplements on Growth Kinetics of Insect Cells in Culture. AIChE Annual Meeting (November 27–December 2), Washington, DC, 1988.
42. Goldblum, S., Bae, Y.-K., Hink, W.F., and Chalmers, J. Biotechnol. Prog. 6:383–390, 1990.
43. Bavarian, F., Fan, L.S., and Chalmers, J.J. Biotechnol. Prog. 7:140–150, 1991.
44. Tramper, J., Joustra, D., and Vlak, J.M. In: Plant and Animal Cells: Process Possibilities (Webb C., and Mavituna, F., eds.). Ellis Horwood, Chichester, England, 1987.
45. Wu, J., King, G., Daugulis, A.J., Faulkner, P., Bone, D.H., and Goosen, M.F.A. Appl. Microbiol. Biotechnol. 32:249–255, 1989.
46. Murhammer, D.W., and Goochee, C.F. Biotechnol. Prog. 6:142–148, 1990.
47. Hink, W.F., and Strauss, E.M. In: Invertebrate Systems In Vitro (Kurstak, E., Maramorosch, K., and Duebendorfer, A., eds.). Elsevier/North Holland Biomedical Press, Amsterdam, 1980.
48. Miltenburger, H.G., and David, P. Dev. Biol. Stand. 46:183–186, 1980.
49. Eberhard, U., and Schugerl, K. Dev. Biol. Stand. 66:324–330, 1987.
50. Graf, H., and Schugerl, K. Biotechnol. Tech. 5:91–94, 1991.
51. Inlow, D., Harano, D., and Maiorella, B. Large-Scale Insect Cell for Recombinant Protein Production. American Chemical Society National Meeting, New Orleans, LA, 1987.
52. Schopf, B., Howaldt, M.W., and Bailey, J.E. J. Biotechnol. 15:169–186, 1990.
53. Kamen, A.A., Thom, R.L., Caron, A.W., Chavarie, C., Massie, B., and Archambault, J. Biotechnol. Bioeng. 38:619–628, 1991.
54. King, G.A., Daugulis, A.J., Faulkner, P., and Goosen, M.F.A. Biotechnol. Prog. 8:567–571, 1992.
55. Streett, D.A., and Hink, W.F. J. Invertebr. Pathol. 32:112–113, 1978.
56. Boraston, R., Thompson, P.W., Garland, S., and Birch, J.R. Dev. Biol. Stand. 55:103–111, 1984.
57. Shirai, Y., Hashimoto, K., Yamaji, H., and Kawahara, H. Appl. Microbiol. Biotechnol. 29:113–118, 1988.
58. Wohlpart, D., Kirwan, D., and Garner, J. Biotechnol. Bioeng. 36:630–635, 1990.
59. Glacken, M.W., Flesichaker, R.J., and Sinskey, A.J. Trends Biotechnol. 1:102–108, 1983.
60. Taticek, R.A., Moo-Young, M., and Legge, R.L. Appl. Microbiol. Biotechnol. 33:280–286, 1990.
61. Fowler, M.W. Biotechnol. Genet. Eng. Rev. 2:41–67, 1984.
62. Weiss, S.A., Orr, T., Smith, G.C., Kalter, S.S., Vaughn, J.L., and Dougherty, E.M. Biotechnol. Bioeng. 24:1145–1154, 1982.
63. Wang, M.-Y., Vakharia, V., and Bentley, W.E. Biotechnol. Bioeng. 42:240–246, 1993.
64. Reuveny, S., Kim, Y.J., Kemp, C.W., and Shiloach, J. Appl. Microbiol. Biotechnol. 38:619–623, 1993.
65. Scott, R.I., Blanchard, J.H., and Ferguson, C.H.R. Enzyme Microb. Technol. 14:798–804, 1992.
66. Gardiner, G.R., Priston, A.J., and Stockdale, H. Studies on the Production of Baculoviruses in Insect Tissue Culture. International Colloquium on Invertebrate Pathology, Queen's University, Kingston, Canada, 1976.
67. Licari, P., and Bailey, J.E. Biotechnol. Bioeng. 37:238–246, 1991.
68. Licari, P., and Bailey, J.E. Biotechnol. Bioeng. 39:432–441, 1992.
69. Wu, S.-C., Dale, B.E., and Liao, J.C. Biotechnol. Bioeng. 41:104–110, 1993.
70. Kool, M., Voncken, J.W., van Lier, F.L.J., Tramper, J., and Vlak, J.M. Virol. 183:739–746, 1991.
71. Wickham, T.J., Davis, T., Granados, R.R., Hammer, D.A., Shuler, M.L., and Wood, H.A. Biotechnol. Lett. 13:483–488, 1991.

72. MacKinnon, E.A., Henderson, J.F., Stoltz, D.B., and Faulkner, P. J. Ultrastruct. Res. 49:419–435, 1974.

73. Kool, M., van Lier, F.L.J., Vlak, J.M., and Tramper, J. In: Agricultural Biotechnology in Focus in the Netherlands (Dekkers, J.J., van der Plas, H.C., and Vuijk, D.H., eds.). Pudoc, Wageningen, The Netherlands, 1990.

74. de Gooijer, C.D., Koken, R.H.M., van Lier, F.L.J., Kool, M., Vlak, J.M., and Tramper, J. Biotechnol. Bioeng. 40:537–548, 1992.

75. Hink, W.F., and Strauss, E.M. J. Invertebr. Pathol. 27:49–55, 1976.

76. Ramoska, W.A., and Hink, W.F. J. Invertebr. Pathol. 23:197–201, 1974.

77. Potter, K.N., Jaques, R.P. and Faulkner, P. Intervirology 9:76–85, 1978.

78. Volkman, L.E., and Summers, M.D. J. Virol. 16:1630–1637, 1975.

79. Hink, W.F., Strauss, E.M., and Ramoska, W.A. J. Invertebr. Pathol. 30:185–191, 1977.

80. Stockdale, H., and Gardiner, G.R. J. Invertebr. Pathol. 30:330–336, 1977.

81. Wood, H.A., Johnston, L.B., and Burand, J.P. Virology 119:245–254, 1982.

82. Lynn, D.E., and Hink, W.F. J. Invertebr. Pathol. 32:1–5, 1978.

83. Lindsay, D.A., and Betenbaugh, M.J. Biotechnol. Bioeng. 39:614–618, 1992.

84. Reuveny, S., Kim, Y.J., Kemp, C.W., and Shiloach, J. Biotechnol. Bioeng. 42:235–239, 1993.

85. Wang, M.-Y., Kwong, S., and Bentley, W.E. Biotechnol. Prog. 9:355–361, 1993.

86. Wickham, T.J., Shuler, M.L., Hammer, D.A., Granados, R.R., and Wood, H.A. J. Gen. Virol. 73:3185–3194, 1992.

87. de Gooijer, C.D., van Lier, F.L.J., van den End, E.J., Vlak, J.M., and Tramper, J. Appl. Microbiol. Biotechnol. 30:497–501, 1989.

88. Power, J., Greenfield, P.F., Nielsen, L., and Reid, S. Cytotechnology 9:149–155, 1992.

89. Kumar, A. A Model of a Split-Flow Air Lift Bioreactor for Attachment-Dependent Baculovirus Infected Insect Cells. M.S. Thesis, Cornell University, Ithaca, NY, 1994.

90. Forestell, S.P., Kalogerakis, N., Behie, L.A., and Gerson, D.F. Biotechnol. Bioeng. 39:305–313, 1992.

91. Forestell, S.P., Milne, B.J., Kalogerakis, N.E., and Behie, L.A. Chem. Eng. Sci. 47:2381–2386, 1992.

92. Weiss, S.A., Smith, G.C., Kalter, S.S., Vaughn, J.L., and Dougherty, E. Intervirology 15:213–222, 1981.

93. Finter, N.B. Lab. Technol. March/April:157, 1984.

94. Lubiniecki, A. In: Advances in Animal Cell Biology and Technology for Bioprocesses (Spier, R.E., Griffiths, J.B., Stephenne, J., and Crooy, J.J., eds.). Butterworths, Guildford, 1989.

95. Weiss, S.A., Belisle, B.W., DeGiovanni, A., Godwin, G., Kohler, J., and Summers, M.D. Dev. Biol. Stand. 70:271–279, 1989.

96. Weiss, S.A., Gorfien, S., Fike, R., DiSorbo, D., and Jayme, D. In: Biotechnology: The Science and the Business. Ninth Australian Biotechnology Conference, Gold Coast, Queensland, Australia. (September 24–27), 1990.

97. Weiss, S.A., Whitford, W.G., and Godwin, G. Improved Production of Recombinant Proteins in High Density Insect Cell Culture. Eighth International Conference on Invertebrate and Fish Tissue Culture, Anaheim, CA, 1991.

98. Barkhem, T., Carlsson, B., Danielsson, A., Frieberg, H., and Ohman, L. Production in a 100 L Stirred Tank Reactor of Functional Full Length Human Thyroid Receptor β_1 in Sf-9 Cells Using a Recombinant Baculovirus. Baculovirus and Recombinant Protein Production Workshop, Interlaken, Switzerland, 1992.

99. Weiss, S.A., Belisle, B.W., Chiarello, R.H., DeGiovanni, A.M., Godwin, G.P., and Kuhler, J.P. Large-Scale Cultivation of Insect Cells. Conference on Biotechnology, Biological Pesticides and

Novel Plant–Pest Resistance for Insect Pest Management. Boyce Thompson Institute for Plant Research, Ithaca, NY, 1988.

100. Chung, I.S., Shuler, M.L. Biotechnol. Lett. 15:1007–1012, 1993.

101. Jager, V., Grabenhorst, E., Kobold, A., Deutschmann, S.M., and Conradt, H.S. High Density Perfusion Culture of Insect Cells for the Production of Recombinant Glycoproteins Baculovirus and Recombinant Protein Production Workshop, Interlaken, Switzerland, 1992.

102. Cavegn, C., Bertrand, M., Payton, M.A., and Bernard, A.R. Optimization of High Density Insect Cell Cultures for the Production of Recombinant Proteins. AIChE Annual Meeting, Abstract 152h, Miami, FL, November 1–6, 1992.

103. Massie, B., Tom, R., and Caron, A.W. Scale-Up of a Baculovirus Expression System: Production of Recombinant Protein in Perfused, High-Density Sf9 Cell Cultures. Baculovirus and Recombinant Protein Production Workshop, Interlaken, Switzerland, 1992.

104. Guillaume, J.M., Couteault, N., Hurwitz, D.R., and Crespo, A. High Density Insect Cell Homogeneous Perfusion Culture for High Level Recombinant Protein Production. Baculovirus and Recombinant Protein Production Workshop, Interlaken, Switzerland, 1992.

105. Kim, H.R., Oh, J.H., Yang, J.M., Kang, S.K., Yoon, H.H., and Chung, I.S. In: Biochemical Engineering for 2001: Proceedings of Asia-Pacific Biochemical Engineering Conference 1992 (Furusaki, S., Endo, I., and Matsuno, R., eds.). Springer-Verlag, Tokyo, 1992.

106. Zhang, J., Kalogerakis, N., Behie, L.A., and Iatrou, K. Biotechnol. Bioeng. 42:357–366, 1993.

107. Kloppinger, M., Fertig, G., Fraune, E., and Miltenburger, H.G. Cytotechnol. 4:271–278, 1990.

108. Kompier, R., Tramper, J., and Vlak, J.M. Biotechnol. Lett. 10:849–854, 1988.

109. van Lier, F.L.J., van den End, E.J., de Gooijer, C.D., Vlak, J.M., and Tramper, J. Appl. Microbiol. Biotechnol. 33:43–47, 1990.

110. van Lier, F.L.J., Kool, M., van den End, E.J., de Gooiger, C.D., Usmany, M., Vlak, J.M., and Tramper, J. Ann. N.Y. Acad. Sci. 613:183–190, 1990.

111. Shapiro, M. In: Microbial and Viral Pesticides (Kurstak, E., ed.). Marcel Dekker, New York, 1982.

112. Polland, R., and Khosrovi, B. Process Biochem. May:31–37, 1978.

113. Lery, X., and Fediere, G. J. Invertebr. Pathol. 55:342–349, 1990.

114. Karel, S.F., Libicki, S.B., and Robertson, C.R. Chem. Eng. Sci. 40:1321–1354, 1985.

115. Scheirer, W. Animal Cell Immobilization. 8th International Biotechnology Symposium, Paris, France, 1988.

116. Mattiasson, B. In: Immobilized Cells and Organelles (Mattiasson, B., ed.). CRC Press, Boca Raton, FL, 1983.

117. Radovich, J.M. Enzyme Microb. Technol. 7:2–10, 1985.

118. Lazar, A., Silberstein, L., Reuveny, S., and Mizrahi, A. Dev. Biol. Stand. 66:315–323, 1987.

119. Cherry, R.S., and Papoutsakis, E.T. Biotechnol. Bioeng. 32:1001–1014, 1988.

120. Croughan, M.S., and Wang, D.I.C. Biotechnol. Bioeng. 33:731–744, 1989.

121. Weiss, S.A., Smith, G.C., Kalter, S.S., and Vaughn, J.L. In Vitro 17:495–502, 1981.

122. Wickham, T.J., and Nemerow, G.R. Biotechnol. Prog. 9:25–30, 1992.

123. King, G.A., Daugulis, A.J., Faulkner, P., Bayly, D., and Goosen, M.F.A. Biotechnol. Lett. 10:683–688, 1988.

124. Morita, K., and Igarashi, A. J. Tissue Culture Methods 12:35–37, 1989.

125. Posillico, E.G. Bio/Technology 4:114, 1986.

126. Rupp, R.G. In: Large-Scale Mammalian Cell Culture (Feder, J., and Tolbert, W.R., eds.). Academic Press, New York, 1985.

127. King, G.A., Daugulis, A.J., Goosen, M.F.A., Faulkner, P., and Bayly, D. Biotechnol. Bioeng. 34:1085–1091, 1989.

128. Kompier, R., Kislev, N., Segal, I., and Kadouri, A. Enzyme Microb. Technol. 13:822–827, 1991.
129. Wickham, T.J., Davis, T., Granados, R.R., Shuler, M.L., and Wood, H.A. Biotechnol. Prog. 8:391–396, 1992.
130. Rees, D.A., Lloyd, C.W., and Thom, D. Nature 267:124–128, 1977.
131. Lotz, M.M., Burdsal, C.A., Erickson, H.P., and McClay, D.R. J. Cell Biol. 109:1795–1805, 1989.
132. Ward, M.D., and Hammer, D.A. Biophys. J. 64:936–959, 1993.
133. Volkman, L.E., Goldsmith, P.A., and Hess, R.T. Virol. 156:32–39, 1987.
134. Volkman, L.E. Virology 163:547–553, 1988.
135. Hess, R.T., Goldsmith, P.A., and Volkman, L.E. J. Invertebr. Pathol. 53:169–182, 1989.
136. Volkman, L.E., and Zaal, K.J.M. Virology 175:292–302, 1990.
137. Hortsch, M., and Goodman, C.S. Annu. Rev. Cell Biol. 7:505–557, 1991.
138. Bunch, T.A., and Brower, D.L. Development 116:239–247, 1992.
139. Oldshue, J.Y. Biotechnol. Bioeng. 8:3, 1966.
140. Bailey, J.E., and Ollis, D.F. Biochemical Engineering Fundamentals. McGraw-Hill, New York, 1986.

9

The Effect of Hydrodynamic Forces on Insect Cells

Jeffrey J. Chalmers

Department of Chemical Engineering, The Ohio State University, Columbus, Ohio 43210

INTRODUCTION

The successful commercialization of the production of biopesticides and recombinant proteins from the insect cell–baculovirus expression system requires the successful and economical production of high-density, large-scale cultures of insect cells. By *large-scale,* I mean working volumes greater than 1 liter. This goal, as in animal cell culture, is inhibited by several problems, not the least of which is the susceptibility of the suspended cell to damage/death as a result of hydrodynamic forces, commonly called the *shear sensitivity* of the cell.

Over the last 10 years, a variety of unique, and in most cases complicated, bioreactors have been developed to avoid this problem; however, only recently have the actual forces responsible for this damage begun to be understood. With a fundamental understanding of these forces, and a corresponding understanding of how to prevent this cell damage as a result of these forces, more classical stirred tank and airlift bioreactors can be routinely used to culture insect cells. It is one of the goals of this chapter to discuss both the forces responsible and how the damage can be prevented.

The first reported case of the large-scale culture of insect cells in suspension was of the *Antherae* cell line by Vaughn [1] in 1967. Fruit flies (*Drosophila melanogaster*) and mosquitoes (*Aedes albospictus*) have also been grown in suspension culture [2,3]. More recently, Hink [4] grew *Trichoplusia ni,* TN368, in large-scale culture to determine the feasibility of producing baculoviruses for bioinsecticide purposes. In 1988, Maiorella et al. [5] reported the large-scale growth of *Spodoptera frugiperda,* SF9, cells in 21 liter airlift bioreactors to produce human macrophage colony-stimulating factor.

While, as has been reported by these authors, large-scale growth of these cells is possible, they and others have reported that without the addition of "protective"

Baculovirus Expression Systems and Biopesticides, pages 175–204

additives, or modified operating conditions, the cells can be killed by hydrodynamic forces. These results parallel observations by researchers working with animal cells.

In addition to these lethal effects, many reports exist in the biomedical engineering literature that sublethal levels of hydrodynamic forces can have significant morphological and physiological effects on cells. Whether these or similar effects are observed with suspended insect cells remains to be seen.

In this chapter, a review of the research conducted on the hydrodynamics within a bioreactor, the effects of one well-defined hydrodynamic force—laminar shear stress—on cells, cell adhesion to interfaces, and mechanisms of additive protection is given. As mentioned above, much of the work conducted on insect cells has paralleled the work conducted on animal cells, and those results will also be included. When there are actual or possible differences between insect and animal cells, the results and conclusions will be so noted.

HYDRODYNAMICS WITHIN A BIOREACTOR

Overview

The question of defining the specific hydrodynamics forces within a bioreactor, especially in light of how it affects suspended cells, is exceedingly complex due to the turbulent nature of the system. Despite this complexity, animal cells have been grown in suspension culture for over 35 years, one of the earliest reported cases being that of Elsworth et al. [6] in 1956. Contained in the early reports are accounts of how the suspended cells could be damaged, especially when sparged with air. In addition, several researchers observed that this damage could be reduced through the addition of "protective additives" [7,8]; however, not until 1968 did Kilburn and Webb [9] suggest a possible mechanism for this protection.

Despite this early work, a commonly held belief is that the mechanical mixing damaged the cells. Other mechanisms that have been suggested to cause cell damage are forces in the bubble injection region, forces resulting from bubbles rising through the medium, and forces created as the bubbles rupture at the air–medium interface. Each of these possible mechanisms is discussed below.

Mechanical Mixing

A substantial number of studies have been conducted to determine the effects of mechanical mixing, with and without sparging, on animal and insect cells [10–14]. Some of these reports claim that high agitation intensity is not a problem, while other reports claim that cell damage begins at agitation rates as low as 150 rpm. As a result of the variety of cell types, types of bioreactors, and varieties of operating conditions used, it is not possible to compare the different and seemingly contradictory observations.

In addition to damage as a result of hydrodynamic forces on suspended cells, several research groups have studied the effects of these forces on anchorage-dependent animal cells attached to microcarrier beads suspended in a well-mixed vessel. While a complete review of this work is beyond the scope of this chapter, a short overview, including its application to suspend cells, is given.

As previously mentioned, the flow field in a agitated bioreactor is turbulent. This turbulence is characterized by seemingly random variations of fluid direction and velocity, and this motion has been called *turbulent eddies*. Croughan et al. [15–17] and Cherry and Papoutsakis [18,19] suggested that, when an eddy is significantly larger than the microcarrier beads, the beads with cells attached will travel within the eddy and not be damaged by the fluid motion. However, when a fluid eddy smaller than the microcarrier interacts with the surface of the bead, large shear forces can be exerted on the cells attached to the bead. They further suggested that these shear forces are sufficient to damage and/or remove cells from the microcarrier.

In an attempt to characterize the sizes and forces associated with these eddies acting on the cells, they applied the dimensional analysis approach of Kolmogorov [20–22]. In this approach, large eddies transfer their energy to smaller eddies, and so forth, until a small eddy is reached at which the energy in the eddy is not transferred to a smaller eddy but is lost as a result of viscous dissipation and a corresponding generation of heat. Kolmogorov suggested that the smallest eddy in which viscous dissipation takes place can be defined as

$$\eta = (\upsilon^3/\epsilon)^{1/4} \tag{1}$$

$$v = (\upsilon\epsilon)^{1/4} \tag{2}$$

where η is the characteristic length of the eddy, υ is the kinematic viscosity, ϵ is the mean energy dissipation rate, and v is the velocity scale.

According to Croughan et al. [15] and Cherry and Papoutsakis [18], when the characteristic lengths of these smallest eddies, which undergo viscous dissipation (according to Kolmogorov approach), approach the sizes of the microcarriers (usually 100–200 mm), cell damage will be observed. With this line of reasoning, in several cases good agreement with experiments were obtained.

However, several words of caution are in order. First, Kolmogorov's approach is a correlation; the actual, physical existence of eddies as imagined above is still a debated topic in the field of turbulent research. Second, as pointed out by Oh et al. [23], considerable disagreement exists regarding the region in which the energy is dissipated. Consequently, the value of ϵ can vary over several orders of magnitude depending on how one determines the "mean energy dissipation rate." Third, Oh et al. [23] note the work of Yianneskis et al. [24], who used laser Doppler anemometry techniques to determine the turbulent Reynolds shear stress levels in the flow close to the Rushton turbine blades and whose observations indicate that the major Reynolds shear stress exists in the trailing vortices behind the impeller blade of the Rushton turbine. Yianneskis et al. [24] developed a relationship

$$\tau_{re} = 0.11\rho\,(\pi ND) \tag{3}$$

where τ_{re} is the Reynolds shear stress, ρ is the fluid density, N is the agitator speed, and D is the diameter of the impeller. Using this relationship, Oh et al. [23]

demonstrated that the shear stress calculated from Eq. 3 is over an order of magnitude greater than that calculated with Kolmogorov's approach.

At this point the reader should be reminded that the work discussed above was for cells attached to 100–200 mm microcarrier beads while the average diameter of a suspended insect cell ranges from 10 to 20 mm. Consequently, as both Oh et al. [23] and Kunus and Papoutsakis [25] point out, much higher levels of energy dissipation, and correspondingly much higher levels of agitation, are needed to damage cells based on Kolmogorov's approach.

Oh et al. [23] report that the growth rate, viability, and metabolic activity of three murine hydridomas is unaffected by agitation speeds from 100 to 450 rpm in bioreactors containing Rushton turbines. Only after gas sparging was introduced was damage observed, and this damage increased with increasing agitation rates.

Kunus and Papoutsakis [25] verified and extended these observations. They were able to show, again for a hybridoma cell line, that agitation rates up to 600 rpm in a Setric Genie bioreactor resulted in no cell damage as long as no gas bubble ruptures occurred at the air–medium interface. This was accomplished by removing the gas head space by completely filling the vessel with medium, thereby preventing the formation of a center, gas–liquid, vortex with the associated bubble entrainment and rupture. They further reported that for values of rpm in the range of 600–800, at which cell damage begins to be observed, the Kolmogorov characteristic eddy length is 12–15 mm, which is approximately the size of a cell. From these results they conclude that it is the rupture of bubbles at the air–medium interface, and not just the presence of air bubbles, that causes cell damage.

All of these results seem to indicate that the primary mechanism of damage of suspended animal and insect cells in sparged reactors is the result of cell–bubble interactions, unless excessively high levels of agitation are used. There exists only one reported case where the effects of mechanical mixing is separated from sparging in insect cell cultures. Tramper et al. [11] studied the effect of impeller rpm on a culture of SF9 cells in a surface aerated, 1 dm^3, Applicon bioreactor with a marine impeller operated as a CSTR. They reported that significant cell damage began to appear at an impeller speed of 510 rpm. However, it is not known whether any bubbles became entrained in the vessel and whether a vortex was formed. It is highly likely that, at the rpm they reported, both were present, which makes it impossible to separate the effect of turbulent eddies from cell–bubble interaction.

At this point, another word of caution is in order. As was mentioned previously, care must be taken in extending the results with animal cells to insect cells. As is shown later (see Comparison of Lethal Levels of Shear Stress for Different Cell Lines, below), insect cells are significantly more sensitive to laminar shear stress than many types of animal cells. Until a well-controlled study is conducted without any gas–liquid interfaces present, the level of agitation at which damage begins with insect cells is not known, and only estimates can be made.

Cell–Bubble Interactions

Just as the hydrodynamics associated with mechanical mixing is exceedingly complex, so are the hydrodynamics, and the associated forces, of gas bubbles interacting with a turbulent liquid medium. In fact, the hydrodynamics of two- and three-phase systems have been studied for many years in fluidization engineering [26,27], and it is still an area of intense research because of a lack of fundamental understanding.

As a result of this complexity, the approach that has been used by most researchers to begin to understand cell damage as a result of cell–bubble interactions has been of a correlational, phenomenological nature. This approach has lead to an understanding that a majority of the damage as a result of sparging occurs in certain regions of the bioreactor. In an attempt to understand this damage from a more fundamental basis, two research groups, Handa et al. [28,29] and Bavarian et al. [30] used microscopic, high-speed videosystems to observe these interactions. An interesting observation made by one of these groups is that cells can adsorb onto the rising bubble. These and other results will be discussed in the following sections.

While a majority of the experimental evidence seems to imply that cell damage takes place at the air–medium interface, three possible regions of damage as a result of cell–bubble interactions have been suggested: the bubble injection region, the freely rising bubble region, and the air–medium interface region [11,30,31]. In the following sections, each of these regions is discussed in light of the experimental and theoretical evidence that exists.

Bubble Injection Region. It has been suggested by several research groups that it is possible that cell damage, as a result of sparging, can take place at the point of air injection [31,32].

Probably the most compelling evidence for this hypothesis is the results of Murhammer and Goochee [32]. Using two similar airlift bioreactors under identical operating conditions, including bubble diameter, they reported large differences in the amount of cell damage. The only major difference between the two systems was the type of air sparger; the reactor that had the greater cell damage had twice the pressure drop as the system with the lower cell damage.

Murhammer and Goochee concluded that this higher pressure drop creates a more severe environment for the cells in the region of bubble formation and that this environment is sufficient to kill the cells. They then hypothesize that this higher pressure affects how the bubbles are formed and creates greater turbulence and concurrent cell death. However, they give no mechanism by which this higher pressure drop causes greater cell death. Obviously, further work is needed to confirm these observations.

Freely Rising Bubble Region. As stated above, the hydrodynamics associated with a rising bubble is very complex. Some of the variables that influence these hydrodynamics are bubble diameter, presence of surface-active agents in the liquid medium, and the number of bubbles introduced. In a conventional bioreactor with bubbles ranging in size from 1 to 5 mm in diameter, the Reynolds number can range from approximately 100 to 1,000. However, the presence of a large amount of surface-active agents and/or surfactants can create a hydrodynamic regime known as *large wake* hydrodynamics, which can further complicate the problem [33]. In this regime the high concentration of surfactants can result in the formation of a boundary layer around a rising bubble that moves to the front or leading edge of the rising bubble.

In an attempt to visualize the interactions of cells with bubbles, Handa et al. [28] developed a microscopic-movie system. They reported that at high gas flow rates cells near rising bubbles could be observed to be channeled into fast-flowing fluid elements and that these channeled flows appeared to be highly shearing. No further observations were given.

To obtain more complete observations, Bavarian et al. [30] developed a "two-dimensional" bubble column to observe cell–bubble interactions. With equipment that provided larger depth of field, field of view, and much higher shutter speeds than the system used by Handa et al., images of rising bubbles and suspended cells were obtained. Figure 1a,b shows two photographs of video images, set at a shutter speed of 1/60 of a second, of a 1.7 mm air bubble rising through a solution of SF9 insect cells. The bubble is traveling from left to right, and in Figure 1a it has just entered the left side of the photograph while in Figure 1b it is in the right half of the photograph. As a result of the low shutter speed, the bubble is blurred; however,

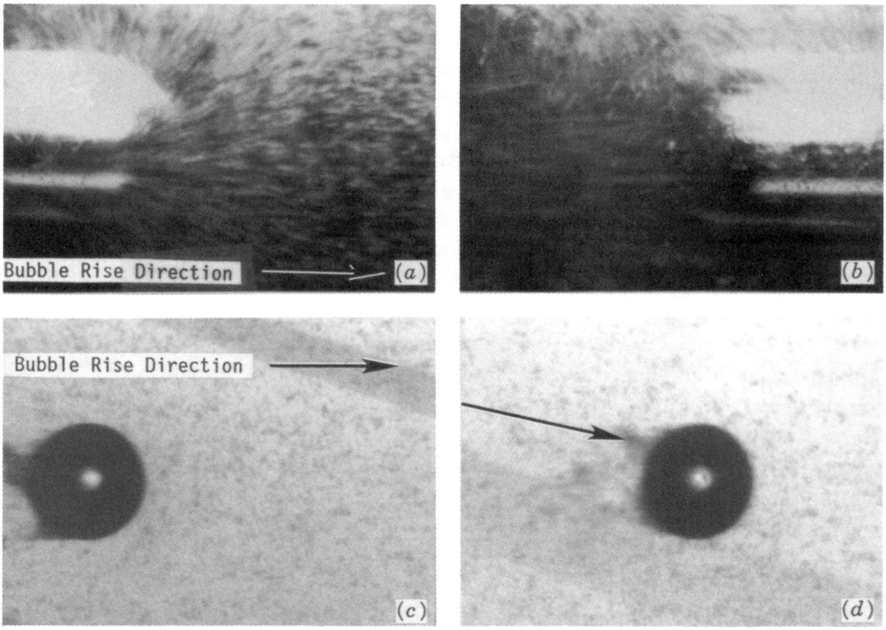

Fig. 1. Photographs of video images of a 1.7 mm air bubble rising through a suspension of TN368 insect cells in TNM-FH medium. The direction of bubble rise is from left to right in the photographs. **a,b:** The cells appear as white spots and white streaks. **c,d:** The cells are black dots. The video camera shutter speed was 1/60 of a second in a and b and 1/10,000 of a second in c and d. In c and d, cells attached to the rising bubble are indicated by the arrow.

the cells, which appear as white spots or streaks, act as tracer particles. These streaks indicate the rapid acceleration and velocities that the cells are subjected to. The vortices within which cells are trapped can be clearly observed as curved streaks of the cells in Figure 1b.

With the video camera set at a higher shutter speed (1/10,000 second), clumps of cells attached to a rising bubble can be observed (Fig. 1c,d). In these images a different lighting angle was used; consequently, the cells appear as dark spots. A significant limitations with these observations was the size of the air bubble relative to the size of a cell. A complete bubble in the field of view results in the cells only appearing as small dots. This size limitation lead Bavarian et al. [30] to use oxygen microbubbles (50–100 mm in diameter) produced through electrolysis of water. These smaller bubbles allowed the use of higher magnification to observe cell–bubble interactions. Figure 2 contains several photographs of cells attached to the rising oxygen bubbles.

These two classes of observations—1) cells subjected to rapid accelerations and shear forces in vortices and 2) cells attached to a rising bubble—suggest two possible mechanisms of cell damage. While the velocity gradients within the vortices have not been quantified, the shear stress on a cell attached to a rising bubble has been estimated. Bavarian et al. [30] reported that the shear stress on the membrane of a cell attached to a 110 mm rising bubble can approach 2.2 dynes/cm^2. This level of shear stress has been shown by Goldblum et al. [34] to result in

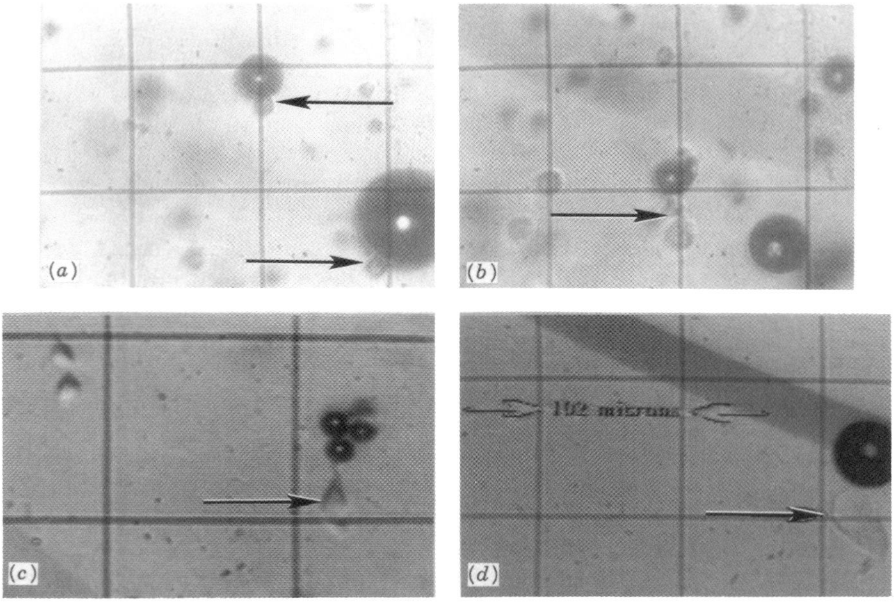

Fig. 2. Photographs of video images of cells, either SF9 or TN368 insect cells, attached to rising oxygen bubbles. The cells are indicated by arrows and appear a lighter gray than the black oxygen bubbles. **d** is twice the magnification of **a–c**.

significant cell lysis for a period of time equal to the time it takes for a bubble to rise in a normal bioreactor. However, as was stated by Bavarian et al. [30], care should be taken in comparing shear stress created in a cone and plate viscometer and the shear stress on a cell attached to a rising bubble. In addition, a majority of the bubbles in a sparged bioreactor are significantly larger (>1 mm) and the hydrodynamics, as discussed earlier, is not a simple Stokes's flow. Further work is necessary to characterize the flow and forces surrounding a rising bubble.

The Air–Medium Interface. Correlation studies of suspended cell damage in bubble columns by Handa et al. [28,29] and Tramper et al. [11,35] identified several variables that increase the rate of cell damage. Initial results seemed to indicate that decreasing bubble diameter and increasing superficial gas velocities contribute to higher cell death rates and that these death rates are first order. Further work by Tramper et al. [35] indicated that bubble diameter is not a factor and that cell death is only a function of a "hypothetical killing volume" associated with each bubble. However, the range of bubble sizes used was small; therefore, the conclusion with regard to bubble diameter is not necessarily true. Evidence is presented (see Bubble Rupture, below) that indicates that bubble diameter is probably important.

Using the assumptions of this "hypothetical killing volume" and a first-order death rate constant, Tramper et al. [35] went on to develop a model that was experimentally validated. Through the use of this model, they demonstrated that by increasing the height of a bubble column while keeping the air–medium interfacial area constant less cell damage will be obtained. In separate work, Handa-Corrigan et al. [29] suggested that cell damage only takes place in a "destructive zone" at the region of bubble disengagement at the air–medium interface. Their hypothesis is consistent with the results of Tramper's model since, as the bubble column increases in height at constant diameter, the ratio of the "destructive" air–medium interface to the medium volume decreases.

In an attempt to understand what the mechanism of destruction is in this "destructive zone" Handa et al. [28,29] used the same microscopic system mentioned earlier in the section on rising bubbles. Based on their observations they suggested the following mechanisms: 1) In medium containing antifoam, cells are damaged by oscillatory disturbances resulting from rapidly bursting bubbles; 2) In medium without antifoam cells are damaged as a result of physical shearing of the cells as they drain in the bubble lamella between and/or around the bubbles; and 3) cells can be lost in the foam.

Bavarian et al. [30], also using the microscopic-video system discussed above in the section on rising bubbles, made observations of cells and bubbles at the air–medium interface. Before bubbles were introduced into the system, cells were observed to be attached to the air–medium interface. Once oxygen bubbles had been introduced into the system, a large number of cells were seen attached to the bubbles at the interface and, after a foam layer formed, a large number of cells could be observed to be trapped within the foam layer. Figure 3 contains photo-

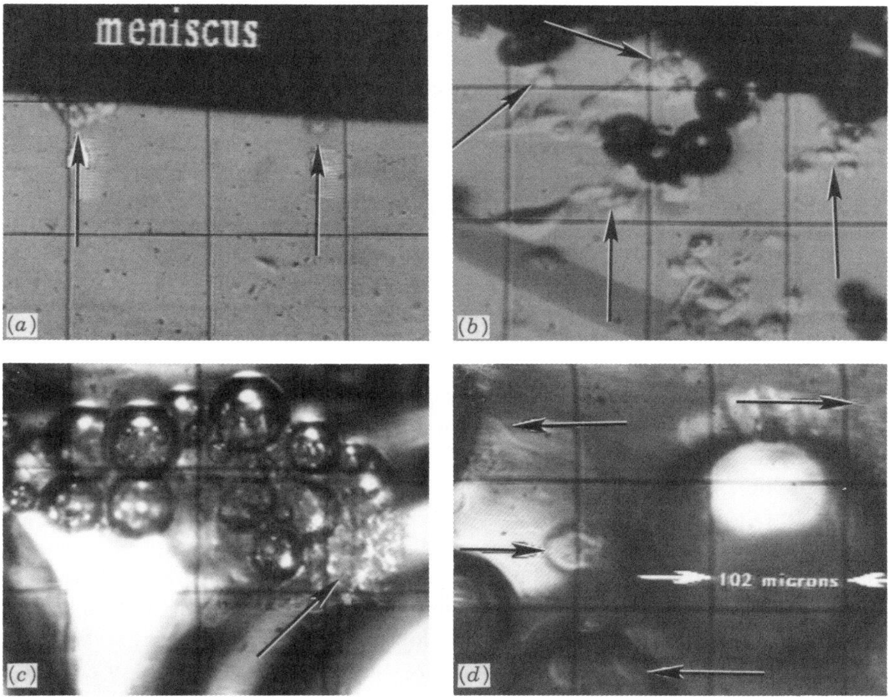

Fig. 3. Photographs of video images of TN368 insect cells interacting with the air–medium interface. **a,b:** Images of cells attached to the meniscus and bubbles at the interface. **c,d:** Images of cells trapped in the foam layer.

graphs of video images of each of these observations. Based on these and other video images, Bavarian et al. [30] suggested that cells become trapped within the foam layer either by being drawn into the foam or by being covered from underneath by rising bubbles.

In addition to the formation of a foam layer, which can trap cells, gas bubbles at an interface can rupture. Since this is a complicated and important hydrodynamic event that can damage cells, a separate section discussing this follows.

Bubble Rupture

Once an air or oxygen bubble is introduced into a bioreactor, two fates await it at the air–medium interface: it will either remain relatively stable and form a foam layer with other bubbles or it will rupture. As has been discussed previously, a decrease in the number of viable cells can result from the physical loss of cells in the foam layer. Handa-Corrigan et al. [29] suggested that the addition of protective agents, such as Pluronic F-68, create a stable foam layer that excludes cells and also prevents cells from being damaged during bubble film drainage. However, the reports of Murhammer and Goochee [14], Kunas and Papoutsakis [25], and others indicate that cells can be damaged even without the formation of a stable foam layer that indicates that bubble rupture can result in cell death.

The mechanism of bubble rupture has been studied for over 100 years, with the first work being conducted by Lord Rayleigh [36] on soap films. More recently, using various techniques including high-speed photography, several researchers have studied the process and dynamics of film and bubble rupture [37–40]. In 1972, MacIntyre [41] reviewed the process of bubble rupture and attempted to quantify the dynamics and forces involved. While the hydrodynamics are far too complicated to obtain an analytical solution, he did try to approximate the types of flow and obtain an idea of the forces involved. This process, as described by MacIntyre, is outlined below.

When a gas bubble reaches the air–medium interface, the bubble comes to rest at the interface, with a thin film cap separating the gas within the bubble from air above the medium. The surface area and shape of this bubble film cap and the shape of the bubble is a function of the size of the bubble and has been extensively studied by Medrow [42].

The length of time the bubble stays at the surface before rupture is a function of many variables, not the lest of which is the presence and type of surface active agents. This bubble film thins with time until it becomes sufficiently thin, ranging in size from 500 Å to 1 μm, that a hole develops. Once this hole develops, the bubble film begins to role up to form a toroidal ring (basically a donut shape) of constantly increasing diameter (Fig. 4).

The velocity of this expanding toroidal ring can be determined by conducting a force and energy balance. Upon rupture, the potential energy associated with the surface tension of the film cap is converted into two energies: the kinetic energy of the expanding torus and viscous dissipation as a result of inelastic acceleration of the stationary film to the velocity of the torus [43]. This expanding velocity, which is constant, can be determined from

$$v = (2\sigma/\rho h)^{1/2} \tag{4}$$

where v is the torus velocity, σ is the surface tension, ρ is the density of the liquid, and h is the film thickness. As Eq. 4 indicates, this velocity is a function of the surface tension of the liquid and the film thickness. Even though both of these values have been reported to vary over a bubble film, reasonable agreement with experimental results have been obtained [39,41,43]. For a bubble film thickness of 2 mm in seawater, this velocity has been reported to be 8 m/sec [41].

As a result of this high expanding velocity, the film outside of this rapidly expanding torus is stationary until reached by the expanding torus. As a result, the distance over which the liquid in the film is accelerated when the torus reaches it is very short. Culick [43] showed that this distance is on the order of

$$\delta \sim h\,(v\mu/\sigma) \tag{5}$$

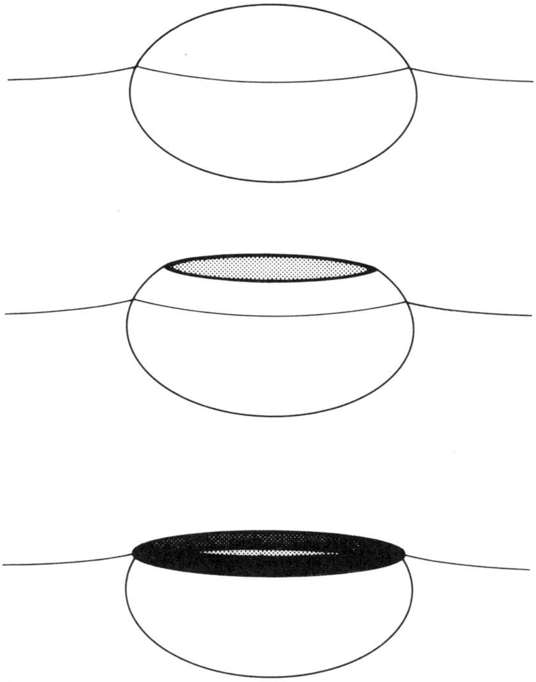

Fig. 4. A drawing of the process of bubble film rupture. Once a hole in the film develops, the film rolls up into a torus, indicated by the dark ring in the middle and bottom figures.

where δ is the acceleration distance, h is the film thickness, μ is the fluid viscosity, and σ is the surface tension. For the same bubble in seawater, the acceleration that the liquid in the film would experience is approximately 2.8×10^8 m/sec^2.

Chalmers and Bavarian [44] and Garcia-Briones and Chalmers [45], using a high shutter speed video camera and a high depth of field lens, reported that SF9 and TN368 cells, grown in TNM-FH medium without protective additives, become attached to these bubble film caps as a bubble interacts with an air–medium interface. A photograph of one of these observations is given in Figure 5a. As can be observed, a large number of cells is present on the film of the 3 mm bubble. Chalmers and Bavarian [44] further suggest that the forces associated with film rupture and the expanding torus is sufficient to kill cells. However, further work, including high-speed photographic techniques, are needed to verify this suggestion. In addition to these observations, they noted that the addition of 0.2% Pluronic F-68 to the growth medium results in the prevention of cells attaching to or being contained within the film cap (Fig. 5b). A discussion of this observation is given below (see Mechanism(s) of Protection).

MacIntyre [41] states that once the expanding torus reaches the edge of the

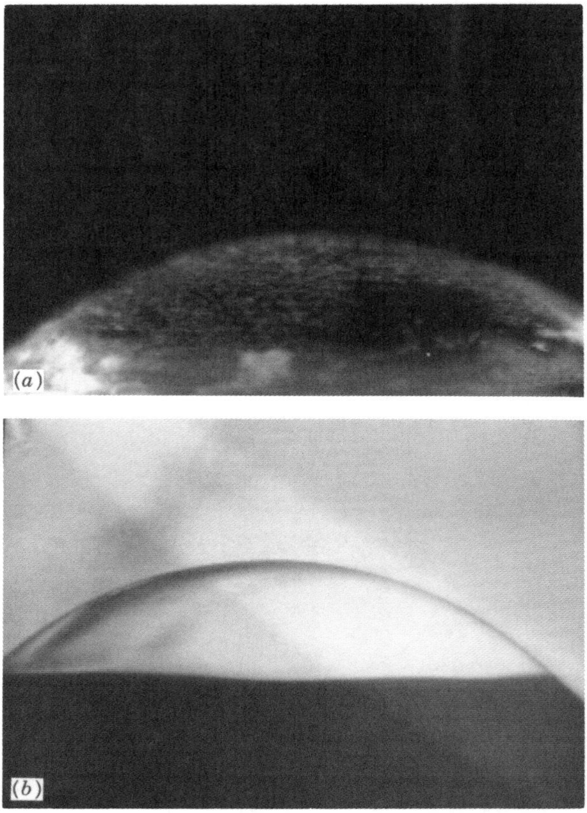

Fig. 5. Photographs of video images of 3 mm air bubbles resting at the air–medium interface in medium containing SF9 insect cells in midexponential stage of growth. In both examples TNM-FH medium containing 10% FBS was used, except in b the medium also contained 0.2% Pluronic F-68. In **a,** cells (white spots) can be seen on and in the bubble film; **b,** no cells are observed.

bubble cavity, liquid, partially originating in the bubble film, flows down the bubble cavity wall in a boundary layer type flow. Once this liquid reaches the bottom of the cavity, a stagnation point occurs and the liquid forms two jets: one that flows upward through the cavity and into the air above the interface, and one that flows downward into the liquid below the bubble cavity. MacIntyre approximated this liquid flow into the cavity as a boundary layer type flow, and Chalmers and Bavarian [44] suggested that the shear stress in this boundary layer flow, which is very high, is sufficient to damage or kill insect cells. Figure 6 is an approximate scale drawing of a 1.7 mm bubble rupturing. Each line corresponds to the successive positions of the air–liquid interface at 6.67×10^{-4} second intervals. The figure inset is a 10× enlargement of the air–liquid interface 1.0×10^{-3} seconds after the film ruptured. The dark circles correspond to 20 μm cells.

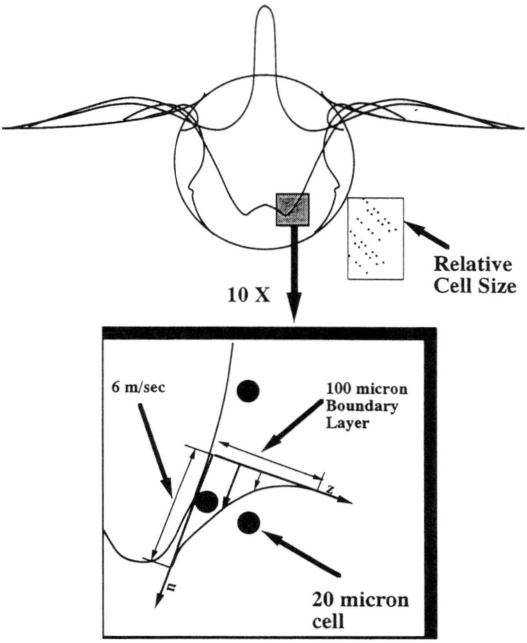

Fig. 6. A drawing of the process of a bubble rupture. The various dotted lines correspond to the air–medium interface at 6.67 × 10 second intervals.

As further evidence that this jet formation is sufficient to damage cells, Garcia-Briones and Chalmers [45] were able either to capture drops that originate in this rising jet or sample the top of the jet. They discovered that these drops contained insect cells at higher concentrations than in the bulk medium surrounding the bubble. This concentration effect was first observed by Blanchard and Syzdek [46,47] when they observed concentrations of bacteria from 10 to 1,000 times greater in the jet drops than in the bulk medium surrounding the bubble. Other researchers [48,49] concluded that this concentration effect was the result of cells, and in one case latex particles, that were adsorbing to or within a thin liquid layer surrounding the gas–liquid interface of the bubble.

Furthering the work of Garcia-Briones and Chalmers, Trinh et al. [50] quantified the number of insect cells killed per bubble rupture in well-controlled experiments. On average, 1,200 SF9 cells are killed per bubble rupture when suspended in TNM-FH medium without any protective additives. However, when 0.1% Pluronic F-68 is present, no cells are killed. They suggest that this number of cells killed per bubble rupture is sufficient to account for most if not all of the cell death observed in suspended cell culture.

Trinh et al. [50] was also able to quantify the number of dead cells in samples of the upward jet. On average 75% of the cells that are ejected in the upward jet are killed by the process, and approximately the same number of cells are killed in the

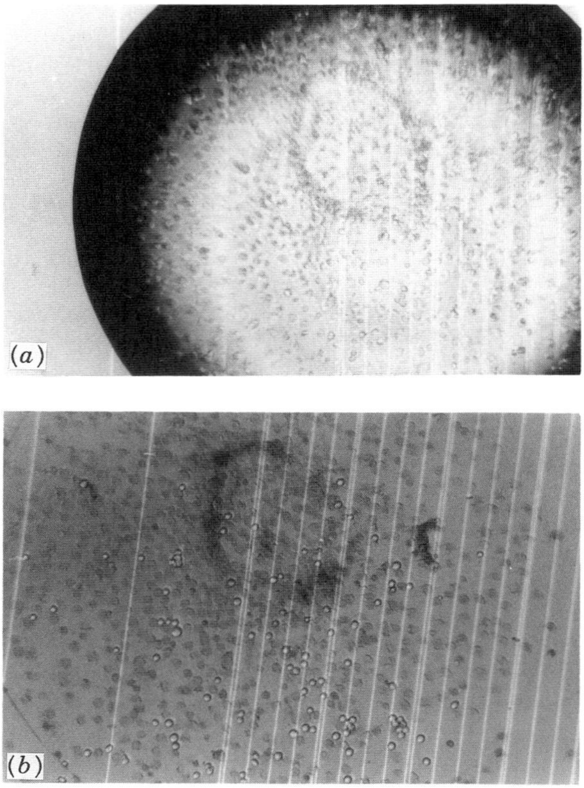

Fig. 7. Photographs of a jet drop that was collected from a bubble rupture. **a:** Immediately after the bubble was captured. **b:** After an equal volume of a Trypan blue solution was added to the drop. Cells can be observed in a, while in b a majority appear dark, which indicates that the cells are dead.

upward jet as are killed by the rupture itself. Trinh et al. suggested that, for the 3.5 mm bubbles used in their studies, there is a very weak or nonexistent downward jet; consequently, the cell death is only the result of the film rupture and upward cavity collapse. Computer simulations of bubble ruptures performed by Garcia-Briones and Chalmers [51] confirms this observation that downward jets are only present for small bubbles. These simulations also confirm that small bubbles produce much greater hydrodynamic forces than larger bubbles. The maximum shear stress for a 0.7 mm bubble rupture is 2,900 dynes/cm^2, while for a 6.5 mm bubble the maximum shear stress is 100 dynes/cm^2. These computer simulations have also been confirmed by the experimental results of both Blanchard [52] and MacIntyre [41], who reported that the acceleration of the liquid into the bubble cavity and the velocity of the upward jet increase significantly with decreasing bubble diameter.

Summary

Based on the picture that is emerging, the primary cause of cell death in sparged bioreactors is the result of either cells killed when bubbles rupture at the air–medium interface or, if large amounts of foam forms, cells trapped in the foam layer. The primary mechanism by which additives, such as Pluronic F-68, protect cells seems to be by preventing the cells from adhering to the bubble. While the hydrodynamic forces associated with bubbles rupturing is high, these forces are highly localized at the air–medium interface and when cells are prevented from adhering to that interface they are excluded from the regions of highest hydrodynamic forces. Finally, it is interesting to note that several researches have reported that small bubbles (approximately 1 mm) diameter are more damaging to cells than bigger bubbles. Based on the computer simulations of Garcia-Briones and Chalmers, this is not surprising since the hydrodynamic forces associated with a bubble rupturing increase greatly as the bubble diameter decreases.

EFFECT OF WELL-DEFINED FLOW-LAMINAR SHEAR STRESS

Overview

As stated earlier, the hydrodynamics within a bioreactor are turbulent and, by definition, very complex. Consequently, it is difficult at the current level of understanding to characterize accurately the type and magnitude of forces that a cell in a bioreactor will experience. Over the last several years a number of researchers have suggested that these forces might have nonlethal effects on the function of cells of biotechnological interest, in addition to the lethal effects described earlier. Evidence for these suggestions exists in the biomedical engineering literature, where the effect of one well-defined hydrodynamic force—laminar shear stress—on cells has been extensively studied. The two most commonly studied cell types are endothelial cells (cells that line arteries and veins) and blood cells.

Laminar shear stress, τ, is a tangential stress that results from the product of the velocity gradient times the liquid viscosity, $\acute{\eta}$:

$$\tau = \eta \, (dv/dx) \tag{6}$$

and acts on the surface in question parallel to the direction of flow.

In addition to studying nonlethal effects of hydrodynamic forces, studying cells in a well-defined hydrodynamic environment allows the determination of the "strength" or, alternately, the "sensitivity" of the cells to these forces. It also provides a means to determine if the cells' resistance to these forces can be altered through environmental or physiological means. Finally, it provides a means to compare the relative "strengths" of different cell lines and cell types.

This section contains a brief review of the nonlethal effects of hydrodynamic forces on cells of medical interest and comparison of the "strength" of different cell lines.

The Effect of Laminar Shear Stress on Cells of Medical Interest

Anchorage-Dependent Cells. A majority of the studies on the effect of laminar shear stress have been on anchorage-dependent cells, which means that the cells will only grow when attached to a surface. This allows two relatively simple apparatuses to be used to create a constant shear stress on the surface of a cell (Figure 8a,b). With these two types of devices, laminar shear stress has been shown to affect the morphology, the metabolism, the physiology, and the rates of uptake and release of compounds from these anchored cells.

Dramatic changes in the shape and orientation of endothelial cells in response to flow have been reported by several researchers [53–55]. These changes involve the change of shape from a random, cobblestone orientation to an ellipsoidal shape in the direction of flow. Depending on the level of shear stress this change in orientation can become apparent in as little as 1 hour. While their observations were not based on shape changes in response to shear stress, Folkman and Moscona [56] have reported that DNA synthesis is dependent on cell shape, and Ben-Ze'ev et al. [57] reported that shape modulation of epithelial cells and chondrocytes significantly changes the type of proteins that the cells synthesize. It is interesting to note that Diamond et al. [58] have shown that at least one protein, tissue plasminogen activator (TPA), is specifically produced by human umbilical vein endothelial (HUVE) cells in response to shear stress.

Mammalian and insect cells contain a filamentous internal network of proteins, the cytoskeleton, that is an integral part of the shape of the cell and is associated with DNA, RNA, and protein synthesis. As can be imaged, the dramatic change in shape in anchored endothelial cells also results in an alignment of the cytoskeleton in the direction of flow as well as the formation of stress fibers [59,60]. In addition, associated with the change in shape of the cell is an increase in the mechanical stiffness of the cell membrane that is believed to be related to the changes in the cytoskeleton [55].

While the structure and physiology of the cell can change in response to laminar shear stress, the rates of both the release and the uptake of compounds by a cell is also affected. The release of at least two compounds, urokinase from human embryonic kidney cells and prostacyclin from HUVE cells, has been documented [61,62]. The rate of fluid-phase endocytosis (pinocytosis) on bovine aortic cells has also been shown to be affected by shear stress [54].

While a majority of the work cited above was on endothelial cells that, in their native environment, are subjected to shear stress, it is significant that one cell type, human embryonic kidney cells, are not subjected to shear stress in their native environment. However, significant responses were observed.

Suspended Cells. A large amount of work has also been conducted on the effect of shear stress on cells suspended in blood such as blood and tumor cells. Much of this work was motivated by the desire to understand the interactions of blood-material with artificial devices. And again, it is beyond the scope of this chapter to present a

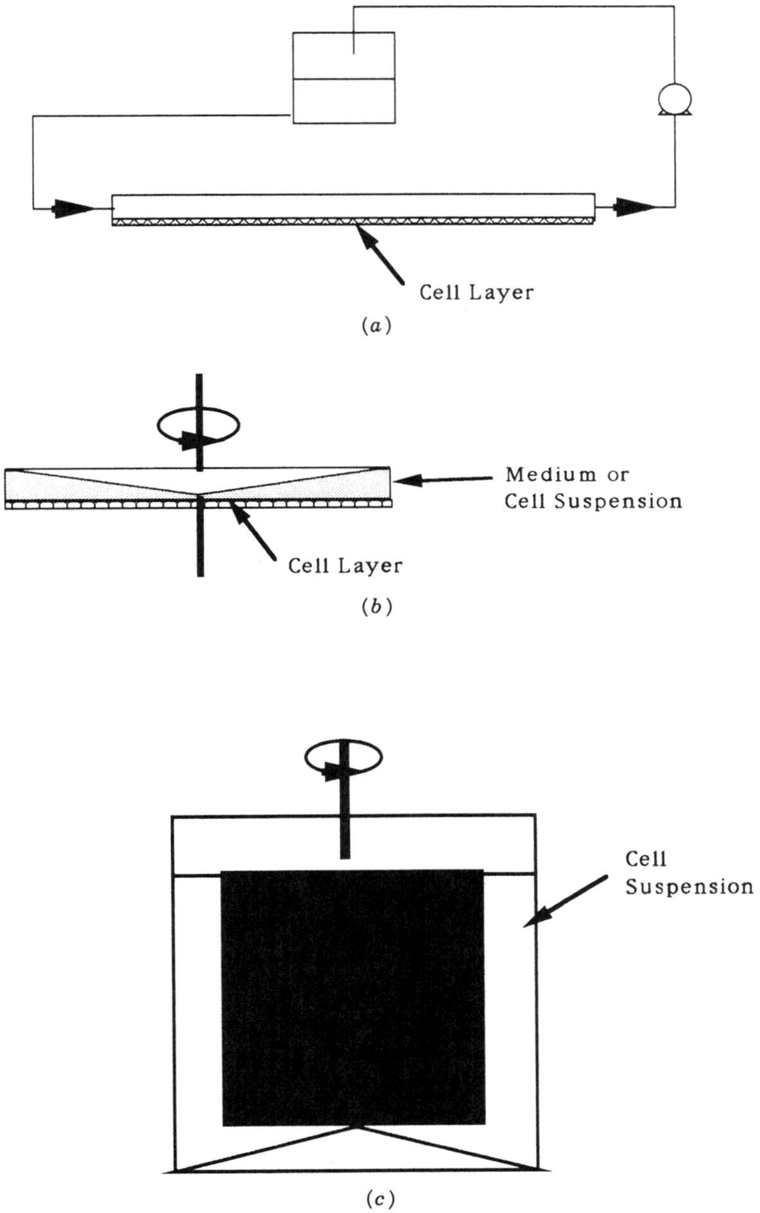

Fig. 8. Drawings of various types of systems used to subject cells to laminar shear stress. **a:** Parallel plate device used for anchorage-dependent cells. **b:** Cone and plate device that can be used for both anchorage-dependent or suspended cells. **c:** Modified Couette viscometer for suspended cells.

complete review of this work. However, shear stress on the order of magnitude that is observed in these experimental devices can result in platelet aggregation, decreased deformability of red blood cells, and decreased ability of polymorphonuclear leukocytes (PMNL) to stage a phagocytic response to a foreign organism [63–65].

Chittur et al. [66] have also studied the effect of shear stress on human T lymphocytes. They observed that high (100–200 dynes/cm^2), yet sublytic, levels of shear stress will significantly reduce the ability of these cells to proliferate and that this inhibition is concentration dependent.

While several devices have been used to subject these cells to shear stress, two of the most commonly used are depicted in Figure 8b,c. In both of these devices, the bulk shear stress is constant throughout the liquid within the vessel. However, the stress on the cell membrane is osculatory (as a result of the rotation of the cell in shear field) of different magnitude and far more complicated (see Martin et al. [65] for further explanation and quantification of this effect).

Comparison of Lethal Levels of Shear Stress for Different Cell Lines

A commonly held assumption is that insect cells are more shear sensitive than other cell lines. The problem with this assumption is, what does *shear sensitivity* mean? As was shown previously, the most probable explanation for suspended cell damage in bioreactors is that the damage results from gas bubble rupture at the air–medium interface. Since the forces that result from this rupture are so large, it is believed that any cell type, except for bacteria, will be killed in the process. Therefore, the "sensitivity" of a cell line is probably more of a function of whether a cell will attach to a rupturing bubble and not the actual "strength" of the cell. Consequently, the phrase *shear sensitivity* is somewhat misleading.

However, under well-defined flow conditions, such as constant laminar shear stress, a large variability in the magnitude of the level of shear stress needed to kill cells has been reported. Table 1 is an attempt to collect all of the reported studies on the levels of shear stress reported to have morphological, physiological, and lethal effects on different cell types. Several words of caution are needed, however. First, these reports originate in different laboratories using different types of apparatuses. Second, as discussed previously, the bulk shear stress on suspended cells is not necessarily the shear stress that the cell membrane will experience [30,65]. Third, other factors, such as medium ingredients, that can have a significant effect on the strength of the cell might not be properly accounted for in this collection (see Protective Additives, below).

When one is subjecting suspended cells to laminar shear stress three questions arise: Is it the shear rate or the shear stress that is the cause of damage? Are normal stresses of importance? What is the relationship between the magnitude of shear stress and the length of time at the given shear stress? The first question was addressed many years ago in the biomedical engineering literature and shown not to be important [67]. It is currently not known whether normal stresses are important; however, research is currently being conducted on this question. The third question, with regard to insect cells, was recently addressed by Goldblum et al. [34]. They demonstrated that the rate of cell lysis is only a function of shear stress and not a

TABLE 1. Comparison of the Effects of Shear Stress on Different Cell Types

Cell type	Method of cell growth	Magnitude of shear stress at which an effect is observed (dynes/cm^2)		References
		Nonlethal	Lethal	
Insect				
TN368	Stationary		1.2	Goldblum et al. [34]
SF9	Stationary		0.59	Goldblum et al. [34]
SF21	Suspension		10	Tramper et al. [11, 35]
Hybridoma				
CLR-8018	Suspension		10–20	Petersen et al. [69]
P-3X63-Ag8	Suspension		−10	Schurch et al. [68]
B45	Suspension		6.84	Smith et al. [70]
HDP-1	Suspension		50	Abu-Reesh and Karg [71]
Endothelial				
Bovine	Anchorage	5–10	>50	Dewey et al. [53]
Human Umbilical	Anchorage	6–24	>50	Diamond et al. [58]
Human embryonic kidney	Anchorage	2–7	26	Stathopoulos and Hellums [62]
Human blood cells				
Polymorphonuclear leukocytes	Suspension	150	50–500[a]	Martin et al. [65]
T cells	Suspension	100	>200	Chittur et al. [66]

[a]The effect is also a function of time at the different levels of shear stress.

function of time. In their case they determined the rate of lysis by calculating the percent of the original amount of intracellular lactate dehydrogenase released (an intracellular enzyme) per minute. The same type of relationship was observed by Schurch et al. [68] with hybridoma cells.

As can be observed in Table 1, a large range of lethal levels of shear stress has been observed for different cell types, and insect cells appear to be the most sensitive. This observation is not necessarily surprising, since blood cells and endothelial cells in their natural state are subjected to shear stress. However, the reported strength of human embryonic kidney cells is surprising. Based on this variability between cell types, a significant question that should be studied is what structural features provided some cell types with increased mechanical strength relative to other cell lines?

Nonlethal Effects of Shear Stress on Cells of Biotechnological Interest

The exposure of cells (bacterial, plant, insect, and animal) to stress can result in a metabolic/physiological response by the cell. An example of this type of response is the "heat shock" response in which a set of "heat shock" proteins are synthesized as a result of an increase in temperature. This response has been shown to occur in many prokaryotic and eukaryotic cell lines. Other stress that can invoke a reaction by the cell are DNA damage and energy limitation.

It has been suggested that hydrodynamic forces might elicit a stress response in suspended cells [72,73]. However, very few studies have been conducted on the nonlethal effects of shear stress or hydrodynamic forces on suspended cells of biotechnological interest, and no studies have been published on insect cells. In one of the few studies in this area, Passini and Goochee [73] studied the effect that agitation had on protein synthesis in a mouse hybridoma cell line. Without the addition of Pluronic F-68 as a protective agent, no cell growth was obtained. However, with 0.2% Pluronic F-68 added to the growth medium, the pattern of protein synthesis had no noticeable difference when compared with cells grown in stationary cultures. Private discussions with researchers in industries that deal with large-scale culturing of suspended animal cells also indicate that no noticeable metabolic or physiological effects are observed.

While these initial studies seem to indicate that there is not a "stress response" in suspended cells as a result of hydrodynamic forces within a bioreactor, one should be careful in using results obtained based on taking the "mean." Isolated, important responses could and probably do take place in the bioreactor. For example, in our laboratory we have observed significant morphological changes in TN368 insect cells, under well-defined flow conditions, at levels of shear stress and time scales that would be commonly encountered in bioreactors (unpublished results). As has been discussed above, changes in cell shape imply significant changes within the cell. Therefore, until further experiments are conducted in well-defined flow conditions, one should be careful in stating that there are no physiological or metabolic effects of agitation on suspended cells.

CELL ADHESION

Overview

While it has been observed for some time that insect cells can aggregate together to form large clumps [74], it was only recently observed that insect cells can attach to gas–liquid interfaces [30,44]. While as far as I know there exist no other reported cases of insect or animal cells adsorbing to gas–liquid interfaces, there do exist many reported cases of bacteria attaching to these interfaces. In this section, cell–cell adhesion and cell–gas adhesion are discussed.

Cell–Cell Adhesion

Again, as in other sections in this chapter, a review of cell aggregation is beyond the scope of this chapter; consequently, only cell–cell interactions with regard to insect cells is discussed. The first reported cases of insect cell aggregation was by Hink and Strauss [74]. They reported that TN368 cells grown in spinner flasks tended to clump together, and this seemed to decrease the viability of the cells. They also reported that the addition of 0.3% Dow 65HG Methocel would reduce this clumping.

In 1988, Chalmers et al. [75] reported that low levels of shear stress will result in aggregation of TN368 insect cells. Using a cone and a plate viscometer in which a viewing port had been added for microscopic observations, they were able to determine that low levels of shear stress, 0.04 dynes/cm^2 for 3 minutes will result in very large cell clumps of greater than 1,000 cells. The same sample subjected to 0.1 dynes/cm^2 for 3 additional minutes resulted in smaller clumps. Finally, after 3 minutes at 5 dynes/cm^2 no cell clumps were observed, but noticeable morphological changes, indicating cell damage, were observed. Representative photographs of cells before and after being subjected to each of these shear stresses are shown in Figure 9. These magnitudes of shear stress are levels that a cell could experience within a typical bioreactor.

Chalmers et al. [75] also reported that the addition of 0.5% Dow 65HG Methocel decreased the amount of aggregation and 0.5% Pluronic F-68 completely prevented cell aggregation in well-defined flow systems. In more recent work, Goldblum et al. [34] indicated that the amount of cell aggregation is a function of the molecular weight of the Methocel, with higher molecular weight forms of Methocel actually increasing the amount of cell aggregation relative to growth medium without Methocel present. They also reported that in medium without any additives the amount of cell aggregation was greater with TN368 than with SF9 insect cells.

The observation that the amount of cell aggregation is a function of the presence and types of surface active agent (Pluronic F-68 and Methocel) seems to indicate that this aggregation is the result of nonspecific hydrophobic–hydrophilic cell–cell interactions. This is discussed further below.

Cell–Gas Interface Adhesion

A large amount of work has been conducted on the adhesion of bacteria to surfaces and gas–liquid interfaces. According to several researchers, the most probable mechanism of this adhesion is the result of hydrophobic–hydrophilic interactions [76–78]. However, other mechanisms, such as the interactions of cells with lipids adsorbed on the air–medium interface, may also play a part [79].

To begin to quantify these interactions, Absolom et al. [80,81] defined a change in free energy, ΔF, with the process of bringing a cell and an interface together from an infinite separation. For the adhesion of a bacterial cell to a surface, this thermodynamic equation takes the form of

$$\Delta F^{adh} = \gamma_{CS} - \gamma_{CL} - \gamma_{SL} \tag{7}$$

where γ is the interfacial tension and C, S, and L correspond to the cell, solid, and liquid surface, respectively.

When ΔF^{adh} is negative, cell–solid adhesion is thermodynamically favorable.

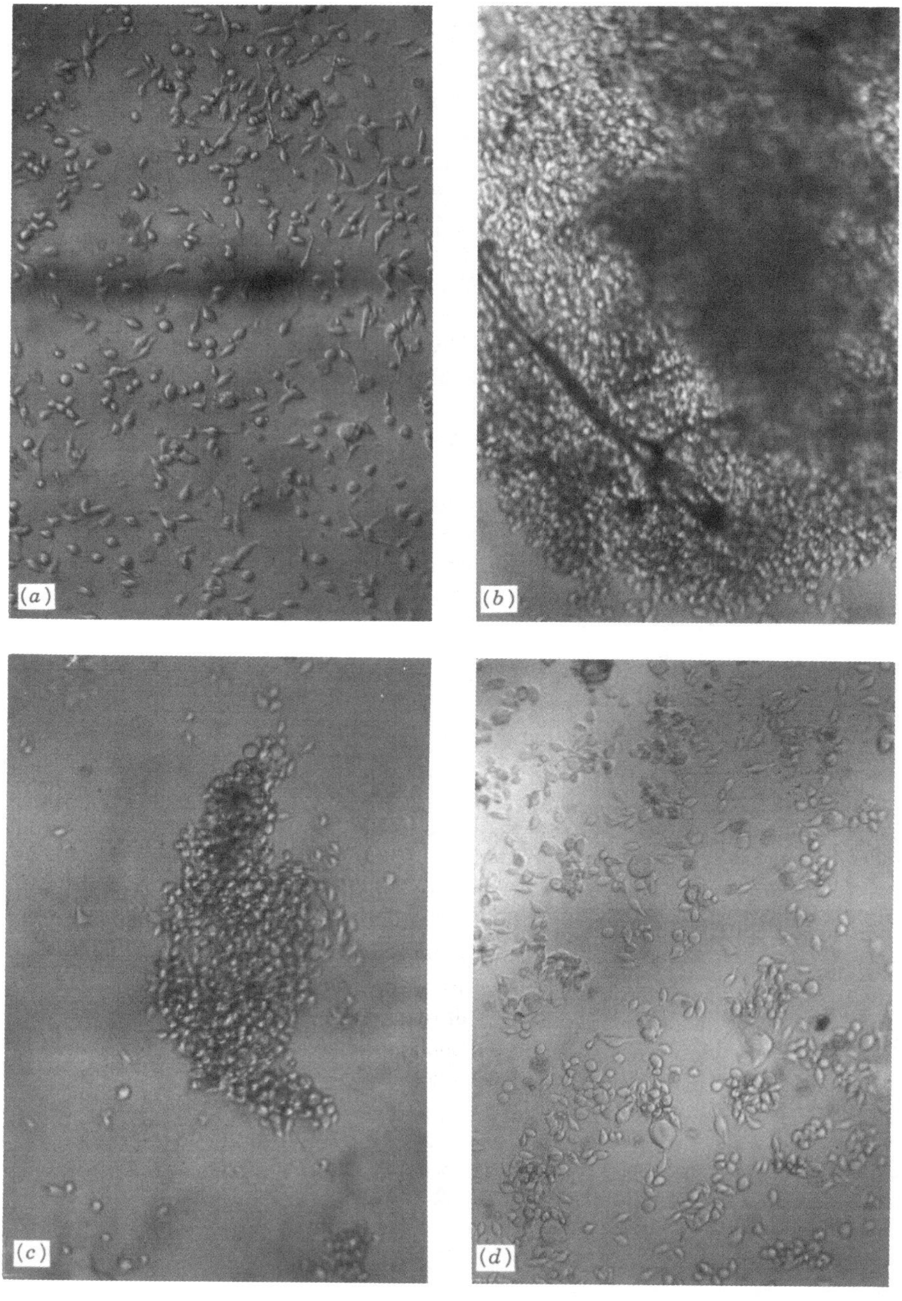

Fig. 9. The clumping of TN368 insect cells after low levels of shear stress. **a:** Photograph before the cells were subjected to shear stress. **b:** Photograph of cell clumps after being subjected to 0.04 dynes/cm^2 for 3 minutes. **c:** After 3 minutes at 0.1 dynes/cm^2. **d:** After 3 minutes at 5.0 dynes/cm^2.

While it is not possible to measure directly some of these interfacial tensions, Neumann et al. [82] developed an empirical equation of state approach that allows the determination of the interfacial tension of interest by using appropriate measurable quantities and relationships. With this approach, the surface tension of over 15 strains of bacteria and over 9 strains of mammalian cells have been determined.

Using Eq. 7, and the empirical equation of state approach discussed above, Absolom et al. [80] developed a theoretical model to predict whether a cell will adhere to a surface. This model has been experimentally verified for a variety of systems and cell types with various types of bacteria, erythrocytes, and plant cells [83–85]. In addition, this approach has been used to predict a wide range of adhesion and phagocytosis of various bacteria, of different surface tensions, by human neutrophils [81]. It has also been reported that while the relationship in Eq. 7 was written for cells adhering to a solid surface, the relationship is true for any three phases that would include gas–liquid–cell [80]. Written in terms of these three phases, one obtains

$$\Delta F^{adh} = \gamma_{CV} - \gamma_{CL} - \gamma_{VL} \tag{8}$$

where C, V, and L now correspond to cell, vapor, and liquid surface. Now, by directly changing the surface tension of the medium, γ_{LV}, it should be possible to change ΔF^{adh} directly. It is interesting to note that some of the additives that have been reported to protect animal and insect cells in suspension culture are surface active agents and have been shown to decrease the surface tension of the growth medium [28,86,87]. Therefore, it is entirely possible that the protective action of these additives is to raise ΔF^{adh} and thereby prevent cells from adhering to the gas–liquid interface. Work is being conducted in the author's laboratory to investigate this hypothesis.

PROTECTIVE ADDITIVES

Overview

Over the last 30 years, a variety of polymers have been reported to protect suspended cells from damage. Some of these polymers included methylcellulose (trade name of Methocel), increasing concentrations of serum, carboxymethylcellulose, tryptose phosphate (TPB), Primatone RL (a peptic digest of animal tissue), hydroxyethyl starch (HES), polyvinyl pyrrolidone, bovine serum albumin (BSA), Pluronic F-68, and some dextrans [74,86,88–95]. In an attempt to summarize all of these observations and results, Table 2 lists the cell type, the additive and its concentration, the method of cell growth, and the authors of the study.

Mechanism(s) of Protection

In one of the earliest reported cases of the addition of protective polymers, Runyan and Geyer [8] reported that without the addition of either Pluronic F-68 or meth-

TABLE 2. Summary of the Cell Type, Method of Cultivation, Additive and Concentration, and Studies

Cell type	Method of cultivation	Additive	Optimum concentration (%)	References
Mammalian				
L cells	Reciprocal	Pluronic F-68	0.1	Runyan and Geyer [8]
NCTC 929	Agitated	Methylcellulose	0.05	
Mouse LS	Sparged, agitated	Pluronic F-68	0.02	Kilburn and Webb [9]
Hamster kidney cells	Sparged, agitated	CMC[a]	2.4	Telling and Elsworth [92]
		TPB[b]	6.0	
Human lyphoblastoid	Agitated, sparged	CMC[a]	0.1	Mizarhi [86]
		HES[c]	0.2	
		Pluronic F-68	0.1	
Hybridoma				
Hybridoma, NS1 myeloma	Bubble column	Pluronic F-68	0.1	Handa et al. [28]
Hybridoma	Airlift	BSA	?	Heulscher and Onken [95]
Hybridoma	Agiated, sparged	Serum	>5.0	Kunas and Papoutsakis [25]
Insect				
SF9	Sparged, agitated, airlift	Pluronic F-68	0.2	Murhammer and Goochee [14]
TN368	Sparged, agitated	Methocel 65 HG	0.3	Hink and Strauss [74]
SF9	Airlift	Pluronic F-68	0.1	Maiorella et al. [5]
Other				
Dunaliella	Roux bottle miniloop	CMC[a] agar	0.1 0.045	Silva et al. [91]
Solanum dulcamara (plant cell)	Shaker flask	Pluronic F-68	0.1–1.0	King et al. [88]

[a]Carboxymethylcellulose.
[b]Tryptose phosphate broth.
[c]Hydroxyethyl starch.

ylcellulose to the growth medium L cells (NCTC clone 929) could not be grown in a reciprocally shaken, nonsparged, modified Warburg apparatus. However, no explanation was given for the mechanism of protection.

In 1968, Kilburn and Webb [9] reported that without increasing the concentration of serum from 2% to 10% or by adding 0.2% Pluronic F-68 to the growth medium LS mouse cells grew poorly in suspended, sparged bioreactors. They suggested that this protective effect resulted from the formation of a "highly condensed interfacial structure of adsorbed molecules" at the cell–liquid interface. This structure prevents the surface-active forces at the bubble interface from drastically modifying or stripping a surface layer from the surface of the cell. Unfortunately, Kilburn and Webb [9] provide no experimental evidence for this hypothesis.

More recently, several reports of the protective effect of Pluronic F-68 have been published. Handa et al. [28,29] reported that the addition of 0.1% Pluronic F-68 protected hybridoma and BHK-21 cells from damage as a result of gas sparging. They suggest that the Pluronic acts as a protective agent by creating a stable foam layer that excludes cells from the damaging effect of film drainage and/or bubble bursting.

Murhammer and Goochee [14] reported that the addition of 0.1%–0.2% Pluronic F-68 to suspension cultures of SF9 insect cells in both agitated, sparged bioreactors and in airlift bioreactors completely or almost completely protected the cells. Unlike Handa et al. [28,29], but in an argument similar to that of Kilburn and Webb [9], Murhammer and Goochee [14] suggest that the mechanism of protection is the result of interactions of the polymer with the cell membrane. They also report that Pluronic F-68 protects cells even when a stable foam layer is not present. In addition, they report that the presence of F-68 in the growth medium significantly increases the percentage of viable cells present, based on Trypan blue stain. They suggest that the most probable explanation is that when the Pluronic F-68 interacts with the cell membrane it inhibits dye uptake.

Maiorella et al. [5] also reported that the addition of 0.1% Pluronic F-68 to the growth medium will prevent bubble damage in 21 liter, airlift bioreactor fermentations of SF9 insect cells. In addition to animal and insect cells, preliminary studies have been conducted on plant cells in medium containing Pluronic F-68 [88]. However, while no inhibitory effects were reported, no protective effects were, either.

In addition to the mechanism above, several others have been proposed to explain the protective action of the various additives. These mechanisms can be classified into three groups: 1) a nutritional effect in which cells are strengthened as a result of the metabolism of the additive; 2) a protection resulting from the polymers interacting with or incorporated into the cell membrane; and 3) as was discussed above (see Cell–Gas Interface Adhesion), the additives make the adhesion of the cells to the gas–medium interface thermodynamically unfavorable and therefore prevent cell damage.

Two arguments can be made against the mechanism of protection resulting from a nutritional/metabolic effect. First, it is highly unlikely that a cell would metabolize any of the additives such as Pluronic F-68 and methylcellulose. Second, using radiolabeled CMC and HES, and other quantification techniques for the pluronics, Mizrahi [86] observed that at least 99% of the added polymers remained in the culture supernatant during the course of the experiments. However, it is possible that the polymers facilitate the delivery of membrane constituents, such as cholesterol, to the membrane. Ramirez and Mutharasan [96] reported that increases in the plasma membrane fluidity will result in increases in the cell shear sensitivity as measured in a laminar shear stress field. They further report that increasing the concentration of cholesterol in the cell membrane will lower the membrane fluidity and thereby the "shear sensitivity" of the cell.

It does seem likely that the polymers are interacting with the membrane, and this interaction is strengthening the cell. To determine if some of these additives make cells more resistant to lysis when subjected to one type of well-defined hydrodynamic force, Goldblum et al. [34] subjected TN368 and SF9 insect cells, grown in medium containing different protective polymers, to laminar shear stress. As stated earlier, the rate of lysis of the cell is a linear function of the level of shear stress. This linear relationship allowed Goldblum et al. to compare quantitatively cells grown in medium containing different protective additives. They observed that Pluronic F-68 does in fact strengthen the cells in a concentration-dependent manner, this "strengthening" ranging from a factor of 15 to 42 for concentrations of 0.2%–0.5%, respectively. Even more surprising, 0.5% of E4M Methocel strengthens the cells by a factor of 76. They also observed that this protection was obtained within 5 minutes after the additive was introduced to the medium, which also provides evidence for the fact that this protection is purely physical and not nutritional.

In contrast to the results of Goldblum et al., Papoutsakis [97] reported that Pluronic F-68 does not protect mammalian cells in viscometric studies. It is not known if this is a result of the cell type being studied, the method of the hydrodynamic force, or some other reason. However, based on these observations, and on the work of other researchers with polymers and red blood cells, it can be concluded that these polymers adsorb on the cell membrane and at least in some cases protect the cells.

As further evidence that these polymers are adsorbing and/or interacting with the cell membrane, Murhammer and Goochee [98] studied the structural features of Pluronic F-68 that are responsible for its protective effect. Pluronic F-68 is one of a number of block copolymers, called *Polyols,* of hydrophobic (polyoxypropylene, PPO) and hydrophilic (polyoxyethylene, PEO) constituents manufactured by BASF Corporation (Parsippany, NJ). There are three characteristics that describe these copolymers: 1) the ratio of PPO blocks to PEO blocks, which is defined as the hydrophilic–lipophilic balance (HLB); 2) the molecular weight of the polymer; and 3) the order of the PPO, PEO blocks. To determine which of these structural factors are responsible for cell protection, Murhammer and Goochee studied 19 different Pluronic Polyols with various HLB ratios, molecular weights, and polyol structure. The most significant observation they made with regard to these three variables is that when the HLB ratio was less than 18, which indicates that the Polyol is highly hydrophobic, the cells lysis. This indicates that the Polyol is interacting with the membrane to such an extent that the membrane's permeability is disrupted and there is leakage from the cell. When the HLB ratio is greater, such as in the case of

Pluronic F-68 (HLB >24), there is no cell lysis or inhibitory effect. To determine if the hydrophobic block is necessary for cell protection, Polyols of only polyoxyethylene (hydrophilic) were tested. In airlift bioreactors no (or very little) protection was obtained. This observation indicates that some hydrophobic characteristics, and most probably cell membrane interactions, are necessary for the Polyol's protective characteristic.

The third possible mechanism of polymer protection is probably the most significant. As was discussed in Cell–Gas Interface Adhesion, above, if cell adhesion to gas–liquid interfaces is the result of nonspecific, hydrophobic–hydrophilic interactions, then the addition of compounds that will lower the interfacial tension of the gas–liquid interface will reduce the adhesion of cells to the interface. It is interesting to note that when the hydrophobic portion of the Polyol is removed, which would greatly reduce the surface active nature of the Polyol, Murhammer and Goochee reported that little protection was obtained, giving further evidence for this mechanism.

CONCLUSION AND RECOMMENDATIONS

With the recent reports on cell damage as a result of cell–bubble interactions, a more fundamental understanding of the mechanisms of suspended cell damage in sparged bioreactors is beginning to emerge. Based on the most recent reports, the primary mechanisms for suspended insect cell damage in sparged bioreactors are cells trapped in the foam layer and cells killed in the bubble rupture process.

In vessels without gas sparging, relatively low levels of agitation (<250 rpm) apparently are not lethal to insect cells; however, care should be taken in using this value since the level of intensity of agitation is as much a function of the type of system as the magnitude of the rpm.

The addition of surface active polymers can almost completely prevent cell damage in most types of bioreactors. While the mechanisms of protection have not been conclusively proven, it appears that these polymers prevent cell–bubble interaction and thereby prevent interactions with rupturing bubbles and cell entrapment in the foam layer. Further work needs to be conducted to confirm that this is in fact the mechanism and to develop criteria for the type and amount of protective additive to use. This type of work is currently being conducted in my laboratory.

Whether sublethal levels of hydrodynamic forces affect the metabolism and/or the physiology of insect cells remains to be shown. Based on the results reported in the biomedical engineering literature, there probably is an effect. Further research is also needed in this area.

ACKNOWLEDGMENTS

I express my thanks to Dr. Fred Hink of the Department of Entomology at The Ohio State University and Dr. Robert Brodkey of the Department of Chemical Engineering at Ohio State for their comments on and corrections of the manuscript.

REFERENCES

1. Vaughn, J.L. In: Proc. Int. Colloq. Invertebr. Tissue Cult., 2nd. Tremezzo, Como., 1967.
2. Lengyel, J., Spradling, A., and Penman, S. In Prescott, D.M. (ed.): Methods in Cell Biology, Vol. 10. Academic Press, New York, 1975.
3. Spradling, A., Singer, R.H., Lengyel, J., and Penman S. In Prescott, D.M. (ed.): Methods in Cell Biology. Vol. 10. Academic Press, New York, 1975.
4. Hink, W.F. In Kurstak, E. (ed.): Microbial and Viral Pesticides. Marcel Dekker, New York, 1982.
5. Maiorella, B., Inlow, D., Shauger, A., and Harano, D. Bio/Technology 6:1406–1410, 1988.
6. Elsworth, R., Meakin, L.R.P., Pirt, S.J., and Capell, G.H.J. Appl. Bacteriol. 19:264, 1956.
7. Swim, H.E., and Parker, R.F. Proc. Soc. Exp. Biol. Med. 103:252–254, 1960.
8. Runyan, W.S., and Geyer, R.P. Proc. Soc. Exp. Biol. Med. 112:1027–1030, 1963.
9. Kilburn, D.G., and Webb, F.C. Biotechnol. Bioeng. 10:801–814, 1968.
10. Backer, M.P., Metzger, L.S., Slaber, P.L., Nevitt, K.L., and Boder, G.B. Biotechnol. Bioeng. 32:993–1000, 1988.
11. Tramper, J., Williams, J.B., and Joustra, D. Enzyme Microb. Technol. 8:33–36, 1986.
12. Dodge, T.C., and Hue, W.S. Biotechnol. Lett. 8:683–686, 1986.
13. Lee, G.M., Huard, T.K., Kaminski, M.S., and Palsson, B.O. Biotechnol. Lett. 10:625–628, 1988.
14. Murhammer, D.W., and Goochee, C.F. Bio/Technology 6:1411–1418, 1988.
15. Croughan, M.S., Hamel, J.F., and Wang, D.I.C. Biotechnol. Bioeng. 29:130–141, 1987.
16. Croughan, M.S., Hamel, J.F., and Wang, D.I.C. Biotechnol. Bioeng. 32:975–982, 1988.
17. Croughan, M.S., Sayre, E.S., and Wang, D.I.C. Biotechnol. Bioeng. 33:862–872, 1989.
18. Cherry, R.S., and Papoutsakis, E.T. Bioproc. Eng. 1:29, 1986.
19. Cherry, R.S., and Papoutsakis, E.T. Biotechnol. Bioeng. 32:1001–1014, 1988.
20. Kolmogororff, A.N. C. R. (Doklady) Acad. Sci. URSS 30:301, 1941.
21. Kolmogororff, A.N. C. R. (Doklady) Acad. Sci. URSS 31:538, 1941.
22. Kolmogororff, A.N. C.R. (Doklady) Acad. Sci. URSS 32:16, 1941.
23. Oh, S.K.W., Nienow, A.W., Al-Rubeai, M., and Emery, A.N. J. Biotechnol. 12:45–62, 1989.
24. Yianneskis, M., Popiolek, M., and Whitelaw, J.H.J. Fluid Mech. 175:537–555, 1987.
25. Kunas, K.T., and Papoutsakis, E.T. Biotechnol. Bioeng. 36:476–483, 1990.
26. Fan, L.S. Gas–Liquid–Solid Fluidization Engineering. Butterworth, London, 1989.
27. Fan, L.S., and Tsuchiya, K. Bubble Wake Dynamics in Liquids and Liquid–Solid Suspensions. Butterworth, London, 1990.
28. Handa, A., Emery, A.N., and Spier, R.E. Dev. Biol. Standard. 66:241–253, 1987.
29. Handa-Corrigan, A., Emery A.N., and Spier, R.E. Enzyme Microb. Technol. 11:230–235, 1989.
30. Bavarian, F., Fan, L.S., and Chalmers, J.J. Biotechnol. Progr. 7:140–150, 1991.
31. Tramper, J., Joustra, D., and Valk, J.M. In Webb, C., and Maaituro F. (eds.): Plant and Animal Cell Cultures: Process Possibilities, Ellis Horwood, Chicester, 1987.
32. Murhammer, D.W., and Goochee, C.F. Biotechnol. Prog. 6:391–397, 1990.
33. Andrews, G.F., Fike, R., and Wong, S. Chem. Eng. Sci. 43:1467–1477, 1988.
34. Goldblum, S., Bae, Y.K., Hink, W.F. and Chalmers, J. J. Biotechnol. Progr. 6:383–390, 1990.
35. Tramper, J., Smith, D., Straatman, J., and Valk, J.M. Bioprocess Eng. 3:37–41, 1988.
36. Rayleigh, Lord. Nature 44:249–254, 1891.
37. Woodcock, A.H., Kientzler, C.F., Arons, A.B., and Blanchard, D.C. Nature 172:1144–1145, 1953.

38. Kientzler, C.F., Arons, A.B., Blachard, D.C., and Woodcock, A.H. Tellus 6:1–7, 1954.
39. Ranz, W.E. J. Appl. Phys. 30:1950–1955, 1959.
40. Day, J.A. Q. J. R. Meteorol. Soc. 90:72–78, 1964.
41. MacIntyre, F.J. Geophysical Res. 77:5211–5228, 1972.
42. Medrow, R.A. Ph.D. Thesis. University of Illinois, 1968.
43. Culick, F.E. J. Appl. Phys. 31:1128–1130, 1960.
44. Chalmers, J.J., and Bavarian, F. Biotechnol. Progr. 7:151–158, 1991.
45. Garcia-Briones, M.A., and Chalmers, J.J. Ann. N.Y. Acad. Sci. 665:219–229, 1992.
46. Blanchard, D.C., and Syzdek, L.D. Science 170:626–628, 1970.
47. Blanchard, D.C., and Syzdek, L.D. J. Geophysical Res. 77:5087–5099, 1972.
48. Bezdek, H.F., and Carlucci, A.F. Limnol. Oceanogr. 17:566–569, 1972.
49. Quinn, J.A., Steinbrook, R.A., and Anderson, J.L. Chem. Eng. Sci. 30:1177–1184, 1975.
50. Trinh, K., Garcia-Briones, M., Hink, F., Chalmers, J.J. Biotechnol. Bioeng. 43:37–45, 1994.
51. Garcia-Briones, M., Chalmers, J. Chemical Eng. Sci. 49:2301–2320, 1994.
52. Blanchard, D.C. In Sears, M. (ed.): Progress in Oceanography, Vol. 1. Macmillan, New York, 1963.
53. Dewey, C.F., Bussolari, S.R., Gimbrone, M.A., and Davis, P.F. J. Biomech. Eng. 103:177–185, 1981.
54. Davies, P.F., Dewey, C.F., Bussolari, S.R., Gordon, E.J., and Gimbrone, M.A. J. Clin. Invest. 73:1121–1129, 1984.
55. Sato, M., Levelsque, M., and Nerum, R. Atherosclerosis 7:276–286, 1987.
56. Folkman, J., and Moscona, A. Nature 273:345–349, 1978.
57. Ben Ze'ev, A., Framer, S., and Penman, S. Cell 14:931–939, 1980.
58. Diamond, S.L., Eskin, S.G., and McIntire, L.V. Science 243:1483–1485, 1989.
59. Franke, R.P., Graefe, M., Schnittler, H., Seiffge, D., and Dreckhahn, D. Nature 307:648–649, 1984.
60. Wechezak, A.R., Viggers, R.F., and Sauvage, L.R. Lab. Invest. 53:639–647, 1984.
61. Frangos, J.A., Eskin, S., McIntire, L.V., and Ives, C. Nature 273:345–349, 1985.
62. Stathopoulos, N.A., and Hellums, J.D. Biotechnol. Bioeng. 27:1021–1026, 1985.
63. Jen, C.J., and McIntire, L.V. J. Lab. Clin. Med. 103:115–124, 1984.
64. O'Rear, E.A., Udden, M.M., McIntire, L.V., and Lynch, E.C. Biochim. Biophys. Acta 691:678–681, 1982.
65. Martin, R.R., Dewitz, T.S., and McIntire, L.V. In Hwang, N.H.C., Gross, D.R., and Patel, D.J. (eds.): Quantitative Cardiovascular Studies: Clinical and Research Applications of Engineering Principles. University Press, 1979.
66. Chittur, K.K., McIntire, L.V., and Rich R.R. Biotechnol. Progr. 4:89–96, 1988.
67. Cherry, R.S., and Papoutsakis, E.T. In Spier, R.E., and Griffiths, J.B. (eds.): Animal Cell Biotechnology. Vol. 4. Academic Press, New York, 1990.
68. Schurch, U., Kramer, H., Einsele, A., Widmer, F., and Eppenberger, H.M. J. Biotechnol. 7:179–184, 1988.
69. Petersen, J.F., McIntire, L.V., and Papoutsakis, E.T. J. Biotechnol. 7:229–246, 1988.
70. Smith, C.G., Greenfield, P.F. and Randerson, D.H. Biotechnol. Techniques 1:39–44, 1987.
71. Abu-Reesh, I., and Kargi, F. J. Biotechnol. 9:167–177, 1989.
72. Goochee, C.F., Passini, C.A. Biotechnol. Progr. 4:189–201, 1988.
73. Passini, C.A. Goochee, C.F. Biotechnol. Progr. 5:175–188, 1989.

74. Hink, W.F., and Strauss, E.M. J. Tissue Culture Methods 5:1023–1025, 1979.
75. Chalmers, J.J., Bae, Y.K., and Brodkey, R.S. AIChE Annual Meeting, Washington, D.C., 1988.
76. Kjelleberg, S. In Marshall, K.C. (ed.): Microbial Adhesion and Aggregation. Springer-Verlag, New York, 1984.
77. Kjellelberg, S. In Savage, D.C., and Fletcher, M. (eds.): Bacterial Adhesion: Mechanisms and Physiological Significance. Plenum, New York, 1985.
78. Hermansson, M., and Dahlback, B. Microb. Ecol. 9:317–328, 1983.
79. Kjelleberg, S., and Stenstrom T.A. J. Gen. Microsc. 116:417–423, 1980.
80. Absolom, D.R., Lamberti, F.V., Policova, Z., Zingg, W., van Oss, C.J., and Neumann, W. Appl. Environ. Microbiol. 46:90–97, 1983.
81. Absolom, D.R. Methods Enzymol. 132:16–95, 1986.
82. Neumann, A.W., Good, R.J., Hope, C.J., and Sejpal, M. J. Colloid Interface Sci. 49:291–304, 1974.
83. Absolom, D.R., Zingg, W., Thomson, C., Policova, Z., Van Oss, C.J., and Neumann, A.W. J. Colloid Interface Sci. 104:51–59, 1985.
84. Facchini, P.J., Radvanyi, L.G., Giguere, Y., and Dicosmo, F. Biochim Biophys. Acta. (in press).
85. Ludwicka, A., Jansen, B., Wadstrom, T., Switalski, L.M., Peters, G., and Pulverer, G. In Shalaby, S.W., Hoffman, A.S., Ratner, B.D., Horbett, T.A. (eds.): Polymers and Biomaterials. Plenum Press, New York, 1984.
86. Mizrahi, A. Dev. Biol. Standard. 55:93–102, 1984.
87. Sarkar, N. Polymer 25:481–486, 1984.
88. King, A.T., Davey, M.R., Mulligan, B.J., and Lowe, K.C. Biotechnol. Lett. 12:29–32, 1990.
89. Bryant, J.C. Ann. N.Y. Acad. Sci. 139:143–161, 1966.
90. McQueen, A., and Bailey, J. Biotechnol. Lett. 11:531–536, 1989.
91. Silva, H.J., Cortinas, T., and Ertola, R. J. Chem. Tech. Biotechnol. 40:41–49, 1987.
92. Telling, R.C., and Ellsworth, R. Biotechnol. Bioeng. 7:417–434, 1965.
93. Mizarhi, A., and Moore, G.E. Appl. Microbiol. 19:906–910, 1970.
94. Mizarhi, A. Biotechnol. Bioeng. 19:1557–1561, 1977.
95. Heulscher, M., and Onken, U. Biotechnol. Lett. 10:689–694, 1988.
96. Ramirez, O.T., and Mutharasan, R. Biotechnol. Bioeng. 36:911–920, 1990.
97. Papoutsakis, E.T., TIBECH 9:316–324, 1991.
98. Murhammer, D.W., and Goochee, C.F. Biotechnol. Progr. 6:142–148, 1990.

10

Baculovirus-Mediated Production of Proteins in Insect Cells: Examples of Scale-Up and Product Recovery

Melvin Silberklang, Kripashankar Ramasubramanyan, Sandra L. Gould, Albert B. Lenny, T. Craig Seamans, Shiping Wang, George R. Hunt, Beth Junker, Kathryn E. Mazina, Michael R. Tota, Oksana Palyha, and Deepak Jain

Enzon, Inc., Piscataway, New Jersey 08854 (M.S.); Merck Research Laboratories, Rahway, New Jersey 07065 (S.L.G., A.B.L., T.C.S., S.W., G.R.H., B.J., K.E.M., M.R.T., O.P.); Merck Manufacturing Division, West Point, Pennsylvania 19486 (K.R.); Biotechnology Division, R.W. Johnson Pharmaceutical Research Institute, Raritan, New Jersey 08869 (D.J.)

INTRODUCTION

Despite the wide acceptance of baculovirus vector–mediated recombinant gene expression in insect cells by the molecular biology community [1–5], this system has as yet seen only limited exploitation, beyond basic research, by the pharmaceutical or biotechnological industries. With a few notable exceptions, such as expression of macrophage colony-stimulating factors (M-CSF) [6] or candidate vaccines such as HIV gp160 [7–11], use of the baculovirus expression vector system, especially in the United States, has generally been limited to small-scale production of proteins for research applications. This implicit reluctance to develop the baculovirus system more extensively for the production of candidate human therapeutic agents may be attributed to two prevalent apprehensions, neither of which has yet been substantiated by experience: 1) that the less complex secretory or surface protein glycosylation of insect cells may render human proteins immunogenic to the average patient and 2) that virus removal from the final product may present serious purification difficulties. The most frequently touted exception, production of HIV gp160 as an AIDS vaccine candidate by MicroGeneSys [10],

Baculovirus Expression Systems and Biopesticides, pages 205–231

circumvents both of these objections in that gp160 is not a human, but a foreign, protein intended to be immunogenic, and its clinical use to date has been primarily in terminally ill HIV-positive adults, a unique patient cohort for whom U.S. Food and Drug Administration guidelines [12] are given their most relaxed interpretation. Nevertheless, as clinical experience with baculovirus-produced material increases, the emerging database on product safety and FDA acceptability may pave the way for wider acceptance of the system for development of human therapeutics.

At Merck, a cautious hesitation to consider the baculovirus expression system where therapeutic candidates are concerned has been no exception to this trend. Notwithstanding such discretion, when a particular peptide anticoagulant, antistasin [13], was found to express best in the baculovirus system [14], there was no reluctance, in our research laboratories, to apply a biochemical engineering perspective to developing a scalable insect cell production process for this material so as to facilitate extensive animal efficacy studies. One fortuitous unique property of this particular 119 amino acid protein, which contributed to its perceived suitability for insect cell expression, is its lack of any glycosylation. Nevertheless, the demonstration of process scalability and the experience gained in the production of antistasin became a persuasive influence toward a more general application of the baculovirus expression vector system for insect cell foreign protein expression throughout our research laboratories; this, in turn, provided us with additional opportunities to refine our insect cell culture scale-up processes. In this chapter, we discuss production of one secreted protein, antistasin, and three cell-associated proteins, herpes simplex virus VP16, human interleukin-1β converting enzyme, and hamster β_2-adrenergic receptor, representing nuclear, cytoplasmic, and membrane-associated subcellular localizations, respectively (Table 1). The last three examples underscore the general utility of baculovirus-mediated expression as a rapid and efficient means to produce the large quantities of recombinant proteins critical to modern drug discovery programs.

GENERAL ISSUES

Effective scale-up of baculovirus infection of insect cell cultures involves optimization of four processes: expansion of a viral seed stock, expansion of an insect cell seed stock, viral infection of the expanded cell stock, and product harvest. While each of these processes is necessarily closely related to the others, we have found it simpler to address each individually prior to its integration into an overall process. Some general considerations apply to production of any product and are discussed here, as well as referred to again below by specific example.

Cell Culture

We have used the *Spodoptera frugiperda* cell line SF9 [15], obtained from the American Type Culture Collection (#CRL1711), for recombinant protein expression. Cells were routinely maintained at 27°–28°C in TNM-FH medium (Grace's

TABLE 1. Data and References for the Four Recombinant-Derived Proteins Discussed

Gene	Species of origin	Localization		Molecular Weight (kd)		References
		Natural	Recombinant[a]	Natural	Recombinant[a]	
Antistasin	*Haementeria officinalis* (leech)	Secreted	Secreted	15	15	[13,35,36]
VP16	Herpes simplex virus	Viral tegument +cell nucleus	S100[b]	64	64	[46–48]
Interleukin-1β converting enzyme	Human	Cytoplasmic (S100)[b]	S100[b]	45, 20 + 10	45, 20 + 10	[44]
β_2-Adrenergic receptor	Hamster	Plasma membrane	Plasma and internal membranes	68	55	[39,40,51]

[a]Expressed in insect cells with the baculovirus expression vector system.

[b]Supernatant fraction, primarily but not exclusively cytoplasmic, obtained upon centrifugation of a cell extract at 100,000*g*.

insect medium supplemented, according to Hink [16], with 3.3 g/liter each of yeastolate and lactalbumin hydrolysate [Gibco, Grand Island, NY, Cat. #350-1605]) containing 10% fetal bovine serum (FBS), a now classic medium that is still commonly used for transfection of cells with DNA as well as for production of viral stocks. For larger scale culture and protein production, we have used either IPL-41 medium [17] supplemented with 2% FBS, 3.3 g/liter each of yeastolate and lactalbumin hydrolysate (Gibco, or Difco, Detroit, MI), and 0.1% w/v Pluronic-F68 (Gibco) or serum-free Ex-Cell 400 medium (JRH Biosciences, Lenexa, KS). Cells were adapted to high-shear suspension growth conditions by serial subculture in Techne (Princeton, NJ) stir-rod cell culture flasks, followed by Bellco (Vineland, NJ) paddle-impeller microcarrier-type spinner flasks, with the stir velocity ramped up gradually from 40 to 65 rpm. Subsequently, the cells were also adapted to serum-free medium by gradual medium dilution at each subculture. Master cell banks were prepared with cells specifically adapted to each of the three media routinely used in this work. The master cell bank of suspension-adapted serum-free cells was cryopreserved in Ex-Cell 400/20% FBS/10% dimethylsulfoxide (DMSO), and a fresh vial of cells was thawed and expanded at 2–3 month intervals to ensure uniform cell quality.

Viral Stocks

Viral stocks were expanded by infecting SF9 cells grown in spinner flask suspension at a multiplicity of ≤ 0.1 and were titered by endpoint dilution [15]. Virus can be prepared in either serum-containing or serum-free media. In our hands, the titer of recombinant baculovirus has proven somewhat superior in 10% FBS, so we have generally preferred TNM-FH/10% FBS for viral preps. Two modes of infection have been employed: 1) simple addition of titered virus to a vigorously stirred cell suspension or 2) prior 10× concentration of a portion of the cells by centrifugation and adsorption of the virus with gentle shaking (see below). The latter procedure is especially well suited to recombinant baculoviruses having reduced infectivity, but can be applied to any virus. We have found gentle viral infection conditions to be of notable benefit, for example, in the expansion of virus expressing the hamster β_2-adrenergic receptor (see below). Under scale-up production conditions, the same phenomenon of reduced viral infectivity is manifested as a requirement for a somewhat higher multiplicity of infection (MOI) to achieve optimal product titers. When this situation occurs, the choice of MOI for production runs becomes a compromise between optimal product yield and the relative difficulty of scaling up an inefficient viral seed train or of infecting the production culture with large volumes of low-titer virus stock in pursuit of this optimum.

Infection With Baculovirus and Protein Production

One of the advantages of the baculovirus expression vector system is that mildly toxic foreign proteins, which might inhibit chromosome replication or mitosis in a dividing mammalian cell, can be expressed without untoward consequences in a terminally infected insect cell. Recombinant baculoviruses remain lytic to the host cell and tend to subvert host macromolecular processes within hours after infection [18,19]. The infected insect cell should be regarded as a sick and, ultimately, a dying cell. Occasional reports of vigorous host cell growth continuing in the face of baculovirus infection [20] must be interpreted as presumptive examples of inefficient initiation of infection.

The consequences of this perspective for process development are that infection parameters, as well as maintenance of the more fragile, infected cell, can be critical optimization issues for each new process. As discussed above, it is possible that membrane-associated expressed proteins may get passively incorporated into the viral envelope and may thereby affect the efficiency of viral infection. Other expressed proteins, while not fatally toxic, may exert subtle biological effects on the host cell that influence the course of infection and protein accumulation. Both MOI and harvest time, therefore, tend to optimize somewhat differently for each recombinant baculovirus vector, and we have found it necessary to perform specific MOI and product accumulation kinetics experiments for each new protein expressed. For this purpose, reliable scale-down process models are essential.

DEVELOPING A SCALABLE PROCESS: FROM T-FLASK TO BIOREACTOR

Because we use stirred tanks for production, we have tended to use 100–500 ml paddle-impeller microcarrier-type spinner flasks as scale-down models. In a routine batch culture run, spinner flasks exposed to gas exchange with ambient air by loosening their side-arm caps will support cell growth in Ex-Cell 400 medium up to a plateau of $4–5 \times 10^6$/ml, with the exponential growth phase (essential for optimal baculovirus infection [15,21,22]) continuing through at least 2×10^6/ml. However, to avoid oxygen and/or nutrient limitation artifacts during viral infection under these conditions, either the working volume of the flask should be restricted to $^1/_2$–$^2/_3$ its nominal volume or cells should be infected at a concentration of no more than $1–1.2 \times 10^6$/ml. In our experience, the addition of forced head-space gassing [cf. 22,23] allows sufficient additional oxygenation to accommodate infection at up to $1.5–2.0 \times 10^6$/ml. In the culture media used in the reported work, infection at higher cell densities leads to reduced, or at best equal, product titers, irrespective of the adequacy of the dissolved oxygen level, presumably due to nutrient depletion (unpublished observations) [24]. In addition, because standard insect cell media are just barely rich enough to maintain adequate cell nutrition throughout a 2–4 day baculovirus infection [20,22], it is important to optimize cell seeding densities so as to avoid nutrient-depleting lag phases in cell growth. For this reason, we generally seed fermentors with $\geq 3 \times 10^5$ cells/ml. Recent results imply that newer enriched commercial serum-free media may accommodate the nutritional demands of cell growth and subsequent baculovirus infection at substantially higher batch cell density (Lenny and Silberklang, unpublished observations) [25]. Moreover, as will be discussed below, efficient infection at up to 10^7 cells/ml is possible with a perfusion bioreactor system, where nutrients are constantly replenished.

When we first attempted baculovirus production scale-up 5 years ago, it was not clear whether direct sparging would be acceptable to maintain dissolved oxygen levels or whether serum-free media would perform as well as serum-containing media. Much effort was devoted to validating tank and media performance [26,27]. An example of relative cell growth in a head-space gassed 250 ml spinner flask as compared with two stirred tanks (in IPL-41/2% FBS medium) is illustrated in Figure 1; one tank was oxygenated through bubble-free silicone rubber tubing and the other by intermittent oxygen microsparging. Note that, in the presence of 0.1% w/v Pluronic-F68 [6,28], direct gas sparging does not appear to be harmful to the cells, and cell growth is essentially identical to growth in the absence of sparging (i.e., using silicone tubing). Actual specific growth rates measured in this experiment were between 0.024 and 0.026 h^{-1} (27–28 hour doubling time). With hindsight, we now appreciate that a key factor ensuring such process tolerance was use of a cell population fully suspension and shear adapted and fully adapted to the particular media used. This required passaging the cells serially for up to 6–8 weeks in each medium in spinner flasks at ≥60 rpm prior to cryopreserving master cell banks. Figure 2 illustrates serum-free cell growth curves in a spinner flask, an 18 liter microsparged Sulzer-MBR (now merged with B. Braun, Allentown, PA) Spinferm

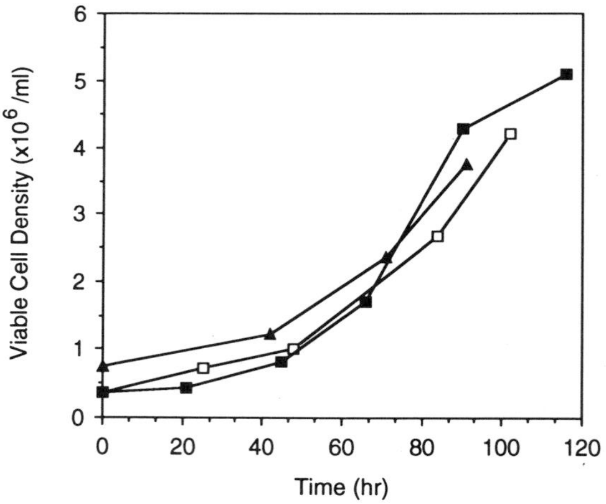

Fig. 1. SF9 cell growth under different gassing systems. SF9 cells were grown in batch culture in IPL-41/2%FBS/0.1% Pluronic-F68 in a 250 ml spinner flask operated with enriched oxygen continuous forced head-space gassing (■), a Braun 8 liter Biostat E fermentor fitted with coils of oxygen-permeable silicone rubber tubing for bubble-free gassing (▲), and an 18 liter Sulzer-MBR Spinferm fermentor operated under intermittent direct oxygen sparging through a 2 μm pore stainless steel microsparger (□). The level of DO in the spinner flask was controlled by manual adjustment of gas composition and/or flow rate as needed, based on off-line measurement, to maintain between 50% and 100% air saturation; the fermentors were automatically controlled at 65% DO based on Ingold probe measurement. Cell viability remained above 95% throughout in all cases. (Reproduced from Jain et al. [26], with permission of the publisher.)

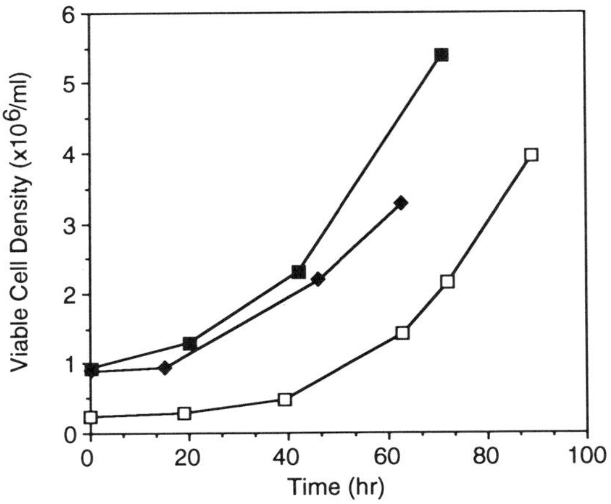

Fig. 2. SF9 cell growth in fermentors of different scales. Cells were grown in serum-free Ex-Cell 400 medium in a head-space gassed 500 ml spinner flask (□), a microsparged (oxygen) 18 liter Sulzer-MBR Spinferm fermentor (■), and a 70 liter direct open pipe air-sparged Chemap fermentor (◆). (Reprinted from Jain et al. [27], with permission of the publisher.)

fermentor and a 70 liter open-pipe (5 mm diameter) sparged Chemap fermentor equipped with hydrofoil impellers. It can be seen that growth is also similar across the three systems, the lag in spinner-flask growth being attributable to a lower seeding density.

Harvest of a recombinant baculovirus culture is ideally based on the attainment of peak product titer. As off-line product assays can be quite time consuming, however, it is useful to correlate a surrogate indicator with peak titer at the spinner flask scale and then determine the harvest time at full scale by monitoring the surrogate indicator. We have generally relied on either percent viability or modal cell diameter as markers for the progress of baculovirus infections (Fig. 3) (Gould et al., unpublished observations) [29]. Modal cell diameter can be monitored with a Coulter Counter (Coulter Instruments, Hialeah, FL) and is found to increase by 30%–40% during the course of a typical infection, although different recombinant viruses may differ in this parameter. Rapid cell death ensues shortly after peak modal diameter is reached. In our experience, most product titers peak when the infected culture average viability has dropped to between 75% and 50%, which occurs 12–24 hours after peak modal cell diameter (Fig. 3). The postinfection kinetics to reach these points its both product and MOI dependent, however, and is modulated by culture conditions, so it must be determined specifically for each process.

As a consequence of the fact that the peak titer for most recombinant proteins is reached when the infected cell viability has declined to approximately 50%, expressed proteins tend to be exposed to the lysosomal and other proteases released by the lysing cells. Product degradation is a constant danger, and earlier harvest times

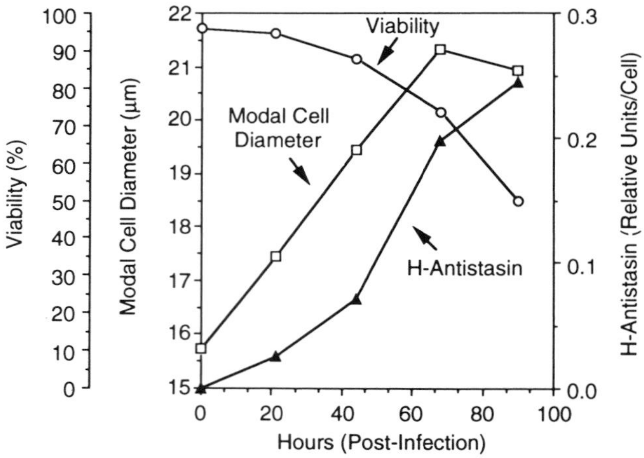

Fig. 3. Relationship between modal cell diameter, percent viability, and production. Changes in modal cell diameter (□) and viability (○) during production of half-antistasin (▲) by a recombinant baculovirus-infected (MOI = 0.1) SF9 batch culture in the 18 liter Sulzer-MBR Spinferm fermentor. This experiment was conducted in IPL-41/2% FBS/0.1% Pluronic-F68 medium. (Reprinted from Jain et al. [26], with permission of the publisher.)

at higher cell viability are sometimes chosen to minimize this problem. Alternatively (or additionally), protease inhibitors may be added at or near the time of harvest. This strategy is illustrated below for the harvests of β_2-adrenergic receptor and IL-1β converting enzyme.

CELL AND VIRAL SEED TRAINS

Insect cells were adapted fully to the culture medium and culture modality for which they were intended (e.g., scale-up in serum-free high-shear [fermentor] suspension) and were then cryopreserved as a master cell bank at 10^7 cells/ml in 1 ml vials and stored in the vapor phase of liquid nitrogen. A thawed vial was diluted into 15–30 ml fresh medium and used to seed either a T-75 flask or a Techne spinner flask culture (30 ml culture in a 100 ml capacity flask), and, in general, seed cultures were divided into multiple scale-up runs. For critical runs, a separate cell seed train from a single vial was established for each fermentation. Optimal growth from a vial was achieved when the cells were centrifuged after initial dilution and resuspended in fresh medium to remove traces of DMSO, but satisfactory results could be obtained without this precaution.

In our hands, SF9 suspension doubling times at 27°–28°C in either IPL-41/2% FBS or Ex-Cell 400 varied with medium lot in the range of 26–48 hours. We attempted to pre-screen commercial media lots for growth promotion so as to obtain more consistent growth kinetics, but, due to limited availability (especially of serum-free formulations), we were not always successful in doing so. To bring a 1 ml vial containing 10^7 cells up to 20 liters at $\geq 3 \times 10^5$/ml (i.e., approximately 7×10^9 cells) therefore required from 10 to 20 days. Bellco paddle-impeller microcarrier spinner flasks in various sizes (100, 250, and 500 ml and 1, 3, 8, and 36 liters) were used throughout. Larger flasks (3 liters and above) were filled to only approximately one-half to one-third of their rated volumetric capacity in order to allow for adequate gas exchange [30]. Cells were split when they reached 1.0 to 1.5×10^6/ml so that they would always be in vigorous exponential growth. Gas exchange with the ambient atmosphere was passive, at the seed train stage, through loosened side-arm caps. Controlled head-space gassing to accommodate higher cell densities was introduced only in production modeling experiments, or in 20 liter spinner flask cultures as an adjunct to baculovirus infection.

Viral seed trains were generally brought up in SF9 suspension cultures grown in TNM-FH/10% FBS. As discussed above, we found this medium to generate slightly higher titers of recombinant viral vectors, as well as to provide excellent long-term viral stability at 4°C. Virus recovered at passage 1 or 2 was cryopreserved as a master viral seed in culture supernatant at −85°C. Each passage (P-2 to P-5) beyond the first T-25 flask (P-1) was accomplished by infecting cells at 10^6/ml with an MOI of ≤ 0.1. This ensured adequate cell nutrition during an immediate postinfection partial population doubling and the subsequent viral infection process. Cultures were generally harvested at 72–96 hours postinfection or when cell viability had dropped below 50%, and recovered virus was titered by endpoint dilution. Cell-free supernatants containing virus were stored at 4°C and were used for up to 12 months. Titers of recombinant virus were found to be partially dependent on the recombinant gene they expressed, presumably due to physiological effects of the expressed proteins, and ranged between 3×10^7/ml and 5×10^8/ml for the viral vectors discussed below. The viral stocks used for scale-up were generally at P-3 to P-5; higher passages were avoided because of the risk of accumulation of defective interfering viral particles [31,32].

One special case worth noting, as discussed above, is that of the β_2-adrenergic receptor (β_2AR) baculovirus vectors. The standard infection conditions described, where virus at an MOI of 0.1 was added to a spinner flask culture of cells at 10^6/ml and stirred at 60 rpm, was found to yield viral titers that were two- to fivefold lower than titers obtained in static T-flask cultures (unpublished observation). This might have been due either to a lower viral infectivity (e.g., poor attachment to cellular receptors, which one might speculate was due to passive incorporation of β_2AR into the viral envelope) or to fragility of the extracellular virus under stirred conditions. To distinguish these two possibilities, cells were infected by transiently putting them into stationary culture. The cells were resuspended after 1 hour, and the infection was allowed to run its course. It was found that titers, under these conditions, matched or exceeded titers obtained in stationary cultures. It became clear that, once produced, the initial burst of virus was of sufficient multiplicity and sufficiently stable to stirring to infect the remaining cells, but that the primary infection was sensitive to stirring. To address this problem in viral seed trains of this vector, we devised a more gentle primary infection process, based on that of Summers and

Smith [15]. Suspension cultures were infected with recombinant β_2AR virus at an effective MOI of $\leq$0.1 by preadsorption of virus onto a $\leq$ 1/10 aliquot of the culture as follows. When the suspension culture reached a density of 1×10^6 cells/ml, one-tenth of the total volume (or proportionally less if the MOI was to be less than 0.1) was removed and centrifuged very gently at 100*g* for 5–10 minutes at room temperature. After centrifugation, the supernatant was discarded and the cell pellet was gently resuspended in one-tenth of its original volume of fresh medium to achieve a cell density of 1×10^7/ml. An appropriate volume of viral stock was then added to this concentrated suspension for an MOI of 1.0. The cells and virus were mixed gently and then allowed to incubate at 27°–28°C for 1 hour, with occasional gentle shaking to prevent settling of the cells. At the end of 1 hour, the infected cells were diluted with fresh medium back to the original volume that had been removed from the suspension culture and then returned to the main culture. This resulted in an effective MOI in the main culture of $\leq$0.1. Among 20 independent infections of suspension cultures with one specific recombinant virus stock expressing β_2AR, 10 that were performed by direct addition of the virus to the culture resulted in viral titers of from 3.1×10^6 to 1.4×10^7 pfu/ml, while 10 subsequent infections using the preadsorption method resulted in viral titers of between 4.0×10^7 and 1.2×10^8 pfu/ml. It should be noted that viral titration by the simple endpoint dilution method [15] is subject to a coefficient of variation of up to 45% [33], so the above values must be considered approximate.

BATCH CULTURE PRODUCTION: EXAMPLES

Protein production with the baculovirus expression vector system is easily scaled up as a batch culture process, keeping in mind that cell density must be kept low in order to avoid nutrient limitation. We have initiated all new processes in batch mode and have accumulated the most experience with this paradigm. Three examples are illustrated below: processes for antistasin and half-antistasin, β-adrenergic receptor, and interleukin-1β converting enzyme.

Antistasin and Half-Antistasin

Antistasin is a 15 kd anticoagulant protein found in the salivary gland of the Mexican leech *Haementeria officinalis,* which has been shown to be a highly selective factor Xa inhibitor [34,35]. The mature protein has 119 amino acid residues, 20 of which are cysteines, and cDNA sequence analysis [13] has revealed an additional 17 amino acid signal peptide, which is cleaved post-translationally. The natural protein exists in several allelic sequence variations in a wild population of leeches [13,35]. Both peptide and cDNA sequence analyses showed a distinct twofold internal sequence homology between the N- and C-terminal halves of the molecule, with the factor Xa-reactive site residing in the N-terminal half [36]. We have expressed two variants of full-length antistasin in the baculovirus system [13,14] and the best yielding one was chosen for scale-up. We have also expressed the

N-terminal half-domain, which we called "half-antistasin," independently in the baculovirus system [37] and scaled it up as well. The signal peptide cleavage of antistasin appears to be accomplished with good fidelity by the insect cells (J. Jacobs, personal communication), and essentially all of the product appears to be secreted. Both full and half forms of the protein were produced in batch processes. The two processes scaled very similarly and are used interchangeably as examples below.

In our first process-oriented experiments with a full-length antistasin AcMNPV expression vector, we found, in static cultures infected at a cell density of 1.6×10^6/ml, that product yield was maximal at a low MOI of 0.1 (Fig. 4a). This trend was subsequently reproduced, albeit less dramatically, with a baculovirus vector expressing a second natural variant of antistasin [15], which was chosen for scale-up because of its higher yield (not shown). The use of a low MOI also proved scalable in spinner flasks. We determined that incomplete infection of the whole cell population, due to the low MOI, was allowing a partial population doubling to occur before the first burst of released virus infected the rest of the culture; this resulted in a somewhat higher average cell density during the productive phase of the infected culture (Fig. 4b). It therefore came as a surprise that half-antistasin production proved relatively independent of MOI when infected at a similar cell density (Fig. 4c). Nevertheless, to conserve viral stocks, all further work with either virus was done at an MOI of 0.1.

Scale-up from spinner flasks was tried in two commercial fermentors, an 8 liter "BioStat E" fermenter from B. Braun (Allentown, PA), utilizing a bubble-free silicone rubber tube gassing system of (cf. Fig. 1), and an 18 liter Sulzer-MBR Spinferm fermenter using a 2 μm (pore size) stainless steel microsparger for direct sparging of oxygen; in each, the level of dissolved oxygen (DO) was monitored and controlled based on Ingold probes. Both fermentors were compared to 250 ml Bellco microcarrier spinner flasks that were head-space gassed at approximately 0.1 vol/vol/min.; in this case, air was used as the starting gas, followed by a mixture of air and oxygen adjusted to maintain the DO level at ≥50% air saturation as measured off-line by a Radiometer (Copenhagen, Denmark) blood–gas analyzer. Figure 5 compares the production of half-antistasin in these formats. Cells were grown to 1.5×10^6/ml and infected at an MOI of 0.1. It is apparent that the process scaled well in either tank format. The slight advantage of the spinner flask over the fermenters correlated with a slightly higher cell density and was not found to be significant in repeat runs (data not shown). In a separate experiment, three different DO levels were compared sequentially in the Braun 8 liter fermentor, 10%, 65%, and 110%; 65% DO was found best [26] and was adopted for all further batch culture work. The scalability of this process up to 70 liters is illustrated in Figure 6. Note that the higher relative activity of antistasin (Fig. 6) produced compared with half-antistasin (Fig. 5) is due to antistasin's intrinsically higher specific activity [37].

When cell viability declined to approximately 50% (cf. Fig. 3), the tanks were harvested. Removal of cells was either by centrifugation, for smaller volumes, or by trangential flow filtration in a Prostak unit (Millipore, Bedford, MA) for larger

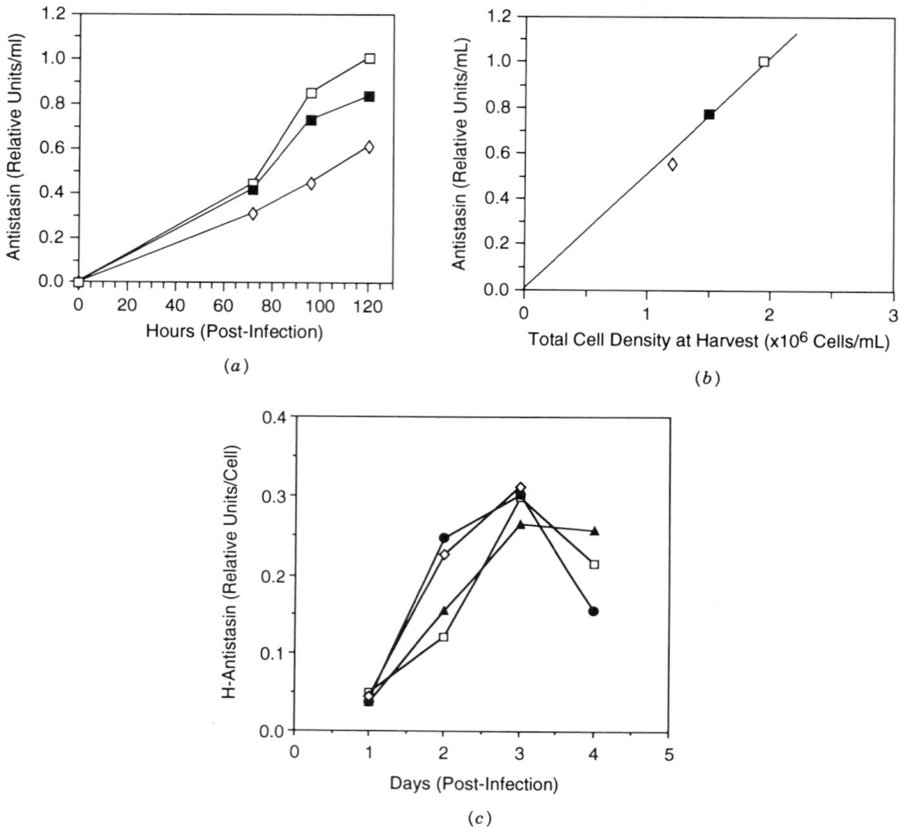

Fig. 4. **a:** The effect of MOI on antistasin production. Cells were grown in suspension in 250 ml spinner flasks in TNM-FH/10% FBS medium, harvested, and infected with recombinant baculovirus, expressing a natural variant of antistasin, at various MOIs in one-tenth their original volume according to Summers and Smith [15]; after 1 hour, the cells were diluted with fresh medium to 1.6×10^6/ml, and the infection was allowed to proceed in static culture in six-well cluster plates. MOIs shown are 5 (◇), 1 (■), and 0.1 (□). **b:** The effect of MOI on total cell density compared with antistasin expression. Cells were grown in suspension in TNM-FH/10% FBS in a 20 liter spinner flask culture. When they reached a density of 10^6/ml, aliquots were withdrawn and 100 ml satellite cultures were set up and infected at 1.2×10^6 cells/ml by direct inoculation of virus at different multiplicities. MOIs shown are 5 (◇), 1 (■), and 0.1 (□). **c:** The effect of MOI on half-antistasin production. Cells were grown in 250 ml spinner flasks in IPL-41/2%FBS/0.1% Pluronic-F68 medium to a density of 1.5×10^6/ml and infected by direct inoculation of specific volumes of virus seed chosen to achieve different nominal multiplicities. MOIs shown are 5 (◇), 2 (●), 0.5 (▲), and 0.1 (□). (Reprinted from Jain et al. [26], with permission of the publisher.)

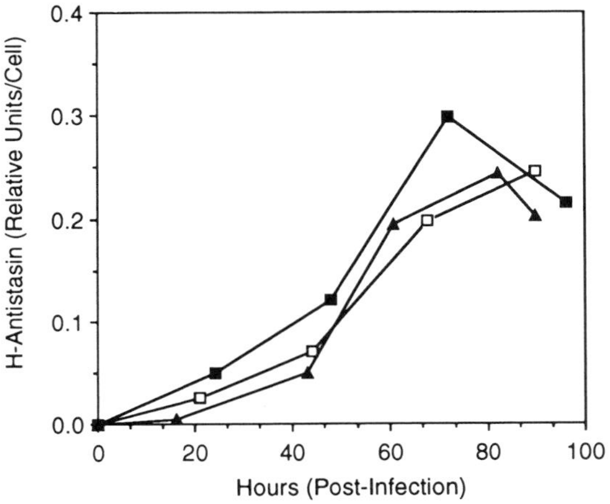

Fig. 5. Half-antistasin production under different gassing systems. SF9 cells in batch culture in IPL-41/2%FBS/0.1% Pluronic-F68 in a 250 ml spinner flask (■), an 8 liter Braun Biostat E fermentor operating with a silicone tube gassing system (▲), and a microsparged (oxygen) 18 liter Sulzer-MBR Spinferm fermentor (□) were infected with a recombinant baculovirus expressing half-antistasin at an MOI of 0.1. (Reprinted from Jain et al. [26], with permission of the publisher.)

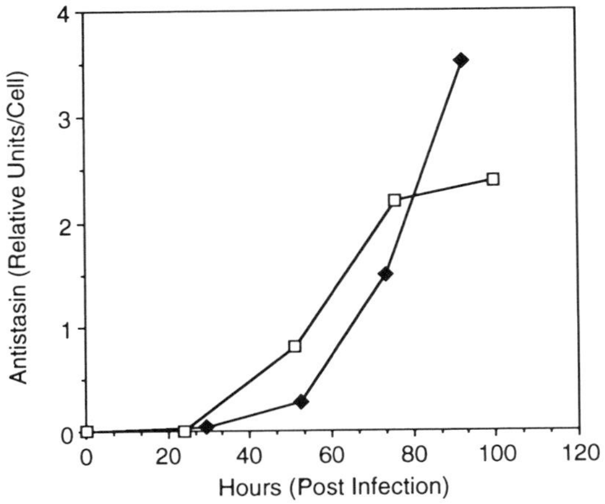

Fig. 6. Antistasin production at two scales, a microsparged (oxygen) 18 liter Sulzer-MBR Spinferm fermentor (□) and a 70 liter direct air-sparged Chemap fermentor (◆), in Ex-Cell 400 serum-free medium. (Reprinted from Jain et al. [27], with permission of the publisher).

volumes. Cell-free supernatants could be stored for 2–4 days at 4°C or frozen. Recovery of antistasin [38] is described below (see Downstream Processing and Purification).

β_2-Adrenergic Receptor (Batch)

Hamster β_2-adrenergic receptor is a glycosylated 68 kd plasma membrane glycoprotein receptor, of the seven transmembrane domain type, that is coupled to an intracellular G-protein [39–41]. Its signal transduction response to β-adrenergic agonists involves stimulation of cyclic AMP production by adenylyl cyclase. The baculovirus system was chosen for expression of this receptor because of the relatively high yields obtained compared with expression in mammalian cells (unpublished observations) [42]. Subsequent to its initial cloning in one of Summers and Smith's original baculovirus vectors, pAc373 [15], the gene was recloned into a higher producing vector, pJVP-10Z [43]. The β_2AR produced in insect cells was fully active in binding β-antagonists, but only about 40%–50% reached the plasma membrane surface (unpublished observations); the rest was also membrane associated, presumably localized to intracellular membranes.

Larger scale production of β_2AR was undertaken to produce material for structural studies. We first optimized MOI in spinner flasks and found, in contrast to antistasin, that a higher MOI of ≥ 1 was necessary to obtain optimal final yields; Figure 7 illustrates the β_2AR product yield response to MOI in 100 ml spinner flasks; in this and other experiments (not shown), the MOI optimum was 1–2. As discussed above, the observed MOI dependence was provisionally attributed to impaired virus–cell interaction, possibly due to the presence of overexpressed β_2AR in the viral envelope. Initially, the yield of recombinant β_2AR, as assayed by radioactive ligand (^{125}I-cyanopindolol [39]) binding, was found to peak at about

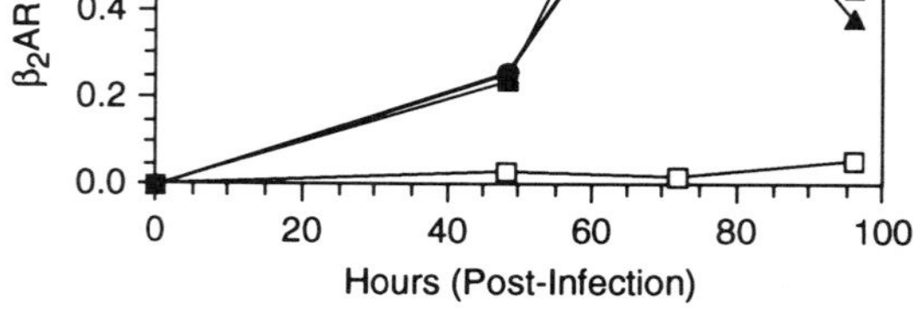

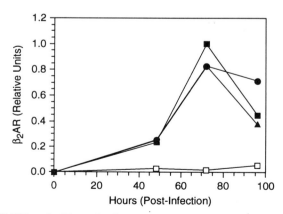

Fig. 7. Effect of MOI on β_2AR production. Experiments were performed in 100 ml spinner flasks in IPL-41/2% FBS/0.1% Pluronic-F68 medium. MOIs shown are 2 (●), 1 (■), 0.5 (▲), and 0.1 (□). No protease inhibitors were used in this experiment.

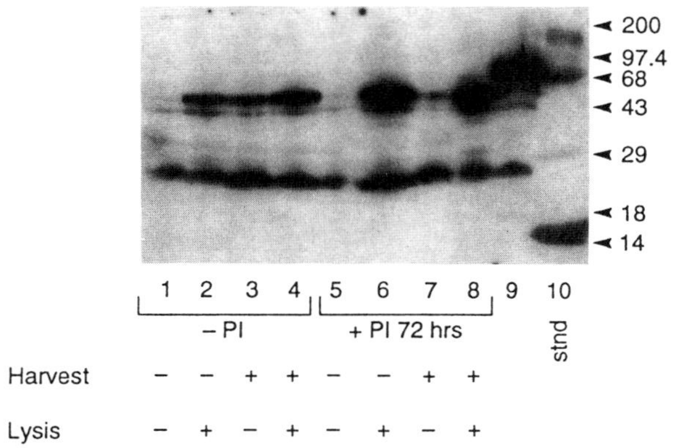

Fig. 8. Effect of protease inhibitors on β_2AR recovery. Immunoblot of SDS-PAGE analysis of recovered cells from parallel spinner flasks cultured, harvested, and/or lysed in the presence of protease inhibitors (PI). Equivalent amounts of total protein were run in each of the insect cell lanes (1–8). The blot was probed with an anti-β_2AR polyclonal serum. Lanes 1–4, PI (+) present at harvest and/or lysis; lanes 5–8, PI (+) present during the last 24 hours of culture as well as at harvest and/or lysis; lane 9, β_2AR standard (made by transient transfection of COS monkey kidney cells [40]); lane 10, molecular weight standards. (Reprinted from Jain et al. [27], with permission of the publisher.)

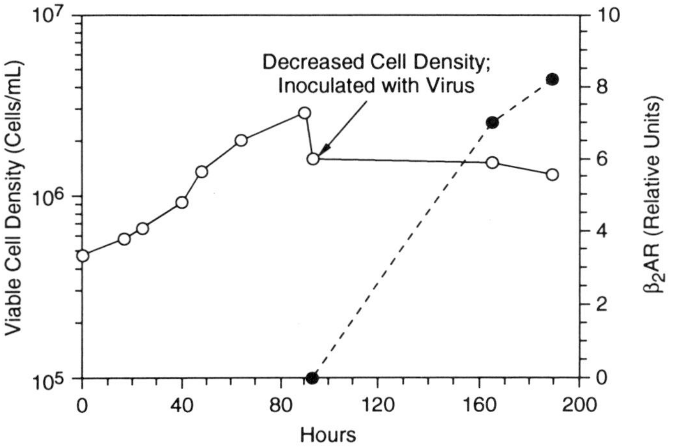

Fig. 9. Cell growth, cutback, and β_2AR production in a 1.25 liter fermentor. A 1.25 liter New Brunswick Scientific Celligen fermentor was seeded with SF9 cells in IPL-41/2% FBS/0.1% Pluronic-F68 and allowed to grow to high density; cells (○) were then cut back by culture fluid withdrawal and replacement with fresh medium, and recombinant baculovirus was added at an MOI of 2. Production of β_2AR (●) was followed by ^{125}I-cyanopindolol binding [40].

72–96 hours postinfection, corresponding to a cell viability of 50%, and then essentially to plateau for another 24–48 hours (data not shown). Upon closer examination, the fully active protein, especially at later times, was found to be protease nicked (M. Candelore and C. Strader, personal communication). This problem was overcome by adding a protease inhibitor cocktail consisting of soybean trypsin inhibitor, aprotinin, and leupeptin to the culture 24 hours prior to harvest, and a cocktail containing, in addition, ethylenediamine tetraacetate (EDTA), phenylmethylsulfonyfluoride (PMSF), and benzamidine in all cell wash and lysis buffers (see Downstream Processing and Purification, below). The protective effect of protease inhibitors can be seen in Figure 8, which represents a Western blot of β-receptor harvested with or without such inhibitors in the culture medium, cell wash, and lysis buffers. Note that maximal yield of intact receptor was obtained with protease inhibitors present throughout late culture times, harvest, and lysis. A typical β_2AR batch production run in a 1.25 liter Celligen bioreactor (New Brunswick Scientific, Edison, NJ) is illustrated in Figure 9. In this particular case, the cells overgrew slightly, but were still in late log phase. Cell density was therefore cut back to 1.6×10^6/ml and the cells were infected at an MOI of 2. Harvest was at approximately 96 hours postinfection, when cell viability had declined to approximately 75%. The application of perfusion culture technology to increase the production yield of β_2AR and its purification are discussed below.

Interleukin-1β Converting Enzyme

Interleukin-1β converting enzyme (ICE) is a cysteinyl protease that processes the interleukin-1β precursor (31–33 kd) to its mature secreted form (17.5 kd), which represents a key step in the activation of this important mediator of inflammation. Studies on the cloned ICE gene and its expression have indicated that the enzyme is composed of two nonidentical 20 and 10 kd subunits that are derived by proteolytic cleavage from a single 45 kd proenzyme [44].

The ICE gene was cloned into a commercially available baculovirus expression vector pBlueBac II (Invitrogen, San Diego, CA), and a recombinant AcMNPV viral vector was isolated by standard means, as described by the manufacturer. MOI and harvest time were optimized in 100 ml spinner flasks. It was found that a synchronous infection, initiated with an MOI of 1–2, provided the best product yields. The molecular weight of ICE varied with time of harvest, from primarily 45 kd precursor at 24–40 hours postinfection to primarily 20 + 10 kd processed subunits [44] by 48 hours postinfection (O. Palyha and A. Howard, personal communication). Harvest times for large-scale cultures were chosen based on which form of ICE was preferred for particular studies (generally the processed heterodimer) and tended to be the earliest times among the examples discussed in this chapter. Serum-free Ex-Cell 400 batch cultures were run at the 20 liter scale in 36 liter Bellco top-drive paddle-impeller spinner flasks with oxygen-enriched head-space gassing; DO was controlled at ≥65% [26], based on Ingold probe measurement, by manual adjustment of gas flow rate. Because assayed ICE activity was found to vary with how the packed, harvested cell pellets were handled (A. Howard, personal communication), accurate production kinetics were not obtained for this process.

PERFUSION CULTURE PRODUCTION: EXAMPLES

Because of the obvious nutrient limitations encountered in batch processes, we were interested in determining the feasibility of nutrient refreshment through continuous medium perfusion. This capability is provided by the Sulzer-MBR Spinferm fermentor, which has an independently driven, top-mounted 20 μm stainless steel screen mesh spin filter [45] inside a full-length draft tube. The draft tube provides a vertical flow of suspended cells past the spin filter, which reduces cell loss and filter clogging. The spin filter is run at ≥120 rpm by the top drive, while the impeller is run at 60–90 rpm by the bottom drive. Under these conditions, a medium dilution rate of 1 vol/day results in less than 2% cell loss per day. We describe perfusion processes for HSV-VP16 and β_2AR.

Herpes Simplex Virus VP16

The herpes simplex virus transcriptional transactivator VP16 (also referred to as αTIF or ICP25) is a 64 kd protein encoded by the virus and packaged in the tegument of the virion [46,47]. It activates gene expression of five viral α (early) genes. Though not itself a DNA-binding protein, VP16 interacts with other transcriptional regulatory proteins, including Oct-1 [47,48], to activate transcription from viral promoters bearing specific upstream regulatory sequences. In herpes-infected mammalian cells, VP16 is localized primarily in the nucleus [47]. When overproduced in SF9 cells with a baculovirus vector, its localization is less distinct, and it can be recovered in cell extracts from the postnuclear supernatant fraction after centrifugation at ≥100,000*g* (S. Ludmerer, personal communication) [48]. A recombinant baculovirus clone expressing VP16 was obtained from Kristie et al. [48], from which a viral stock was passaged and cryopreserved as described above, and a P-4 stock was used for scale-up culture.

Preliminary experiments in spinner flasks verified that a synchronous batch infection, using an MOI of 1–2, was acceptable for VP16 yield. However, at the productivity obtained, the total product required for planned studies would have exceeded the volumetric fermentor tank capacity at our disposal. It was therefore decided to attempt to boost cell density in a smaller volume by application of medium perfusion. We used the spin-filter/draft-tube assembly of the 18 liter Sulzer-MBR Spinferm fermentor for this purpose.

Cells were inoculated in a 17 liter working volume of Ex-Cell 400 medium at 6.8×10^5/ml. The DO level was controlled at 65% air saturation using head-space gassing (air, nitrogen, and oxygen) as well as microsparging (oxygen only). When the cell density reached 1.5×10^6/ml, the perfusion flow was initiated at one tank vol/day to avoid a slowdown in exponential growth, which had previously been found to occur, in this particular tank, beyond 2×10^6/ml (Fig. 10); up to this density, the specific growth rate in both batch and perfusion runs was 0.023–0.024 h^{-1} (28–30 hours doubling time). The agitation was initially maintained at 60 rpm and was then stepped up to 90 rpm when the cell density increased to 5×10^6/ml in order to maintain the cells in uniform suspension. Under perfusion, growth remained exponential and a cell density of 1.9×10^7/ml was eventually attained, with a

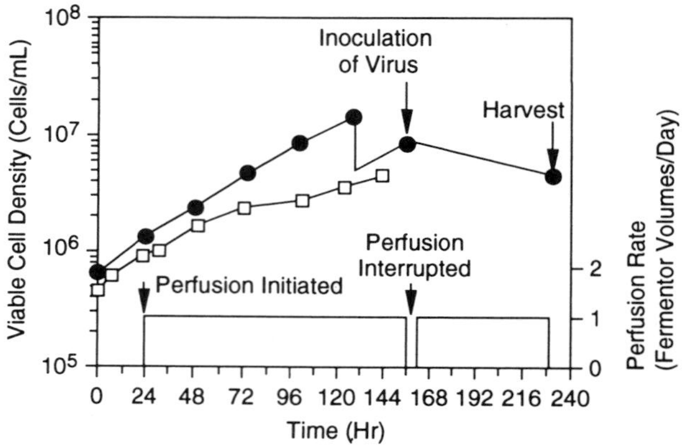

Fig. 10. Perfusion culture of SF9 cells for the production of herpes VP16. SF9 cells were seeded in the Sulzer-MBR Spinferm fermentor in Ex-Cell 400 medium at 4.5×10^5/ml (□, batch) or 6.5×10^5/ml (●, perfusion). In the batch run, no additions or withdrawals of medium were performed other than periodic sampling. In the perfusion run, perfusion was initiated at 1 reactor vol/day at 24 hours; after cell cutback and regrowth (see text), the perfusion was interrupted for 6 hours beginning at 156 hours for viral infection (MOI = 2) and then continued again at 1 reactor vol/day until harvest.

viability of 80% (Fig. 10). Based on our experience with batch cultures, it was decided to infect at no more than 50% the peak density defined by this experiment. The reactor was partially drained and refilled with fresh medium so as to achieve an effective cell dilution to 5×10^6/ml. The cells were then allowed to regrow to 9×10^6/ml, at which time approximately 1 liter of medium was withdrawn and 1 liter of viral stock was added, representing an MOI of 2. The perfusion flow was temporarily interrupted for 6 hours following virus addition to allow uniform viral infection. Perfusion was then resumed at a rate of one tank vol/day. The tank was harvested at 72 hours postinfection, when the cell viability had dropped to 60%. In addition to the higher cell density achieved, it was found that the continual perfusion of fresh medium had maintained the infected cells' specific productivity at the maximum observed in dilute batch cultures (S. Ludmerer, personal communication).

β_2-Adrenergic Receptor (Perfusion)

The first step taken toward improving β_2AR yields over those obtained in batch culture (Fig. 9) was recloning of the gene into a higher expressing vector [43,49]. In batch culture, a second-generation recombinant virus based on this improved vector resulted in a 2.5–3-fold improvement in yield (K. Mazina, unpublished observation). When used to infect a perfusion culture, results proved even more favorable.

The Sulzer-MBR Spinferm fermentor was run at 16.5 liters working volume, with a full-length internal draft tube and 20 μm stainless steel mesh spin-filter as described above. SF9 cells were expanded in spinner flask culture up to the 4 liter scale in Ex-Cell 400 medium and were inoculated into the tank at an initial density of 4×10^5/ml. They were allowed to grow to 1.5×10^6/ml as a batch culture (Fig. 11, 80 hours). Perfusion was then initiated at one tank vol/day and continued for 3.5 more days until the viable cell density (95% viability) reached 7.9×10^6/ml (Fig. 11, 169 hours). Note that the cell growth rate (39 hours doubling time at 27°C) remained exponential from inoculation to peak density. For infection, 2 liters of medium was removed and replaced with 2 liters of virus stock containing 1.2×10^8 PFU/ml, representing an MOI of 2.1. The perfusion was turned off for 6 hours to allow uniform viral infection and then restarted at one vol/day and continued for 66 hours until the tank was harvested at 72 hours postinfection at a viability of 50% (Fig. 11, 241 hours). Note that a protease inhibitor cocktail (aprotinin, soybean trypsin inhibitor, and leupeptin; cf. Fig. 8) was added at approximately 48 hours postinfection (220 hours in Fig. 11). Cells were harvested by batch centrifugation at 200–400*g* in a Beckman J-6M centrifuge, washed once with phosphate-buffered saline solution containing the protease inhibitors EDTA, benzamidine, aprotinin, leupeptin, soybean trypsin inhibitor, and PMSF, and stored at −85°C. In the course of purification (see below), it was determined that the raw yield of β_2AR in the detergent-extracted cell paste was 2.0 mg/liter, which represented a nearly 35-fold volumetric reactor productivity increase and a sevenfold specific productivity increase over batch production described above. The factors contributing to the in-

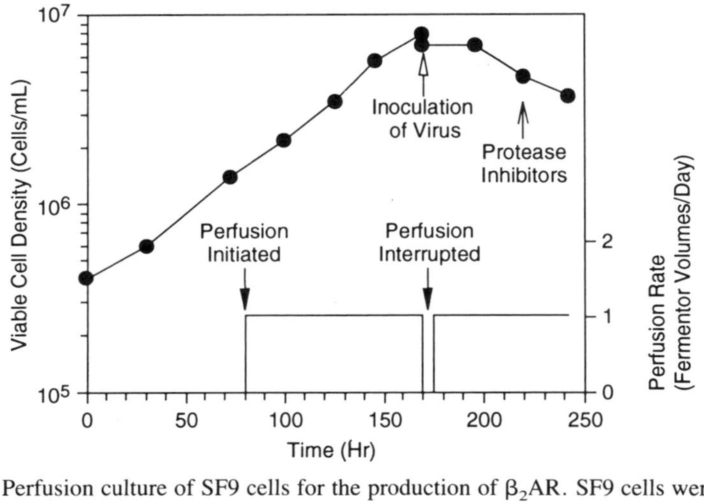

Fig. 11. Perfusion culture of SF9 cells for the production of β_2AR. SF9 cells were seeded in the Sulzer-MBR Spinferm fermentor in Ex-Cell 400 medium 4×10^5/ml, and perfusion at 1 reactor vol/day was initiated at 80 hours; beginning at 159 hours, perfusion was interrupted for 6 hours for viral infection (MOI = 2.1), and then resumed at 1 reactor vol/day until harvest. ●, Viable cell density.

creased specific productivity include the improved vector (ca. 2.5-fold; K. Mazina, unpublished observation) and the benefits of perfusion (ca. 2.5-fold).

DOWNSTREAM PROCESSING AND PURIFICATION

The protein recovery and purification processes described below were intended for preparation of material for biochemical characterization and/or animal studies. These examples illustrate the relative convenience of baculovirus biotechnology, but do not attempt to resolve compliance with U.S. Food and Drug Administration guidelines for Good Manufacturing Practices.

Antistasin

Both antistasin and half-antistasin were recovered chromatographically from harvested batch fermentations. We describe the purification process for (whole) antistasin. When an infected culture declined to about 50% viability, it was processed to remove cells and debris by centrifugation. For larger scale work, harvest was by tangential flow filtration in a Millipore Prostak unit fitted with a 0.65 μm membrane, followed by concentration across a 10,000 molecular-weight-cutoff (MWCO) membrane in a Millipore Pellicon device. The following scheme was developed for purification of antistasin from low-protein serum-free Ex-Cell 400 medium [38]. The cell-free supernatant (4°C) was adjusted up to pH 8.2 with sodium hydroxide, which resulted in a substantial precipitate that included much of the extracellular baculovirus. After centrifugation, the clarified supernatant was adjusted back down to pH 6.8 and fractionated by S-Sepharose fast-flow (Pharmacia-LKB, Piscataway, NJ) cation exchange chromatography, with a net antistasin recovery of 80%. The pooled antistasin peak was desalted by dialysis against double-distilled water using a 3,500 MWCO membrane and further purified by reversed-phase chromatography on a Vydac C_{18} column. The pooled antistasin peak was lyophilized and resuspended in water. Overall antistasin recoveries were 50%–55%, with a final purity of ≥98% [38].

We also examined whether ultrafiltration could be incorporated into the antistasin purification scheme as a potential viral removal step. To our surprise, a substantial fraction of the antistasin in conditioned serum-containing or serum-free medium could be retained by nominal 100,000 (data not shown) or 30,000 MWCO polysulfone ultrafiltration membranes (Table 2); based on similar observations with partially purified preparations (data not shown), we tentatively attributed this anomalous molecular filtration behavior to self-aggregation. As shown in Table 2, this problem could be entirely eliminated by addition of 20% acetonitrile prior to ultrafiltration. The addition of acetonitrile to raw harvest medium caused some precipitation, but antistasin remained in the clarified supernatant. Recently, Millipore Corp. (Viresolve® membranes) [50] and Asahi Chemical Industries, Ltd. (Planova® membranes) developed new lines of validatable ultrafiltration membranes specifically for the purpose of viral removal from process fluids. Although we have not tested these

TABLE 2. Antistasin Recoverable as a 30K MWCO Ultrafiltrate Fraction[a]

Acetonitrile (% v/v)	0	5	10	20	40
Antistasin recovery (%) 30K MWCO	53	51	41	100	100

[a]Ex-Cell 400 cell-free harvest medium containing whole antistasin was subjected to ultafiltration through 30K nominal MWCO polysulfone filters in Millipore Ultrafree-PF filter units; acetonitrile was added prior to filtration at the concentrations indicated.

new membranes ourselves, the results presented in Table 2 predict that similar performance should be obtainable. Ultrafiltration may thus serve as a useful addition to baculovirus downstream processes to help achieve validatable compliance with cGMP viral removal guidelines.

Herpes VP16

The perfused production culture illustrated in Figure 10 was harvested at 60% viability by centrifugation. Cells were resuspended and washed once in phosphate-buffered saline solution and then quick-frozen and stored at −85°C. Cell extracts were prepared from aliquots of the frozen cells by dounce homogenization in a hypotonic buffer containing 10 mM HEPES, pH 7.9, 10 mM KCl, 1.5 mM $MgCl_2$, 0.5 mM dithiothreitol (DTT), and 0.5 mM PMSF (S. Ludmerer, personal communication) and centrifuged at 28,000 rpm for 1 hour in a Beckman (Palo Alto, CA) SW28 rotor to generate a high-speed postnuclear supernatant. This "S100" supernatant fraction was subjected to column chromatography (FPLC) on MonoQ anion exchange resin (Pharmacia-LKB; adapted from Kristie et al. [48]). The final recovered material was judged to be greater than 90% pure by stained SDS-PAGE analysis, and its identity and activity were verified by immunoblotting and in vitro DNA binding complex formation (S. Ludmerer and T. Kristie, personal communication).

IL-1β Converting Enzyme

The production culture was harvested at 60% viability by centrifugation at 200–400*g*, and the cell pellet was washed with cold phosphate-buffered saline solution containing the protease inhibitors aprotinin, leupeptin, pepstatin, PMSF, EDTA, and a proprietary compound (A. Howard, personal communication). ICE was recovered from either freshly washed or frozen cells by first lysing the cell pellet in 25 mM HEPES, pH 7.5, containing 2 mM DTT and 1% CHAPS detergent (Calbiochem, San Diego CA) in addition to the protease inhibitors listed above. A clarified supernatant fraction was prepared by centrifugation at 20,000*g* and was then applied directly to a chromatography column containing an affinity resin based on a potent peptide aldehyde ICE inhibitor, as described [44]. The final purified yield from 8×10^{10} infected cells was estimated at 1 mg of catalytically active ICE, which was nearly homogeneous (A. Howard, personal communication). Because

ICE maturation to subunits in vitro is autocatalytic and can be blocked by an ICE inhibitor, it is assumed that precursor processing in SF9 cells is also primarily autocatalytic. The degree of autoprocessing was found to be influenced to some extent by the handling of the harvested cell pellet (data not shown), so handling was optimized to maximize the yield of active ICE.

β_2-Adrenergic Receptor

Batch cultures of SF9 cells infected with our original hamster β_2AR vector [27] were harvested at approximately 96 hours postinfection (Figs. 7, 9), which corresponded to a cell viability of about 50%, while the perfusion process illustrated in Figure 11, which utilized the second-generation baculovirus vector described above, declined to 50% viability more rapidly and was harvested at 74 hours postinfection. The cells were harvested by gentle centrifugation at 200–400*g*, cooled to 4°C, and the cell pellets were then washed by resuspension in one-fifth their original volume of PBS buffer containing the following protease inhibitor additives: EDTA, benzamidine, leupeptin, aprotinin, soybean trypsin inhibitor, and PMSF. After another 400*g* centrifugation, the pellets were again resuspended in one-fifth volume of PBS containing the same additives and lysed by high-speed centrifugation at 125,000*g* for 1 hour; this was followed by another resuspension in the same buffer and a second high-speed centrifugation at 125,000*g*, which yielded a crude preparation enriched in cellular membranes. β_2AR was extracted from this membrane pellet using a Polytron homogenizer (Brinkman Instruments, Westbury, NY) in a buffer containing 10 mM Tris-HCl, pH 7.4, 100 mM NaCl, the protease inhibitors listed above, and 0.25% (w/v) of the detergent dodecyl-β-D-maltoside (Boehringer Mannheim, Indianapolis, IN) [51]. After 30 minutes on ice, residual high-molecular-weight components were removed by centrifugation for 1 hour at 100,000*g*, and the clarified supernatant was subjected to affinity column chromatography. Two alternative affinity resins were used, alprenolol-Sepharose [24] or an immunoaffinity resin based on a previously described polyclonal rabbit antibody [40]. The alprenolol step was followed by further chromatography on hydroxylapatite [51] and Superose 12 (Pharmacia-LKB), while the immunoaffinity step was followed by lentil lectin agarose chromatography (M. Tota, unpublished data). The purified β_2AR migrated as a 50–55 kd broad band by SDS-PAGE analysis (Fig. 8). This is somewhat smaller than the ca. 68kd mammalian-expressed protein (Fig. 8) [40], with the difference most likely attributable to less complete glycosylation in the SF9 cells. Purified and solubilized β_2AR expressed in insect cells was found suitable for all in vitro biochemical studies undertaken (Tota and Strader, unpublished observations) [51,52].

DISCUSSION

In this chapter, we have summarized our experience with scale-up and production of four proteins expressed in insect cells with baculovirus vectors. These four examples, herpes VP16, interleukin-1β converting enzyme, β_2-adrenergic receptor, and antistasin (including half-antistasin), span the full range of naturally targetted protein expression, from nuclear to cytoplasmic to surface membrane associated to secreted, and illustrate the wide applicability of insect cell expression to proteins of interest to the pharmaceutical industry. These examples also serve to illustrate some of the constants and some of the variables encountered across a variety of recombinant baculovirus vectors, as well as issues particular to scaling up these processes.

We have found that reproducible stepwise expansion of an insect cell seed to pilot production scale, whether in serum-containing or in serum-free medium, is dependent on the process tolerance of the cell stock. Shear and medium tolerance can be ensured by rigorous preadaptation of the cells by continuous passage under simulated process conditions (i.e., higher shear stirring at ≥60 rpm in spinner flasks) prior to cryopreservation of a master cell bank. Cell doubling times and peak cell densities are largely dependent on medium composition; doubling times of 18–28 hours are common in serum-containing media, while 26–48 hour doubling times are characteristic of serum-free Ex-Cell 400 medium, with substantial lot-to-lot variability. In any particular medium, if DO is well controlled at 50%–100%, growth rate is nearly independent of the nature of the gassing system or the reactor configuration, at least across the variety of stirred tanks we have evaluated.

Viral seed expansions vary from recombinant virus to virus, presumably due to the physiological impact of the expressed foreign proteins on the host cells. The recombinant viruses described in these examples have yielded titers ranging from a low of $\leq 10^7$ PFU/ml for the β_2AR virus inoculated without special attention to infection conditions to a high of 5×10^8/ml for the recombinant ICE virus prepared using the gentle cell concentration protocol described above. Overall, we have come to prefer the latter protocol for most viral seed train expansions as, in our experience, it often promotes higher (but never lower) titers. Although we have not conducted a systematic study of viral stability, we have generally found both good titers and good 4°C storage qualities in viral stocks produced in 10% FBS-containing media; we therefore continue to use TNM-FH/10% FBS for most viral seed expansions. It must be acknowledged, however, that use of this medium may pose some potential drawbacks where production of human parenterals (therapeutics or vaccines) is involved.

For production purposes, the minimal MOI needed to achieve optimal product titer has generally been found to be 1. β_2AR yields have proven more reproducible at a higher MOI of 2, while antistasin and half-antistasin are notable in that a lower MOI of only 0.1 resulted in equal or better yields. Use of such a low MOI is very efficient of viral stocks and very economical. Ideally, MOI should be optimized individually for each baculovirus vector.

Harvest time has proven to be one of the more variable parameters between viruses among those surveyed in this report. Although the infectious cycle of a recombinant baculovirus is strongly influenced by the same factors that influence cell doubling time, including dissolved oxygen level [26] and medium composition (and lot), the nature of the expressed protein seems to play a major role as well. Among the seven recombinant baculoviruses described in this report, including those expressing two primary sequence variants of (whole) antistasin, half-antistasin, β_2AR (first-generation and second-generation viruses), VP16, and ICE, optimal harvest times have varied from a relatively early 40–50 hours postinfection for ICE to a relatively late 96 hours postinfection for first-generation β_2AR. Even for a single viral vector, such as second-generation β_2AR, the optimal perfusion harvest time of 72 hours postinfection was somewhat earlier than the optimal batch harvest time (unpublished observation); this may be due to acceleration of the viral life cycle under optimal nutritional conditions. We have therefore come to rely on surrogate variables by which to monitor the progress of large-scale cultures and to decide when to harvest. The most useful marker is percent cell death, with different values in the range between 75% and 50% viability coinciding with peak product titer for each different viral vector. An alternative surrogate marker useful in some cases—among the examples in this report, the secreted proteins antistasin and half-antistasin are most notable—is the cytopathic cell swelling that occurs late in the infectious cycle; this can be quantified as modal cell diameter, with peak product titers typically being reached about 12–24 hours after peak modal cell diameter.

The harvest method we have used up to the 20 liter scale is centrifugation, as it requires no special setup. For larger scales, gentle tangential flow filtration, such as can be accomplished with the Millipore Prostak device, is preferred. The addition of protease inhibitors, to the late phases of the culture process or to the harvest medium and washes, or to both, has proven useful in maintaining the integrity of cell-associated proteins, but has not been found necessary, in our hands, for secreted proteins such as antistasin. Of the examples discussed, β_2AR proved the most sensitive to proteolysis (Fig. 8). It is interesting to note that the possibility of protease sensitivity, in this case, was overlooked at first, because ligand-binding activity was largely unaffected.

We have recently found that product yields may also vary substantially with medium composition. In particular, choosing among the newer commercial serum-free lepidopteran insect cell culture media that are now marketed has provided us with one more variable to survey prior to scale-up (A. Lenny and M. Silberklang, unpublished observations). Finally, while a discussion of alternative insect cell lines for baculovirus infection, such as *Trichoplusia ni,* is beyond the scope of this report, it has been a fruitful subject of recent investigation [53–55].

CONCLUSIONS

We have found processes based on the highly productive baculovirus expression vector system to be routinely scalable to pilot production of recombinant-derived proteins. The recent development of more productive expression vectors for this system, combined with the advent of commercial serum-free media suitable for large-scale insect cell cultivation, have made baculovirus-mediated expression a candidate process alternative for production of proteins not only for research applications but also for use as human vaccines and therapeutics.

While the proteins recovered in the particular examples we have described here were used primarily for in vitro biochemistry or animal efficacy studies, we see no insurmountable barriers to someday developing the same types of processes into FDA-acceptable production processes meeting current Good Manufacturing Practices guidelines. The further progress, through routine FDA review, of several pioneering baculovirus-based processes for biological therapeutic candidates is sure to give more practical credibility to this prediction in the near future and presages the wider acceptance of this system as a process alternative by the pharmaceutical industry.

ACKNOWLEDGMENTS

We gratefully acknowledge our many colleagues, past and present, at the Merck Research Laboratories who generously shared with us their results, reagents, and ideas. In particular, we would like to thank Jack Jacobs, Chris Dunwiddie, Elka Nutt, Bruce Daugherty, Jim DiRenzo, Barry Buckland, Jwu-Sheng Tung, Linda Palladino, Simon Law, Ken Alves, and Greg Cuca for their help and collaboration in the antistasin program; Richard Dixon, Catherine Strader, Mari Candelore, and Rosemary O'Donnell for their help and collaboration in the β-adrenergic receptor program; Steve Ludmerer for collaboration in the VP16 program and for communication of his unpublished results; and Andy Howard for collaboration in the ICE program and for communication of his unpublished results. Finally, we would like to thank Ron Ellis and George Mark, with whose support and guidance this work was undertaken, and Susan Pols for her patience and great care in preparing the manuscript.

REFERENCES

1. Luckow, V.E., and Summers, M.D. Bio/Technology 6:47–55, 1988.
2. Miller, L.K. BioEssays 11:91–95, 1989.
3. Luckow, V.E. In: Recombinant DNA Technology and Applications (Prokop, A., Bajpai, R.K., and Ho, C.S., eds.). McGraw-Hill, New York, pp. 97–152, 1991.
4. Fraser, M.J. Curr. Top. Microbiol. Immunol. 158:131–172, 1992.
5. Luckow, V.E. In: Baculovirus Expression Systems and Biopesticides. Wiley, New York (in press).
6. Maiorella, B., Inlow, D., Shauger, A., and Harano, D. Bio/Technology 6:1406–1410, 1988.
7. Rusche, J.R., Lynn, D.L., Robert-Guroff, M., Langlois, A.J., Lyerly, H.K., Carson, H., Krohn, K., Ranki, A., Gallo, R.C., Bolognesi, D.P., Putney, S.C., and Matthews, T.J. Proc. Natl. Acad. Sci. U.S.A. 84:6924–6928, 1987.
8. Hu, S.-L., Kosowski, S.G., and Schaaf, K.F. J. Virol. 61:3617–3620, 1987.
9. Cochran, M., Ericson, B., Knell, J., and Smith, G. In: Vaccines 1987. (Chanock, R., Lerner, R.A., Brown, F., and Ginsberg, H., eds.). Cold Spring Harbor Laboratories, Cold Spring Harbor, NY, pp. 384–388, 1987.
10. Orentas, R.J., Hildreth, J.E.K., Obah, E., Polydefkis, M., Smith, G.E., Clements, M.L., and Siliciano, R.F. Science 248:1234–1237, 1990.
11. Vlak, J.M., and Keus, R.J.A. Adv. Biotechnol. Process. 14:91–128, 1990.
12. Food and Drug Administration (U.S.), Office of Biologics Research and Review. Points To Consider (PTC) Dockets: 1) PTC in the Production and Testing of new Drugs and Biologicals Produced by Recombinant DNA Technology, 1985; 2) PTC in the Characterization of Cell Lines Used to Produce

Biologicals, 1987; 3) Supplement to the PTC in the Production and Testing of New Drugs and Biologicals Produced by Recombinant DNA Technology: Nucleic Acid Characterization and Genetic Stability, 1992.

13. Han, J.H., Law, S.W., Keller, P.M., Kniskern, P.J., Silberklang, M., Tung, J.-S., Gasic, T.B., Gasic, G.J., Friedman, P.A., and Ellis, R.W. Gene 75:47–57, 1989.
14. Silberklang, M., Lenny, A., Munshi, S., Daugherty, B., Zavodny, S., Law, S., Kniskern, P., Petroski, C., DiRenzo, J., Han, J., Mark, G., and Ellis, R.W. J. Cell Biol. 107:1724, 1988.
15. Summers, M.D., and Smith, G.E. Texas Agricultural Experiment Station Bulletin No. 1555, 1987.
16. Hink, W.F. Nature 226:466–467, 1970.
17. Weiss, S.A., Smith, G.C., Kalter, S.S., and Vaughn, J.L. In Vitro 17:495–502, 1981.
18. Fuchs, L.Y., Woods, M.S., and Weaver, R.F. J. Virol. 48:641–6466, 1983.
19. Volkman, L.E., and Knudson, D.L. In: The Biology of Baculoviruses (Granados, R.R., and Federici, B.A., eds.). Vol. 1. CRC Press, Boca Raton, FL pp. 109–127, 1986.
20. Lazarte, J.E., Tosi, P.-F., and Nicolau, C. Biotechnol. Bioeng. 40:214–217, 1992.
21. Licari, P., and Bailey, J.E. Biotechnol. Bioeng. 37:238–246, 1991.
22. Lindsay, D.A., and Betenbaugh, M.J. Biotechnol. Bioeng. 39:614–618, 1992.
23. Hink, W.F., and Strauss, E.M. In: Invertebrate Tissue Culture—Applications in Medicine, Biology and Agriculture (Kurstak, E., and Maramorosch, K., eds.). Academic Press, New York, pp. 297–300, 1976.
24. Caron, M.G., Srinivasan, Y., Pitha, J., Kociolek, K., and Lefkowitz, R.J. J. Biol. Chem. 254:2923–2927, 1979.
25. Radford, K.M., Reid, S., and Greenfield, P.F. Baculovirus and Recombinant Protein Production Processes (Proceedings of a Workshop, Interlaken, Switzerland, 1992) (Vlak, J.M., Schaleger, E.-J., and Bernard, A.R., eds.). Editiones Roche, Basel, Switzerland, pp. 297–303, 1992.
26. Jain, D., Ramasubramanyan, K., Gould, S., Seamans, C., Wang, S., Lenny, A., and Silberklang, M. In: Expression Systems and Processes for rDNA Products (Hatch, R.T., Goochee, C.F., Moreira, A., and Alroy, Y., eds.). ACS Symposium Series No. 477. American Chemical Society Books, Washington, D.C., pp. 97–110, 1991.
27. Jain, D., Ramasubramanyan, K., Gould, S., Lenny, A., Candelore, M., Tota, M., Strader, C., Alves, K., Cuca, G., Tung, J.-S., Hunt, G., Junker, B., Buckland, B.C., and Silberklang, M. In: Production of Biologicals From Animal Cells in Culture (Spier, R.E., Griffiths, J.B., and Meignier, B., eds.). Proceedings of the European Society for Animal Cell Technology, the 10th Meeting. Butterworth-Henemann, Oxford, England, pp. 345–350, 1991.
28. Murhammer, D.W., and Goochee, C.F. Bio/Technology 6:1411–1418, 1988.
29. Gould, S.L., Wang, S., Seamans, C., Lenny, A., Jain, D., and Silberklang, M. In Vitro 25:47A, 1989.
30. Neutra, R., Levi, B.-Z., and Shoham, Y. Appl. Microbiol. Biotechnol. 37:74–78, 1992.
31. Wickham, T.J., Davis, T., Granados, R.R., Hammer, D.A., Shuler, M.L., and Wood, H.A. Biotechnol. Lett. 13:438–488, 1991.
32. De Gooijer, C.D., Koken, R.H.M., VanLier, F.L.J., Kool, M., Vlak, J.M., and Tramper, J. Biotechnol. Bioeng. 40:537–548, 1992.
33. Nielsen, L.K., Smyth, G.K., and Greenfield, P.F. Cytotechnology 8:231–236, 1992.
34. Tuszynski, G.P., Gasic, T.B., and Gasic, G.J. J. Biol. Chem. 262:9718–9723, 1987.
35. Nutt, E., Gasic, T., Rodkey, J., Gasic, G.J., Jacobs, J.W., Friedman, P.A., and Simpson, E. J. Biol. Chem. 263:10162–10167, 1988.
36. Dunwiddie, C., Thornberry, N.A., Bull, H.G., Sardana, M., Friedman, P.A., Jacobs, J.W., and Simpson, E. J. Biol. Chem. 264:16694–16699, 1989.
37. Palladino, L.O., Tung, J.-S., Dunwiddie, C., Alves, K., Lenny, A.B., Przysiecki, C., Lehman, D., Nutt, E., Cuca, G.C., Law, S.W., Silberklang, M., Ellis, R.W., and Mark, G.E. Protein Expression Purification 2:37–42, 1991.

38. Nutt, E.M., Jain, D., Lenny, A.B., Schaffer, L., Siegl, P.K., and Dunwiddie, C.T. Arch. Biochem. Biophys. 285:37–44, 1991.

39. Dixon, R.A.F., Kobilka, B.K., Strader, D.J., Benovic, J.L., Dohlman, H.G., Frielle, T., Bolanowski, M.A., Bennet, C.D., Rands, C.D., Diehl, R.E., Mumford, R.A., Slater, E.E., Sigal, I.S., Caron, M.G., Lefkowitz, R.J., and Strader, C.D. Nature 321:75–79, 1986.

40. Dixon, R.A.F., Sigal, I.S., Candelore, M.R., Register, R.B., Scattergood, W., Rands E., Strader, C.D. EMBO J. 6:3269–3275, 1987.

41. Dohlman, H.G., Caron, M.G., and Lefkowitz, J.J. Biochemistry 26:2657–2664, 1987.

42. Strader, C.D., Cheung, A.-H., Tsai, A.-M., Gould, S.L., Lenny, A.B., and Silberklang, M. J. Cell Biol. 107:343, 1988.

43. Vialard, J., Lalumière, M., Vernet, T., Briedis, D., Alkhatib, G., Henning, D., Leven, D., and Richardson, C. J. Virol. 64:34–70, 1990.

44. Thornberry, N.A., Bull, H.G., Calaycay, J.R., Chapman, K.T., Howard, A.D., Kostura, M.J., Miller, D.K., Molineaux, S.M., Weidner, J.R., Aunins, J., Elliston, K.O., Ayala, J.M., Casano, F.J., Chin, J., Ding, G.J.-F., Egger, L.A., Gaffney, E.P., Limjuco, G., Palyha, O.C., Raju, S.M., Rolando, A.M., Salley, J.P., Yamin, T.-T., Lee, T.D., Shively, J.E., MacCross, M., Mumford, R.A., Schmidt, J.A., and Tocci, M.J. Nature 356:768–774, 1992.

45. Scheirer, W. In: Animal Cell Biotechnology (Spier, R.E., and Griffiths, J.B., eds.). Vol. 3. Academic Press, London, pp. 263–281, 1988.

46. Batterson, W., and Roizman, B. J. Virol. 46:371–377, 1983.

47. McKnight, J.L.C., Kristie, T.M., and Roizman, B. Proc. Natl. Acad. Sci. U.S.A. 84:7061–7065, 1987.

48. Kristie, T.M., LeBowitz, J.H., and Sharp, P.A. EMBO J. 8:4229–4238, 1989.

49. Luckow, V.E., and Summers, M.D. Virology 167:56–71, 1988.

50. DiLeo, A.J., and Allegrezza, A.E., Jr. Nature 351:420–421, 1991.

51. Tota, M.R., and Strader, C.D. J. Biol. Chem. 265:16891–16897, 1990.

52. Parker, E.M., Kameyama, K. Higashijima, T., and Ross, E.M. J. Biol. Chem. 266:519–527, 1991.

53. Hink, W.F., Thomsen, D.R., Davidson, D.J., Meyer, A.L., Castellino, F.J. Biotechnol. Prog. 7:9–14, 1991.

54. Weiss, S.A., Whitford, W.G., Godwin, G.P., and Reid, S. Baculovirus and Recombinant Protein Production Processes (Proceedings of a Workshop, Interlaken, Switzerland, 1992) (Vlak, J.M., Schlaeger, E.-J., and Bernard, A.R., eds.). Editiones Roche, Basel, Switzerland, pp. 306–315, 1992.

55. Wickham, T.J., Davis, T., Granados, R.R., Shuler, M.L., and Wood, H.A. Biotechnol. Prog. 8:391–396, 1992.

11

Potential Application of Insect Cell-Based Expression Systems in the Bio/Pharmaceutical Industry

Laurie K. Overton and Thomas A. Kost

Department of Molecular Biology, Glaxo, Inc. Research Institute, Research Triangle Park, North Carolina 27709

INTRODUCTION

The past decade has witnessed the rapid emergence of genetic engineering technology coupled with major improvements in the development of large-scale insect cell culture processes. These developments have led to the widespread use of insect cells as a valuable host cell system for the expression of recombinant proteins. Since its development in the early 1980s [1,2], the baculovirus gene expression system has been used extensively for both small- and large-scale production of heterologous proteins [3–5]. Recently, a stable insect cell line expression system, using *Drosophila melanogaster* cells and a plasmid employing an inducible *Drosophila* metallothionein promoter, has also been utilized for the successful production of recombinant proteins [6,7]. This chapter focuses primarily on the potential application of the baculovirus expression system for the production of large quantities of recombinant proteins for use in drug discovery-based research programs and in the development of therapeutic proteins.

ADVANTAGES OF INSECT CELL-BASED GENE EXPRESSION SYSTEMS

The baculovirus expression system possesses numerous advantages for the large-scale production of recombinant proteins. The development of a wide variety of expression vectors [5,8,9] along with advances in recombinant virus isolation and identification procedures [10] have significantly reduced the time and effort required

Baculovirus Expression Systems and Biopesticides, pages 233–242

to generate purified recombinant viruses. Typically, the process requires only 2–3 weeks from the availability of the appropriate gene clone to the isolation and purification of the desired recombinant virus. This time factor may be an important consideration, since significantly more time and effort may be required to develop, clone, and identify high-expressing mammalian cell lines, using a system such as the Chinese hamster ovary (CHO) cell–dihydrofolate reductase (DHFR) gene amplification system [11]. A major trade-off in choosing the baculovirus system is that the viral infection process ultimately leads to cell death versus the generation of potentially stable cell lines. An alternative approach aimed at developing stable insect cell lines expressing heterologous proteins has recently been described [6,7]. *D. melanogaster* Schneider 2 (S2) cells were transfected with a plasmid encoding hygromycin resistance and a plasmid encoding the gene of interest linked to a heavy metal-inducible *Drosophila* metallothionein promoter. After 3 weeks of selection in the presence of hygromycin, a stable polyclonal S2 line was generated containing an average of up to 1,000 gene copies per cell. Thus, S2 cells containing high copy numbers of the gene of interest could be obtained without a time-consuming gene amplification selection procedure. This system may prove an attractive alternative to the baculovirus expression system or currently used mammalian cell-based expression systems.

The relative ease of constructing and isolating recombinant viruses is matched by the utility of insect cell cultures. Recombinant baculoviruses are primarily used to infect cell lines derived from the insect *Spodoptera frugiperda* (Fall armyworm). Typical lines include SF21AE, derived from ovarian tissue [12] and the SF9 line [13], a clonal isolate derived from SF21AE cells. Cell lines, such as TN368 [14] and BTI-TN5B1-4 (established at Boyce Thompson Institute, Ithaca, NY, and commercially available as High Five cells from Invitrogen Corp.) derived from *Trichoplusia ni* (cabbage looper) and those derived from *Mamestra brassicae* (cabbage moth) [15] can also serve as host cells for infecting virus. The SF cell lines and certain derivatives of the *T. ni* lines can be grown attached or in suspension. The cells grow readily between 26° and 28°C in a variety of culture media in the presence or absence of fetal bovine serum and have no CO_2 requirement. A number of excellent review articles, as well as earlier chapters in this book, describe the various cell types available and culture conditions in more detail [16–21].

An important consideration in choosing any gene expression system is the level of expression that can be achieved. Years of experience have proven that expression levels can vary widely depending on the expression system used, the proteins expressed, and the culture conditions employed. A survey of the available literature indicates that expression levels of various baculovirus-expressed foreign proteins have ranged from 1 to >500 mg/liter, with most reports citing values on the order of 1 to 50 mg/liter. Expression level data for the *Drosophila* cell system are limited to a single report describing the production of the human immunodeficiency virus (HIV-1) envelope glycoprotein (gp120) at levels ranging from 5 to 35 mg per liter of culture medium [7]. These values compare favorably with the levels produced in mammalian cell expression systems.

POST-TRANSLATIONAL MODIFICATIONS OF BACULOVIRUS-EXPRESSED RECOMBINANT PROTEINS

The baculovirus expression system has been used successfully to produce a wide variety of foreign proteins in insect cells [4,5]. These studies have generated an extensive database characterizing the post-translational modifications carried out by these cells. It is beyond the scope of this chapter to address this subject in detail; therefore, discussion is limited to a summary of the currently available information. A more detailed description of the post-translational processes can be found in a number of recent reviews [3–5] and in Chapter 4.

In virtually every instance studied, foreign proteins expressed in baculovirus-infected insect cells have been found to be enzymatically, antigenically, and functionally similar to the native protein. The proteins are properly folded, indicating the presence of correct disulfide bond formation, and for the most part native signal peptides are recognized and appropriately cleaved resulting in transport through the endoplasmic reticulum. As a general rule, the proteins are directed to the appropriate subcellular location. In addition, insect cells have been shown to phosphorylate naturally phosphorylated proteins and perform a variety of lipid-based modifications including acylation, α-amidation, and isoprenylation [5].

Glycosylation is a post-translational modification found predominantly associated with membrane-bound and secreted proteins in eukaryotes. Insect cells appear to recognize asparagine-linked *N*-glycosylation sites (Asn-X-Ser/Thr) appropriately; however, the actual pattern of glycosylation at these sites may differ from that of the native protein. In general, glycosylation in insect cells appears to be limited, and conversion to complex *N*-linked oligosaccharides either does not occur or occurs to a lesser degree than in mammalian cells. *O*-linked glycosylation can clearly take place in baculovirus-infected insect cells. The relative importance of the glycosylation state of recombinant proteins produced in insect cells will certainly vary on a case by case basis, depending on the intended use of the protein. For example, in some cases it may actually be advantageous to have limited glycosylation of proteins destined to be crystallized for structural studies, whereas proteins targeted for vaccine or therapeutic development may require a more stringent control of glycosylation.

SCALE-UP OF INSECT CELL CULTURE

The commercial use of the baculovirus system depends on the ability of the system to be readily scaled-up. The most commonly used method for insect cell culture is the batch method; however, insect cells have also been cultured in continuous and semicontinuous culture systems [22–25]. There are a number of reviews on the scale-up of insect cell culture that address issues such as bioreactor design, oxygenation systems, multiplicity of infection (MOI), medium development, and the selection of the appropriate insect cell lines (see Chapter 7) [18,19,26–29]. In the following sections our experiences with scaling up recombinant protein production in 36 liter stirred vessels and the use of *T. ni* High Five cells to increase expression levels of secreted proteins are discussed.

SCALE-UP OF RECOMBINANT PROTEIN PRODUCTION IN 36 LITER STIRRED VESSELS

We initially began the scale-up of insect cell culture (SF9 cells) in 5 liter airlift fermentors [30]. This system functioned very well, but the need existed to increase culture capacity rapidly. We have found that complex systems were not necessary for the scale-up of insect cell culture to meet our protein demands. Increased production capacity was readily accomplished by using banks of 36 liter stirred vessels as shown in Figure 1. This system provided a simple and economical way to scale-up into the hundreds of liters range. The vessels were equipped with medium, cell, and virus inoculation ports. Additional ports were also included for cell sampling and harvesting. The sparger was a ring sparger with 2 mm holes. The assembled units were sterilized by autoclaving. Dissolved oxygen (DO) was monitored and maintained at 50% air saturation by oxygen sparging. Constant temperature can be maintained in a warm room or with an internal stainless steel coil containing 12 feet of tubing that is connected to a circulating water bath set at 28°C.

Figure 2 illustrates a typical growth curve of SF9 cells grown in this system. The cells were seeded into SF900 II medium (GIBCO) at a density of 3.5×10^5 cells/ml. The cells were in log growth until a density of 6×10^6 cells/ml and viability remained high. In a typical production run using a recombinant baculovirus, the cells were infected at 2×10^6 cells/ml at an MOI of 2. This system was employed to produce 500 liters of baculovirus-infected SF9 cells containing

Fig. 1. Thirty-six liter stirred vessels for scale-up of recombinant protein expression using the baculovirus gene expression system. This system provides a simple and economical way to increase culture volumes.

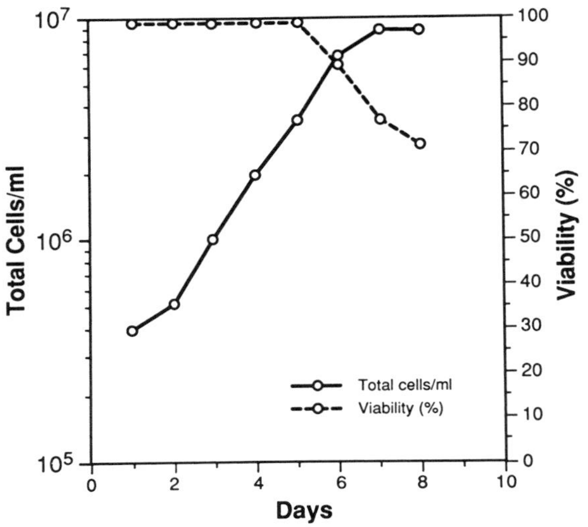

Fig. 2. Growth curve of SF9 cells in a 36 liter stirred vessel. The cells were inoculated into SF900 II medium at a density of 3.5×10^5 cells/ml. DO was maintained at 50% air saturation, pH was not controlled, and temperature was maintained by circulating 28°C water through an internal coil.

human synovial phospholipase A_2 ($hsPLA_2$) in comparable quantities to that produced in either a shake flask or a 5 liter airlift fermentor (data not shown). $hsPLA_2$ was purified by extracting cells with 0.18N H_2SO_4. The supernatant was pH adjusted and loaded onto a Poros II HS cation exchange column and eluted with 0.5–2.0 M NaCl. Fractions were pooled and desalted over a G-25 column. The protein was purified on a C-4 Rainin Dynamax reversed-phase semipreparative column. The purified protein was lyophilized to dryness. The $hsPLA_2$ was >95% pure and was used for structural studies. In Figure 3 is an 8%–16% Tris-glycine-Coomassie–stained gel of the steps in the purification pathway. The expression level of purified $hsPLA_2$ obtained using this system ranged between 1.0 and 1.5 mg/liter.

Harvesting large volumes of cells in a gentle and relatively quick manner was a major obstacle in the processing of large volume batches. After infection, the cells were especially fragile due to increased size and to the nature of the lytic infection. Tangential-flow filtration, hollow fiber, and continuous flow centrifugation systems were evaluated for their suitability for cell harvesting. Cell lysis occurred with all systems except the continuous-flow centrifuge. The continuous-flow centrifugation system that separates cells from medium and directs each into separate harvest vessels provided an effective and contained system for harvesting cells from a wide range of production volumes. With this system there was no decrease in cell viability, as determined by trypan blue staining, throughout the harvest. The cells were only in the separation loop for approximately 10 minutes as compared with being circulated through a pump in the other systems for the duration of the harvest.

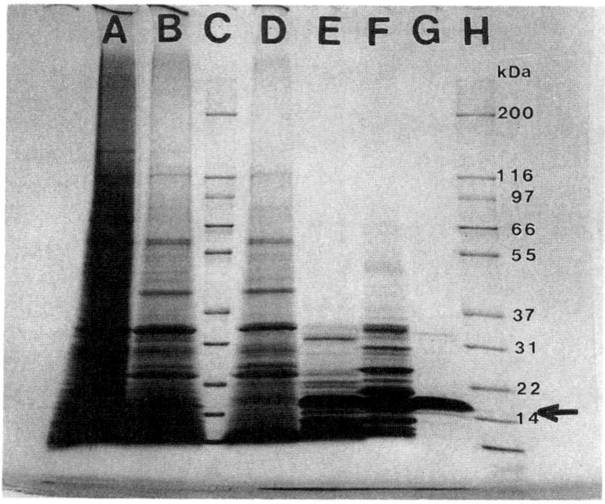

Fig. 3. The steps in the purification protocol for $hsPLA_2$ are shown on an 8%–16% Tris-glycine-Coomassie–stained gel. Lane A, acid extract; lane B, pH adjustment of supernatant; lane C, markers; lane D, S-Poros flow-through; lane E, Poros pool of $hsPLA_2$; lane F, G-25 pool; lane G, Reversed-phase purified $hsPLA_2$; lane H, markers.

ALTERNATIVE CELL LINES FOR RECOMBINANT PROTEIN EXPRESSION

Hink et al. [18] have demonstrated that the choice of host insect cell lines can significantly affect recombinant baculovirus expression levels and that the expression level was highly protein dependent. Recently, a new isolate of *T. ni* cells (High Five or TN5B1-4) have been introduced with the claim that up to a 25-fold increase in the expression level of secreted proteins can be obtained over that of SF9 cells. However, these cells maintained as adherent cultures are not ideal for scale-up. Recently, High Five cells have been adapted to suspension culture. These cells can now be readily scaled-up to large volumes. In suspension, small clumps (two to six cells) may form, but the majority of the culture consists of singlets and doublets.

High Five cells that had been adapted to suspension culture were used to produce the secreted protein tissue inhibitor of metalloproteinases (TIMP). TIMP had previously [30] been produced by recombinant baculovirus-infected SF9 cells at 4–8 mg of purified protein per liter. A production run using the High Five cells in a 5 liter airlift fermentor is shown in Figure 4a. The cells were seeded in Ex-Cell 401 medium (JRH Biosciences) at 7.5×10^5 cells/ml. They were infected at 1.4×10^6 cells/ml at an MOI of 2, and the culture medium containing TIMP was harvested 81 hours postinfection. High Five cells appeared to be more sheer sensitive than SF9 cells because the culture could not be as vigorously sparged. The growth conditions have been optimized in the 36 liter stirred vessels. Figure 4b shows a Coomassie-stained gel comparing the TIMP (MW = 22 kd) produced in adherent cultures of

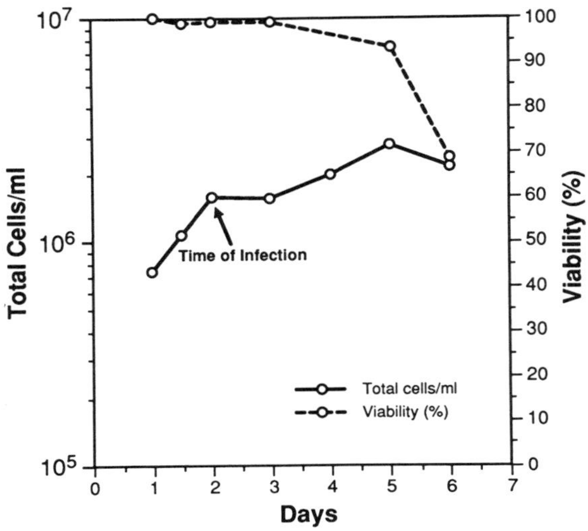

(*a*)

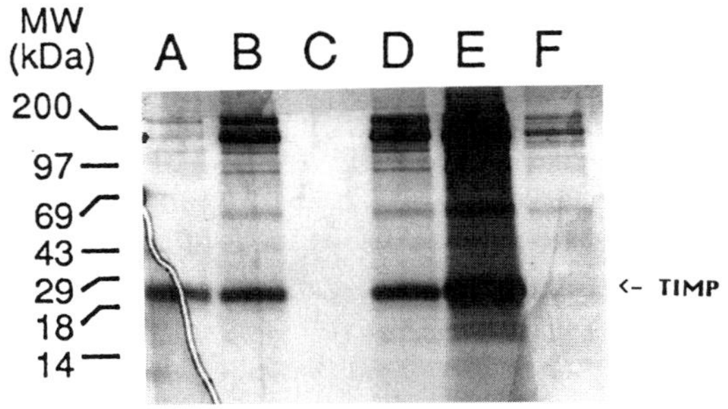

(*b*)

Fig. 4. a: Growth curve of High Five cells in a 5 liter airlift fermentor for the production of recombinant TIMP. The cells were seeded at 7.5×10^5 cells/ml and infected at 1.4×10^6 cells/ml with an MOI of 2. The culture medium containing TIMP was harvested 81 hours postinfection. **b:** Expression levels of TIMP produced by different cell lines are shown on a Coomassie-stained 10%–20% SDS-PAGE gel. Ten microliters of culture supernatant was loaded per lane except in lane D. The position of TIMP is indicated by the arrow. Lane A, TIMP produced by adherent High Five cells; lane B, TIMP produced by High Five cells in a 5 liter airlift fermentor; lane C, permeate after 5× concentration; lane D, 2 μl of 5× concentrated medium; lane E, 10 μl of 5× concentrated medium; lane F, TIMP produced by SF9 cells.

High Five cells, the TIMP produced in High Five cells grown in suspension in a 5 liter airlift fermentor, and the TIMP produced by SF9 cells.

By sodium dodecylsulfate polyacrylamide gel electrophoresis (SDS-PAGE) it appears that equivalent levels of TIMP were produced by the adherent High Five cultures compared with suspension-adapted cells. Ninety milligrams per liter of purified TIMP was obtained from the medium of the infected High Five cells. The increase in TIMP expression by High Five cells represented a 10–20-fold increase compared with that produced by the SF9 cells. As exemplified by our results, the expression of TIMP by High Five cells was in the same range as previously reported for a stable CHO cell line [31].

Recently, we used Ex-Cell 401 modified medium for the growth of High Five cells. In this medium cells grow to a density of approximately 5×10^6 cells/ml, which is two to three times higher than in Ex-Cell 401. Cells grown in this medium and harvested 86 hours postinfection were 95% viable. The level of recombinant protein produced by cells cultured in this medium is now being determined.

SUMMARY

There is a real potential for the use of insect cell culture for the production of therapeutic proteins and vaccines using the baculovirus gene expression system. Other insect cell culture-based expression systems, such as *Drosophila* S2 cells, may also prove to be suitable for this purpose. The tools are now available for the successful growth and infection of insect cells on an industrial scale. After optimization, expression levels can equal or exceed those of mammalian cell-based expression systems. The simple glycosylation carried out by insect cells may or may not be a problem, depending on the protein produced and its intended application. MicroGeneSys has set a precedent with approval by the Food and Drug Administration for clinical testing of their baculovirus-derived HIV subunit vaccine [32]. Issues such as the removal of virus, host cells, and cellular DNA and the purity and stability of the recombinant protein have been addressed and validated for a number of mammalian cell culture-derived therapeutic proteins. The same issues will need to be resolved for insect cell-produced proteins targeted for use as therapeutic agents or vaccines. The baculovirus gene expression system remains an excellent system for producing recombinant proteins for pharmaceutical research programs. The system can be readily scaled-up to produce high levels of recombinant proteins for use in high through-put screens, mechanistic studies, and rational drug design programs.

ACKNOWLEDGMENTS

The authors thank Inder Patel and Bruce Wisely for the construction of the shuttle plasmids, Byron Ellis and Pam DeLacy for purification of $hsPLA_2$, and John Gray and Jeff Robbins for critical review of this manuscript. Also, we are grateful for the gifts of Ex-Cell 401 modified medium and suspension culture-adapted High Five cells from Tana Montgomery at JRH Biosciences.

REFERENCES

1. Smith, G.E., Summers, M.D., and Fraser, M.J. Mol. Cell. Biol. 3:2156–2165, 1983.
2. Pennock, G.D., Shoemaker, C., and Miller, L.K. Mol. Cell. Biol. 4:399–406, 1984.
3. Vlak, J.M., and Keus, R.J.A. In: Viral Vaccines. Wiley-Liss, Inc., New York, 1990.
4. Luckow, V.A. In: Recombinant DNA Technology and Applications. McGraw-Hill, New York, 1991.
5. O'Reilly, D.R., Miller, L.K., and Luckow, V.A. Baculovirus Expression Vectors. A Laboratory Manual. W.H. Freeman, New York, 1992.
6. Van der Straten, A., Johansen, H., Sweet, R., and Rosenberg, M. In: Invertebrate Cell Systems Applications. CRC Press, Boca Raton, FL, 1987.
7. Culp, J.S., Johansen, H., Hellmig, B., Beck, J., Matthews, T.J., Delers, A., and Rosenberg, M. Biotechnology 9:173–177, 1991.
8. Vialard, J., Lalumiere, M., Vernet, T., Briedis, D., Alkhatib, G., Henning, D., Levin, D., and Richardson, C. J. Virol. 64:37–50, 1990.
9. Weyer, U., and Possee, R.D. J. Gen. Virol. 72:2967–2974, 1991.
10. Kitts, P.A., Ayres, M.D., and Possee, R.D. Nucleic Acids Res. 18:5667–5672, 1990.
11. Kaufman, R.J. In: Genetic Engineering Principles and Methods. Vol. 9. Plenum Press, New York, 1987.
12. Vaughn, J.L., Goodwin, R.H., Tompkins, G.J., and McCawley, P. In Vitro 13:213–217, 1977.
13. Summers, M.D., and Smith, G.E. Texas Agricultural Experiment Station Bulletin 1555, 1987.
14. Hink, W.F. Nature 226:466–467, 1970.
15. King, L.A., Mann, S.G., Lawrie, A.M., and Mulshaw, S.H. Virus Res. 19:93–104, 1991.
16. Hink, W.F. In: Invertebrate Systems in Vitro. Elsevier, New York, 1980.
17. Hink, W.F. In: Invertebrate Cell System Applications. CRC Press, Boca Raton, FL, 1989.
18. Hink, W.F., Thomsen, D.R., Davidson, D.J., Meyer, A.L., and Castellino, F.J. Biotechnol. Prog. 7:9–14, 1991.
19. Agathos, S.N. Biotechnol. Adv. 9:51–68, 1991.
20. Murhammer, D.W. Appl. Biochem. Biotechnol. 31:283–310, 1991.
21. Vaughn, J.L., and Weiss, S.A. BioPharmacology, 4:16–19, 1991.
22. de Gooijer, C.D., van Lier, F.L.J., van den End, E.J., Vlak, J.M., and Tramper, J. Appl. Microbiol. Biotechnol. 30:497–501, 1989.
23. Kompier, R., Tramper, J., and Vlak, J.M. Biotechnol. Lett. 10:849–854, 1988.
24. van Lier, F.L.J., van den End, E.J., de Gooijer, C.D., Vlak, J.M., and Tramper. J. Appl. Microbiol. Biotechnol. 33:43–47, 1990.
25. Hink, W.F. In: Microbial and Viral Pesticides. Marcel Dekker, New York, Inc., 1982.
26. Weiss, S.A., and Vaughn, J.L. In: The Biology of Baculoviruses. Vol. 2. CRC Press, Boca Raton, FL, 1986.
27. Maiorella, B., Inlow, D., Shauger, A., and Harano, D. Biotechnology 6:1406–1410, 1988.
28. Vaughn, J.L., and Weiss, S.A. In: Large-Scale Mammalian Cell Culture Technology. Marcel Dekker, New York, 1990.
29. Hu, W.-S., and Peshwa, M.V. Can. J. Chem. Eng. 69:409–420, 1991.

30. Overton, L.K., Patel, I., Becherer, J.D., Chandra, G., and Kost, T.A. In: Baculovirus Expression Protocols. Humana Press (in press).

31. Cockett, M.I., Bebbington, C.R., and Yarranton, G.T. Biotechnology, 8:662–667, 1990.

32. Cochran, M.A., Smith, G.E., and Volvovitz, F. U.S. Patent Application No. 920197, 1986.

Index